The Art and Science of Ultrawideband Antennas

For a complete listing of the *Artech House Antennas and Propagation Library*, turn to the back of this book.

The Art and Science of Ultrawideband Antennas

Hans Schantz

ARTECH
HOUSE

BOSTON | LONDON
artechhouse.com

Library of Congress Cataloging-in-Publication Data
Schantz, Hans.
  The art and science of ultrawideband antennas / Hans Schantz.
    p. cm.
  Includes bibliographical references and index.
  ISBN 1-58053-888-6 (alk. paper)
  1. Ultrawideband antennas.   I. Title.

  TK7871.67.U45S33 2005
  621.384'135—dc22   2005045270

British Library Cataloguing in Publication Data
Schantz, Hans
  The art and science of ultrawideband antennas.—(Artech House antennas and propogation library)
  1. Antennas (Electronics)   2. Ultrawideband devices   I. Title
  621.3'824

  ISBN: 1-58053-888-6

Cover design by Leslie Genser

© 2005 ARTECH HOUSE, INC.
685 Canton Street
Norwood, MA 02062

International Standard Book Number: 1-58053-888-6
Library of Congress Catalog Card Number: 2005045270

10 9 8 7 6 5 4 3 2 1

To Cora Elizabeth and Greta Marielle,
born October 29, 2004
who taught me that there are in fact
more difficult,
more challenging,
more time-consuming,
and more rewarding
endeavors
than writing a book.

Table of Contents

North Alabama was full of Liquor Interests, Big Mules, steel companies, Republicans, professors, and other persons of no background.

Harper Lee, *To Kill a Mockingbird*, 1961

Preface

This book is the culmination of a 15-year journey for me: from graduate school through teaching, independent research, still more school, and finally to the gentle green hills of northern Alabama. Traditionally, a preface explains the scope of the appended book, sets the context for the subject matter, and clarifies the author's intentions. The introductory chapter of this book already tackles those tasks, leaving a preface potentially redundant. Thus, my solution to the preface problem is to provide a brief memoir of my personal experience in the ultrawideband (UWB) world. Readers uninterested in the somewhat idiosyncratic tale of how a theoretical physicist by training became a UWB antenna designer by profession are strongly encouraged to stop now and just skip ahead to Chapter 1.

At the University of Texas, Austin, in 1990, I began my studies in physics by delving into nuclear matter theory. While the equation of state for nuclear matter and the subject of nuclear astrophysics were not without their charms, I had difficulty imagining how I could apply my theoretical background to more practical problems. Accordingly, I gave new meaning to the fable of jumping out of the frying pan and into the fire by deciding to study gravitational theory and general relativity instead. In my defense, my idealistic (if somewhat unrealistic) visions of being able to build antigravity machines were at least slightly more practicable than designing and building my very own neutron stars. Fortunately, my dissertation advisor, George Sudarshan, had the good sense to tell me that although the mathematical descriptions of gravity provided by general relativity are reasonably well verified, no one really understands how gravitation works and how gravitational energy is conveyed from place to place. If I wanted to understand the workings of gravitation and gravitational energy, he advised I should begin by attempting to understand electromagnetic energy transfer. I tackled electromagnetic energy with a vengeance and never looked back.

I discovered that there was indeed a comprehensive theory of electromagnetic energy transfer introduced by John Henry Poynting and Oliver Heaviside in the 1880s. However, the Poynting-Heaviside theory had fallen into a certain disrepute for two reasons, one physical and the other philosophic. First, there were many alleged problems, paradoxes, and ambiguities in its formulation. Second, the Poynting-Heaviside theory implies that the phenomena of electromagnetics involve some more fundamental physical processes and entities—the "whatever-it-is" that stores energy in "free space" and conveys it from place to place in accord with Maxwell's laws and the Poynting-Heaviside theory. Thus, the Poynting-Heaviside theory runs directly counter to over a century's worth of mainstream physical thought which holds that physics is all about mathematical prediction of experimental measurements and observations [1].

The physical hurdle was relatively easy. As I investigated the puzzles of the Poynting-Heaviside theory, I retraced the footsteps of other researchers who had managed to resolve and explain the alleged paradoxes. I was unable to find any evidence against the theory, and, to the contrary, I found that the Poynting-Heaviside theory comprehensively explains many additional phenomena including the momentum and angular momentum of electromagnetic fields.

The philosophic hurdle was the more difficult one. All too many physicists are comfortable speaking of vacuum energy, vacuum polarization, the speed of light in a vacuum, vacuum permitivity, and vacuum permeability. They are comfortable ascribing properties to "vacuum," implicitly saying that nothing is something, and yet they deride those willing to entertain speculation regarding an "aether" or other more fundamental phenomenology underlying electromagnetics.

Of course, the concept of an aether comes with its own philosophical baggage. Aether reeks of musty nineteenth-century theories of cogwheels, gears, jellies, and rotational elastic solids invoked by Maxwell and his contemporaries to try to explain electromagnetics. These archaic models wrongly implicate by association Maxwell's firm (and I believe correct) conviction that electromagnetic energy is stored, somehow, in "free space" [2]. I had the great pleasure and privilege of discussing philosophic issues as they pertain to physics with my qualifying exam coadvisor, John Archibald Wheeler. Although neither of us really persuaded the other of our respective philosophic ideas, Professor Wheeler helped me to understand the thinking and premises behind much of contemporary physical thought. I documented my philosophic analysis in a lengthy essay [3] and pushed ahead in applying the Poynting-Heaviside theory to understand how static electromagnetic energy decouples from an electric dipole and radiates away.

What I found amazed me. I had implicitly accepted the notion, drummed into my head throughout my physics education, that accelerating charges are the source of radiation energy. I had studied quantum electrodynamics and seen how Feynman diagrams allow a convenient and easy (well, relatively so) means for summing terms in the mathematical description of electromagnetic processes. Feynman diagrams generally treat electromagnetic interactions in a way I came to describe as "atoms-and-void electromagnetics": point charges emit and absorb point photons in a wide variety of patterns to give rise to electromagnetic phenomena. Although Feynman diagrams provide a convenient mathematical shorthand, taking them literally leads to physical misapprehensions.

I discovered that under certain circumstances, static electromagnetic energy stored in regions remote and dissociated from accelerating charges decouples, converts to radiation energy, and radiates away. I discovered that electromagnetic energy can be partitioned into causally distinct regions and that cause-and-effect relationships can be understood by finding surfaces of no net energy flow. I dubbed these "causal surfaces," and I used them to understand electromagnetic energy around electric dipoles. I published the results in a technical paper [4] and in my dissertation [5].

After my graduation in 1995, I pursued a career in teaching and practical research. I landed a position at ITT Technical Institute in Austin, Texas, and I supported myself by teaching mathematics, physics, and electronics. I sought a position at a four-year institution only to discover that "proven grant-getting ability" generally trumps teaching ability on faculty-selection committees. I also found that potential clients were somewhat reluctant to hire a freshly minted theoretical physicist as an electromagnetic consultant. Finally, I resolved to go back to school for some more practical training in electrical engineering. Remarkably, I had great difficulty finding an institution willing to accept a physics Ph.D. as a graduate student. I eventually landed at the ElectroScience Lab of the Ohio State University in Columbus, Ohio, enrolling in the summer quarter of 1998.

Reviewing the course curriculum and degree requirements, I found that it was physically possible to obtain a Ph.D. in electrical engineering in two years, so that was what I set out to do. I learned a great deal about practical antenna engineering and gained valuable hands-on laboratory experience. The highlight of my Ohio State tenure was the opportunity to take Professor Ben Munk's course on antennas.

During the summer of my second year (in 1999), my officemate (and all-around guru on analysis and processing of electromagnetic signals), Soumya Nag, flew down to Alabama for a plant trip to a potential employer. A few weeks later, he came into the office we shared and found me browsing the Internet. Soumya asked me what I was doing.

"I'm beginning my job hunt," I explained, "and I'm doing some research on potential employers."

"What kind of jobs are you looking for?" Soumya asked.

With my background in time domain electromagnetics and broadband antenna design, I had resolved to go into UWB antenna design. Having been seduced by the publicity surrounding the Time Domain Corporation [6], they were my top choice.

"I'm looking for a job as a UWB antenna designer," I told Soumya, "preferably at an outfit like Time Domain."

"Time Domain?" Souyma asked incredulously. "You've heard of Time Domain?"

"Why, yes," I replied. "They're one of the leaders in practical applications of UWB technology."

"Oh," said Soumya followed by a long pause. "I have a job offer from them."

His plant trip was in fact to Time Domain, and he hadn't bothered telling me about it because he figured I'd have never heard of them. Soumya put me in touch with some contacts at Time Domain; I presented a paper at a Time Domain–sponsored conference in September 1999, took a plant trip, and finally received a job offer.

I was torn between staying in school and leaving to go to Time Domain. In discussions with the administration of the ElectroScience Lab, however, I learned that although under the officially printed and distributed course requirements one might be able to complete a doctoral degree in two years, there were significant hitherto-undisclosed additional requirements that would extend my tenure at Ohio State to at least three years, possibly more. Accordingly, I bade Ohio State farewell and arrived in Huntsville, Alabama, to begin work at Time Domain in December 1999. I soon discovered that northern Alabama had an appeal sadly underappreciated by Harper Lee [7], particularly as I wooed and wed a delightful southern girl, Miss Barbara McNew.

I had the luxury of being paid to work full-time designing and building UWB antennas fully three years before the February 14, 2002, Report and Order that authorized commercial deployment of UWB technology. This head start gave me the advantage of having overcome many of the problems and misconceptions that plague UWB antenna practice today. Further, it gave me a chance to make many of the same mistakes prevalent in contemporary UWB antenna practice, but with ample time since to reflect upon and correct those missteps and misapprehensions.

I had the privilege of inventing a great many UWB antenna designs—a regrettably small fraction of which were actually novel. In many cases, I subsequently learned that I was not the first hound down the trail sniffing for certain particularly interesting and effective designs. In part, this is the book I wish had been available when I started my work in UWB antennas—a book that would have helped me avoid reinventing the proverbial wheel and would have focused my efforts on advancing rather than rediscovering antenna art.

I also learned a great deal about antennas and the radio frequency (RF) arts in general from the highly talented team of engineers assembled by Time Domain, particularly (in alphabetical order) Mark Barnes, Bob DePierre, David Dickson, Doug Fitzpatrick, Herb Fluhler, Larry Fullerton, Charles Gilbert, Soumya Nag, Kai Siwiak, Dennis Troutman, and Glenn Wolenec—many of whom (if they look carefully) will be able to see their imprint on this book.

I left Time Domain in April 2002 to pursue a career as an independent consultant, and now I wear two hats. My "day job" is as chief scientist with the Q-Track Corporation. Q-Track is pioneering a technology that is the polar opposite of UWB. Q-Track's Near Field Electromagnetic Ranging (NFER™) technology is a low-frequency (typically within the amplitude modulation (AM) broadcast band), narrowband, high-precision, RF tracking technology. NFER™ tracking systems can track up to 70m away with an accuracy of about 30cm outdoors and about 1m indoors [8].

I also work part time for the Huntsville-based Next-RF, Inc., as a consulting engineer and physicist. My consulting work has involved companies as close as Alabama, Texas, Florida, Maryland, and California and has taken me as far away as Finland, Korea, and Japan. My work has exposed me to a wide variety of interesting and challenging problems, and I've attempted to share some of the lessons I've learned along the way in my book.

This book is based on a short course I first presented at the Centre for Wireless Communications in Oulu, Finland, in February 2004. The 2004 IEEE International Symposium on Antennas and Propagation was the venue for a revised and extended version of my short course. I am indebted to my sponsors for their support and to my students for their helpful suggestions and critiques.

I'd be delighted to receive feedback at h.schantz@ieee.org from readers that might help me prepare an even better second edition (if this initial edition generates sufficient interest!). I've tried my best to dig into old patents and antenna articles, but inevitably I will have overlooked any number of interesting UWB antennas in my research. Please feel free to bring to my attention any UWB antenna I have not mentioned that you believe warrants some discussion. Also, what additional topics should be covered? What were the strengths and weakness of this text? In the meanwhile, my latest work and a list of corrections for this book will always be available on my website at www.uwbantenna.com.

Endnotes

[1] There is a joke with an unfortunate kernel of truth: "Engineers think equations are an approximation to reality. Physicists think reality is an approximation to the equations. Mathematicians have never made the connection."

[2] See the Chapter 5 epigraph for a taste of Maxwell's thinking on electromagnetic energy.

[3] Schantz, Hans Gregory, "Aristotle, Plato, and Electromagnetics," Proceedings of the Fifth Texas Objectivist Societies Conference, October 23, 1994.

[4] Schantz, Hans Gregory, "The flow of electromagnetic energy in an electric dipole decay," *American Journal of Physics*, 63, No. 6, June, 1995, pp. 513-520.

[5] Schantz, Hans Gregory, "The Energy Flow and Frequency Spectrum about Electric and Magnetic Dipoles," Doctoral Dissertation, University of Texas, Austin, Austin, TX, August 1995.

[6] USA Today, April 9, 1999 (see pp. 1A, 2A, 1B, 2B).

[7] Lee, Harper, *To Kill a Mockingbird*, New York: Warner Books, 1961, p. 16. In the epigraph at the beginning of the preface, Scout describes northern Alabama in the context of explaining her new teacher, who hails from there.

[8] For more information on Q-Track's NFER™ technology, please see www.q-track.com.

Acknowledgments

I'm indebted to my business partners, Bob DePierre, Jerry Gabig, and David Langford for tolerating my year-long diversion in writing this book. Life in a seed stage startup is an adventure with many challenges and frustrations, but the opportunity to work with so outstanding a team makes it all worthwhile.

I'm grateful to Ian Oppermann for the opportunity to present a rough draft of this material as a short course to the folks at the Centre for Wireless Communications in Oulu, Finland, in February 2004. I'll never forget the Snow Show at Rovaniemi, skiing at Iso-Syote, and the experience of working on the early chapters of this book in the Oulu public library with a backdrop of seemingly eternal dusk shrouding the frozen Baltic. Thanks go also to Ryuji Kohno for support in presenting an invited paper at the International Workshop on Ultra-Wideband Systems (IWUWBS) in Kyoto, Japan, in May 2004 and to Kyung Sup Kwak for support in presenting an invited paper at the second International Workshop on UWB Radio Technology in Incheon, Korea, in June 2005.

I greatly appreciate the early readers of this book: Ben Munk of the Ohio State University; Mark Barnes of the Time Domain Corporation; and Kai Siwiak of Time Derivative, Inc. Your insightful analysis and comments have strengthened this text. A particularly honorable mention goes to Herb Fluhler of Time Domain who not only reviewed the book but also generated the index.

Some of the material in this book previously appeared in my many conference papers and journal articles. I'm grateful to the IEEE, Kluwer Academic Press, and CMP Media for permission to reproduce my work that bears their copyright. Darrel Emerson, Guofeng Lu, and Glenn Smith all graciously allowed me to reproduce some key figures and results from their published work. Finally, Globe Photo and Forbes were particularly helpful in allowing me to reproduce the historical 1945 Forbes cover featuring Lindenblad's famous UWB TV antenna (for a modest fee). I'd also like to thank the many businesses and organizations that assisted and cooperated with me in gathering the material presented in this book. These include Aetherwire and Location, Inc., the Centre for Wireless Communications, the ElectroScience Lab of Ohio State University, Innovative Wireless Technology, LLC., Next-RF, Inc., Pulse~LINK, and Time Derivative, Inc. In particular, I'd like to acknowledge my debt to my former employer, the Time Domain Corporation, which generously allowed me to access and cite from work I performed while in their employ.

A special thanks goes to Glenn Wolenec, who volunteered to construct a couple of the more interesting antennas I discuss in this book. Glenn also helped me focus on designs useful for amateur radio applications.

Most of all, I'd like to thank my wife, Barbara McNew Schantz, for shouldering a disproportionately large share of the baby care responsibilities to allow me to finish this book. I owe you a further debt as my first reader and most detailed editor. Your confidence is my strength, and your love is my inspiration.

Chapter 1

An Introduction to UWB Antennas

Rays of light will not pierce through a wall, nor, as we know only too well, through a London fog. But the electrical vibrations of a yard or more in wave-length of which I have spoken will easily pierce such mediums, which to them will be transparent. Here then, is revealed the bewildering possibility of telegraphy without wires, posts, cables, or any of our present costly appliances. Granted a few reasonable postulates, the whole thing comes well within the realms of possible fulfillment.

William Crookes, 1892

The magic and mystery of radio have captured imaginations from the earliest speculations of William Crookes to the present day. The marvel of radio is taken for granted in a world of pervasive and instantaneous wireless communication. All around us quiver vibrations in the aether conveying data: voices, images, and information. The magic of radio plucks these vibrations out of thin air and recovers the original data. The wand responsible for this wizardry is the antenna.

Of course, neither radios nor antennas are really magic. As Arthur C. Clarke observed, "Any sufficiently advanced technology is indistinguishable from magic" [1]. Radio is merely the technical application of electromagnetic science to communications. The extent to which radio appears magical is the extent to which one has failed to understand the advanced technology that makes radio work (or merely takes it all for granted). By any measure, antennas are among the most mysterious and least understood aspects of radio technology—"magic" by Clarke's definition. And if antennas are regarded as magical, then ultrawideband (UWB) antennas must surely be the most arcane and darkest of black magic.

This book aims to replace the mystery surrounding UWB antennas with hard facts and sound science. The fundamental science behind antennas is electromagnetics, and electromagnetics is a quantitative science. Oliver Heaviside defined mathematics as "reasoning about quantities" [2]. Any detailed analysis of antenna behavior necessarily requires mathematical treatment. Often though, such a detailed analysis is not necessary to make effective use of an antenna or to grasp the relatively simple yet fundamental qualitative principles that govern antenna behavior. Thus, this book strives to explain UWB antennas using a minimum of mathematical analysis.

Working with UWB antennas requires more than science alone. There is a creative side to UWB antenna design by which scientific principles are mixed with a healthy dose of imagination and distilled to yield novel element shapes. Art, according to Aristotle, is the realization in external form of a true idea [3]. By this definition, UWB antenna design is not only a science but also an art. Accordingly, this book is replete with figures and diagrams illustrating various ways antenna designers have taken the true ideas of electromagnetic science and realized them in shaped and curved metallic form.

Although recent years have seen increased interest in UWB antennas, the ancestry of UWB antennas can be traced back to the discovery of radio over a century ago. This chapter begins by tracking the early genealogy of UWB antennas through a historical introduction to the subject.

Then, this chapter asks the fundamental question, What is an antenna? There are at least four answers, depending on the viewpoint of the questioner:

- **Transducer**: For many applications, an antenna may be treated as a transducer, a black box that converts signals on a transmission line to signals through the aether. Chapter 2 addresses this viewpoint.
- **Transformer**: In some cases, however, an antenna should be thought of as a transformer, coupling a transmission line impedance to free space. Chapter 3 addresses this viewpoint.
- **Radiator**: In other cases, the traditional electromagnetic point of view has value; an antenna is a defined structure supporting currents in a particular geometry coupled to associated electromagnetic fields. Chapter 4 addresses this viewpoint.
- **Energy Converter**: A final perspective considers antennas as devices that transform guided energy into radiation energy (and vice versa). Chapter 5 addresses this viewpoint.

Chapter 1 provides an overview of these four viewpoints. The following four chapters (Chapters 2–5) focus on each viewpoint in turn. Next, this initial chapter provides a brief taxonomy of UWB antennas, a topic examined in depth in Chapter 6. Finally, Chapter 1 discusses some of the system considerations relevant to UWB antenna applications. System considerations are the subject of Chapter 7.

The reader should feel under no moral obligation to review closely all eight chapters. A manager, casual UWB technologist, or amateur radio operator might review the first two chapters and skim through Chapters 3 through 5 to arrive at the UWB antenna taxonomy of Chapter 6. A radio frequency (RF) or systems engineer might wish to review the first three chapters, skim through Chapters 4 and 5, and then read more closely the antenna taxonomy of Chapter 6 and the system considerations of Chapter 7. A physicist might enjoy the perspective of electromagnetics and energy transfer presented in Chapters 4 and 5. Hopefully, antenna engineers and designers will find all chapters worth their while!

1.1   HISTORICAL INTRODUCTION

Heinrich Hertz (1857–1894) demonstrated the existence of what we would now call radio waves in 1888 [4]. Within a few years, Crookes and others were speculating on the possibility of wireless telegraphy [5]. Before the decade was out, wireless (or radio) communication was born. Modern UWB radios are the direct descendents of these early systems. This examination of UWB antennas will begin with a brief historical survey of radio technology in general and antennas in particular.

Early radio equipment was primitive by modern standards. Lacking any practical means of generating continuous wave signals of a particular frequency, radio pioneers had to resort to discharges in resonant circuits. Their equipment radiated damped sinusoidal impulses that would be considered UWB by modern standards.

Reviewing the history of radio with the benefit of a modern perspective, it is tempting to think of spark-gap radios as early UWB systems. But the UWB damped sinusoidal impulse signals of spark-gap transmitters were emitted by accident and not by design. These UWB emissions were not due to any deliberate intention toward spectrum spreading. Instead, UWB impulses were an inescapable side effect of the then available means of generating sine waves at radio frequencies. Narrowband operation was the explicit goal of the radio pioneers, and the broad bandwidths of their signals were unintentional emissions. Even Hertz used resonant circuits to feed resonant, narrowband antennas, and he used other resonant narrowband antennas to detect signals.

The narrowband frequency domain mind-set was clear as early as 1892 when Crookes wrote,

> Any two friends living within the radius of sensibility of their receiving instruments, having first decided on their special wavelength and attuned their respective instruments to mutual receptivity, could thus commnicate as long and as often as they pleased by timing the impulses to produce long and short intervals in the ordinary Morse code [5].

The full realization of Crookes' vision would require decades of technical development. Nevertheless, by the time of Crookes' writing, the foundation had been laid:

> This is no mere dream of a visionary philosopher. All the requisites needed to bring it within the grasp of daily life are well within the possibilities of discovery, and are so reasonable and so clearly in the path of researches which are now being actively prosecuted in every capital of Europe that we may any day expect to hear that they have emerged from the realms of speculation into those of sober fact. Even now, indeed, telegraphing without wires is possible within a restricted radius of a few

hundred yards, and some years ago I assisted at experiments where messages were transmitted from one part of a house to another without an intervening wire by almost the identical means here described [5].

Even as Crookes wrote, wireless telegraphy was taking its first steps forward. Although narrowband operation was their clear goal, the pioneers of wireless began by radiating UWB damped impulse signals. Under the circumstances, it is no surprise that some of their early antennas—designed to radiate and receive spark-gap impulses—are also well suited for use in modern UWB systems.

The history of early radio is well studied and widely known. This history has been thoroughly documented in both professional histories [6, 7] and in popular treatments [8]. The development of UWB antennas has been superficially examined elsewhere, but not in similar detail [9]. Antenna designers are not immune from the maxim that those who ignore history are condemned to repeat it. Because the history of UWB antennas is relatively unknown, antenna designers have been prone to devise "novel" UWB antennas that were actually known to earlier generations of designers. Thus, it is worthwhile to begin this examination of UWB antennas with a historical review.

1.1.1 The Spark-Gap Pioneers

James Clerk Maxwell (1831–1879) first published his remarkable equations that describe the behavior of electric and magnetic fields in 1865. The jump from Maxwell's equations to understanding electromagnetic radiation and radio waves was not as obvious as the streamlined and simplified derivations in modern textbooks would make it appear. For over twenty years, there was serious debate over whether Maxwell's equations really implied the existence of electromagnetic radiation [10]. The controversy was not resolved until 1888 when Hertz first published the results of his experiments [see Figure 1.1(a)]. Hertz led the way as the first to generate radio waves and subject them to scientific scrutiny.

A great many innovators like Nikolai Tesla, Carl Braun, Alexander Popov, and Reginald Fessenden contributed key elements to radio technology. Looked at from an antenna perspective, however, three other pioneers were at the forefront. British physicist Oliver Lodge (1851–1940) invented biconical and bow tie antennas and introduced the first practical system for "syntony" [see Figure 1.1(b)]. Syntony is the concept of transmitting and receiving using a particular frequency and tuned circuits. An Indian physicist, Jagadis Chandra Bose (1858–1927), performed pioneering work in millimeter wave systems and invented the horn antenna [see Figure 1.1(c)]. An Italian engineer and businessman, Guglielmo Marconi (1874–1937), first commercialized radio as a means of long-range communication [see Figure 1.1(d)]. This section discusses the contributions of Hertz, Lodge, Bose, and Marconi to the development of antenna technology.

Figure 1.1 (a) Heinrich Hertz (1857–1894), the German physicist who discovered radio waves, (b) Oliver Lodge (1851–1940), the British physicist who invented biconical and bow tie antennas and invented the first practical "syntonic" radio, (c) Jagadis Chandra Bose (1858–1937), the Indian physicist who invented horn antennas and pioneered microwave and millimeter wave technology, and (d) Guglielmo Marconi (1874–1937), the Italian engineer and businessman who first commercialized radio as a means of long-range communication (Courtesy of the Library of Congress and Darrel Emerson).

I have succeeded in producing distinct rays of electric force, and in carrying out with them the elementary experiments which are commonly performed with light and radiant heat.

Heinrich Hertz, 1893

1.1.1.1   Heinrich Hertz

Hertz was not the first to generate what we would now call radio waves. Hertz is credited with the discovery of radio waves because he was the first to generate radio waves deliberately, to measure their velocity, and to subject them to reflection, refraction, and diffraction in such detail as to demonstrate conclusively that his radio or electromagnetic waves were in fact light waves with vastly longer wavelengths and smaller frequencies.

Hertz generated radio waves by releasing the energy stored in Leyden jars (an early capacitor) and discharging it through resonant circuits and antennas. In fact, many of Hertz's antenna designs betray their origins as modifications of the capacitive vanes or elements used in Leyden jars. Hertz knew the inductance (L) and capacitance (C) of his resonant circuits, so he was able to calculate the frequency

$$f = \left(2\pi\sqrt{LC}\right)^{-1} \tag{1.1}$$

Measuring the wavelength (λ) and knowing the frequency (f), Hertz was able to estimate the velocity of his signals

$$v = \lambda f \tag{1.2}$$

Hertz was well aware of the concept of a half-wave dipole and sized antennas to match his operating frequency. Further, he employed end-loading to maximize the charge stored on his antennas and, thus, the amplitude of his transmitted signals. Figure 1.2 shows examples of Hertz's end-loaded antennas, including spherical and square plate end-loading.

As a detector, Hertz employed a resonant loop antenna with a small, adjustable screw gap. By observing the intensity of discharges and knowing the gap distance, Hertz was able to make rough estimates of the received field strength. Hertz's technique was sufficient for him to analyze standing waves. By locating maxima and minima, Hertz could directly measure the wavelength of the standing waves he created.

Hertz also introduced reflectors to focus and concentrate RF energy. Figure 1.3 shows one of Hertz's parabolic cylinder designs [11]. Of Hertz's antenna designs, the parabolic reflector is the only one that can be characterized as UWB.

Even Hertz's reflector antennas used relatively narrowband resonant feeds. Later inventors and engineers extended the scientific foundation laid by Hertz to create practical radio systems and UWB antennas.

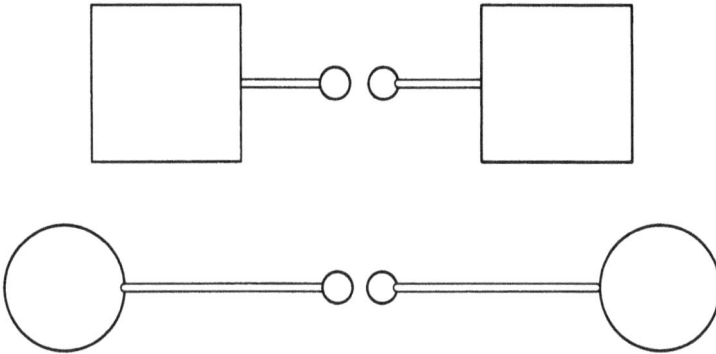

Figure 1.2 Square plate loaded dipole and spherical loaded dipole of Heinrich Hertz.

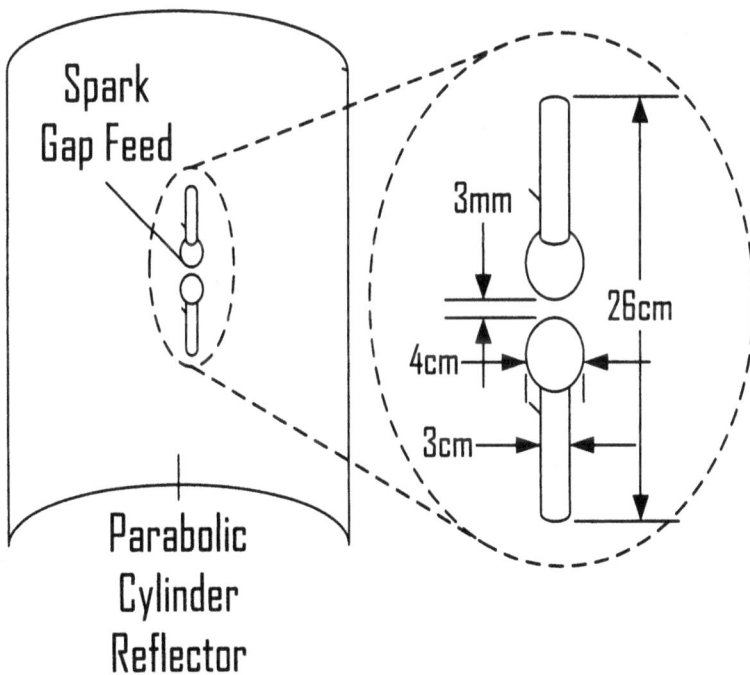

Figure 1.3 One of Hertz's parabolic cylinder antennas (After [11]).

Hertz pioneered both antenna and RF technology, but his accomplishments had much broader scientific implications. Since the time of Isaac Newton, natural philosophers and physicists alike had generally accepted that fundamental physical interactions occurred via "action at a distance." Entities at one location were thought to exert a physical influence at distant locations without having traversed the intervening medium. Michael Faraday challenged this dogma with his concept of "fields" pervading space carrying electric and magnetic influences. Maxwell's theory was in large part aimed at giving a mathematical foundation to Faraday's ideas. By catching electromagnetic energy in the act of moving from one place to another, Hertz and his antennas provided the first, solid, experimental confirmation that Faraday and Maxwell were correct. In others' hands, Hertz's ideas were to become the foundation of commercial radio.

1.1.1.2   Oliver Lodge

Like Hertz, Lodge was also an academic, equally talented as a theoretician and as an experimentalist. Lodge made significant progress in understanding lightning rods. Along with Oliver Heaviside, he was a major contributor to the development of the theory of AC circuits and reactance. Almost simultaneously with Hertz, he made experimental measurements of electromagnetic waves on wires, yet he acknowledged Hertz's priority as the discoverer of free space electromagnetic waves.

After Hertz's discoveries, Lodge remained active in his investigations of radio waves. Spurred by Marconi's successes, Lodge developed the first practical syntonic radio system, one in which a transmitter and a receiver are tuned to the same frequency so as to maximize the received signal. In 1898, Lodge patented a syntonic radio system. Ironically, the very patent that inaugurated this fundamental concept of narrowband frequency domain radio also disclosed some of the first UWB antennas:

> As charged surfaces or capacity areas, spheres or square plates or any other metal surfaces may be employed; but I prefer, for the purpose of combining low resistance with great electrostatic capacity, cones or triangles or other such diverging surfaces with the vertices adjoining and their larger areas spreading out into space; or a single insulated surface may be used in conjunction with the earth, the earth or conductors embedded in the earth constituting the other oppositely-charged surface [12].

In perhaps the most profound and sweeping sentence in the history of antenna technology, Lodge disclosed spherical dipoles, square plate dipoles, biconical dipoles, and triangular, or bow tie, dipoles. He also introduced the concept of a monopole antenna using the Earth as a ground.

In fact, Lodge's patent drawings make very clear his preferred embodiments. Figure 1.4 shows Lodge's second figure, in which triangular, or bow tie, elements are clearly indicated. Figure 1.5 depicts Lodge's fifth figure, in which biconical antennas are unmistakably used in a transmit-receive link.

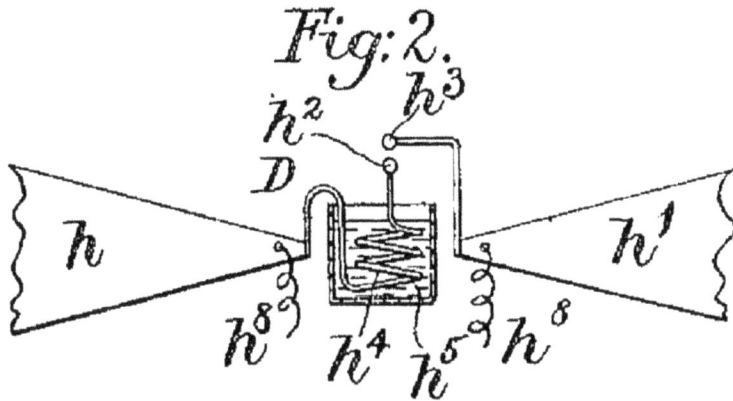

Figure 1.4   Lodge preferred antennas consisting of triangular "capacity areas," a clear precursor to the bow tie antenna [12].

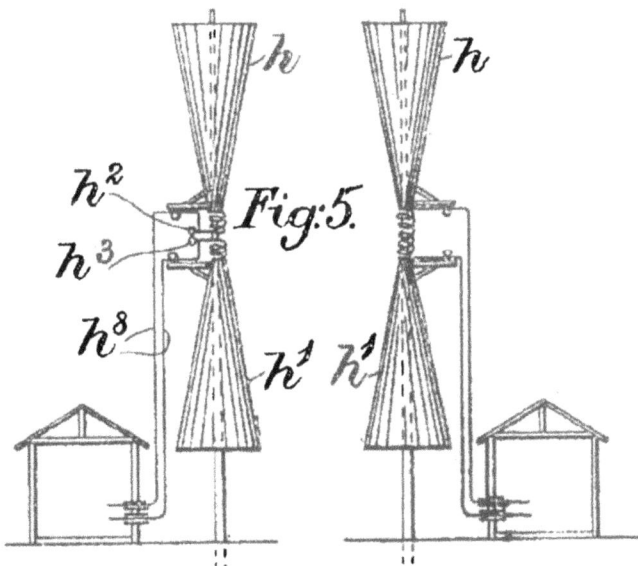

Figure 1.5   Lodge was also the inventor of the biconical antenna [12].

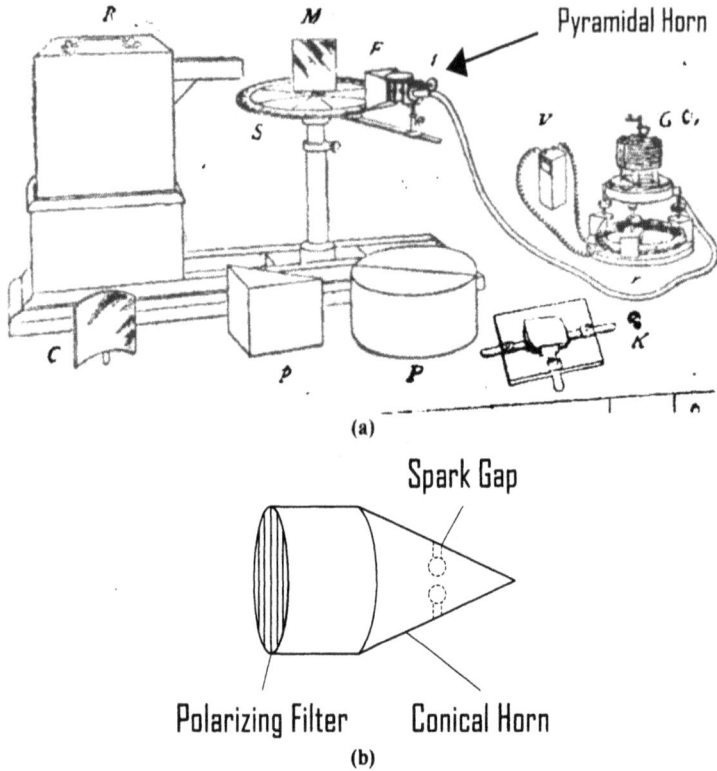

Pyramidal Horn

(a)

Spark Gap

(b)

Polarizing Filter   Conical Horn

Figure 1.6 (a) Bose's 1897 millimeter wave apparatus showing a pyramidal horn antenna ([13];
©1999 IEEE; Courtesy: Darrel Emerson), and (b) sketch of Bose's transmit antenna
showing a spark-gap driven conical horn transmit antenna (After [13]).

1.1.1.3   Jagadis Chandra Bose

In a public demonstration in 1895, Bose used electromagnetic waves to ring a bell
remotely and to detonate gun powder. In 1897, Lord Rayleigh invited Bose to
London to present his research to the Royal Society in London. Bose
demonstrated a complete system for transmitting and receiving radio waves at
millimeter-wave frequencies of around 60GHz. This system included what Bose
dubbed "collecting funnels," the first horn antennas. Bose used a spark-gap
transmitter and a semiconductor junction as a receiver. Figure 1.6(a) reproduces
one of Bose's figures showing a pyramidal horn. Figure 1.6(b) shows a sketch of
one of Bose's conical horn antennas. Darrel Emerson provides an excellent and
detailed description of Bose's pioneering work [13].

> *A short time ago a young Italian electrician came to see me, with a new system, which we immediately tested on Salisbury Plain with excellent results, over a distance of a mile and three quarters. This young man is Signor Guglielmo Marconi....*

Sir William Preece, 1897

1.1.1.4   *Guglielmo Marconi*

Marconi was among the first to realize the possibility of radio communication. Others (including Lodge) had demonstrated the feasibility of signaling via radio waves but were slow to grasp the commercial implications. Marconi was convinced that this scientific curiosity had practical applications: "If we had attributed to the power of light only the possibilities offered by a candle, we would never have built lighthouses and reflectors" [14].

Marconi's original antennas were little changed from those used by Hertz, including parabolic reflectors and square plate antennas (see Figure 1.7) [15]. The most significant difference was that Marconi recognized the value and importance of a good earth ground, essential since he operated at much lower frequencies than those used by Hertz. As Marconi moved to high-power, longer-range systems, he adopted fan and conical monopole antennas [16]. These antennas are shown in Figure 1.8(a, b) respectively.

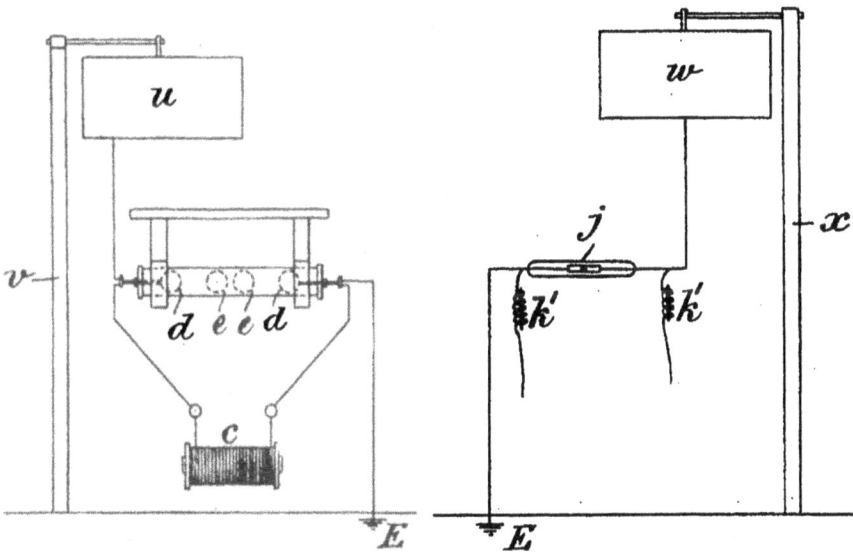

Figure 1.7 Marconi's square plate antenna [15].

(a)

(b)

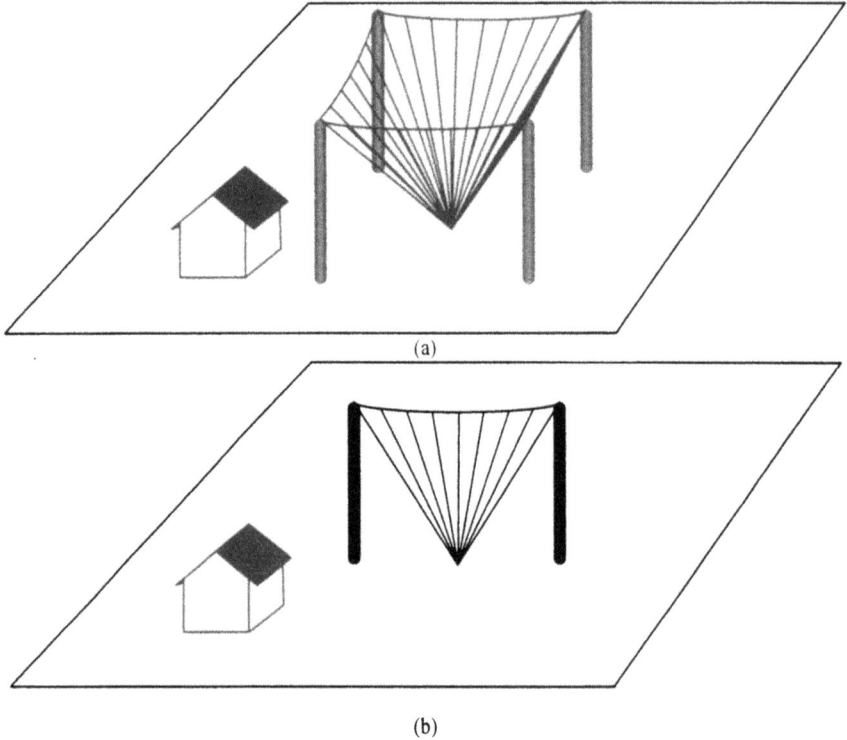

Figure 1.8   (a) Marconi's thin wire conical monopole (or monocone), and (b) Marconi's thin wire fan, or bow tie, monopole (After [16]).

As Marconi and others continued to improve radio technology, they moved beyond early spark-gap devices and implemented continuous wave radio systems. Efficient narrowband transmitters soon became a reality, and there was less incentive to use more complicated and bulky broadband antennas. The compact size and lower cost of thin-wire monopole and loop antennas soon eclipsed the pioneering broadband antenna designs of Lodge, Bose, and Marconi. Their early designs were largely forgotten, awaiting rediscovery by a later generation of antenna engineers.

Figure 1.9 shows a chart of the radio spectrum circa 1930 [17]. The upper limit of the radio spectrum was considered to be around 30,000kc (30MHz), with the band from 30 to 60MHz "not now useful" for commercial purposes and only of interest to amateurs and experimenters. At 30MHz the wavelength of a radio wave is 10m, and a quarter-wavelength is 2.5m. This size begins to permit construction of complicated quarter-wavelength scale antenna structures, as discussed in the following section.

KILOCYCLES

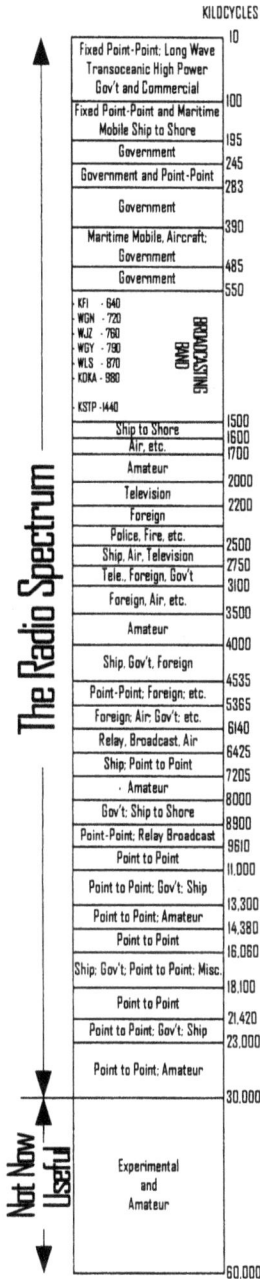

	KILOCYCLES
	10
Fixed Point-Point; Long Wave Transoceanic High Power Gov't and Commercial	
	100
Fixed Point-Point and Maritime Mobile Ship to Shore	
	195
Government	
	245
Government and Point-Point	
	283
Government	
	390
Maritime Mobile, Aircraft; Government	
	485
Government	
	550
KFI - 640 WGN - 720 WJZ - 760 WGY - 790 WLS - 870 KDKA - 980 KSTP -1440	BROADCASTING BAND
	1500
Ship to Shore	1600
Air, etc.	1700
Amateur	
	2000
Television	
	2200
Foreign	
Police, Fire, etc.	2500
Ship, Air, Television	2750
Tele., Foreign, Gov't	3100
Foreign, Air, etc.	
	3500
Amateur	
	4000
Ship, Gov't, Foreign	
	4535
Point-Point; Foreign; etc.	5365
Foreign; Air; Gov't; etc.	6140
Relay, Broadcast, Air	6425
Ship; Point to Point	7205
Amateur	8000
Gov't; Ship to Shore	8900
Point-Point; Relay Broadcast	9610
Point to Point	11,000
Point to Point; Gov't; Ship	13,300
Point to Point; Amateur	14,380
Point to Point	16,060
Ship; Gov't; Point to Point; Misc.	18,100
Point to Point	21,420
Point to Point; Gov't; Ship	23,000
Point to Point; Amateur	30,000
Experimental and Amateur	60,000

The Radio Spectrum

Not Now Useful

Figure 1.9 The radio spectrum circa 1930 (After [17]).

1.1.2   The Shortwave and Television Era

The frontiers of radio science surged past 30MHz in the 1930s, and a new generation of antenna designers tackled the problem of creating broadband antennas. As frequencies used in RF systems increased and wavelengths became shorter, interest in high-performance, broadband antennas grew. Shorter wavelengths made complicated quarter-wavelength-scale antenna elements more practical.

Not only were frequencies higher but also bandwidths were wider. More sophisticated signals placed greater demands on antennas than the traditional 10kHz wide amplitude modulation (AM) of commercial broadcast radio systems. Edwin Howard Armstrong's 1933 invention of wideband frequency modulation (FM) signals demanded bandwidths of 150kHz or more. The advent of television also led to a demand for antennas that could handle the much wider bandwidths associated with video signals, typically on the order of 6MHz wide.

This renewed demand for wideband antennas led to Philip Carter's rediscovery of the biconical antenna and conical monopole in 1939 [see Figure 1.10(a–c)] [18]. Carter also improved upon Lodge's original design by incorporating a tapered feed (see Figures 3.7 and 3.8) [19]. Chapter 3 will examine Carter's taper feed biconical, or "alpine horn," antenna in more detail. Carter was among the first to take the key step of incorporating a broadband transition between a feed line and radiating elements.

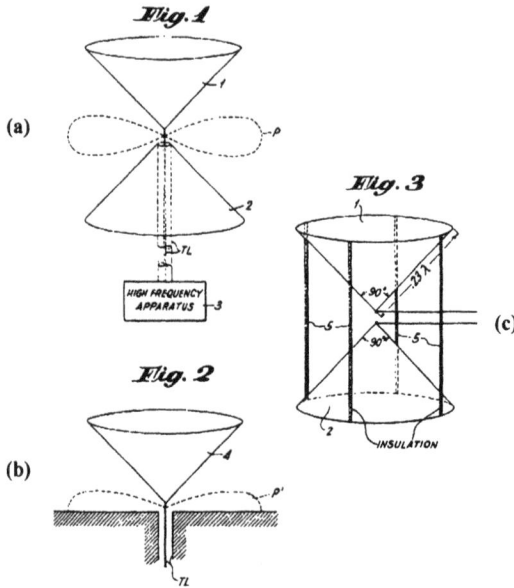

Figure 1.10 (a) Carter's biconical antenna, (b) Carter's conical monopole, and (c) Carter's biconical antenna with supports and dimensions [18].

One of the most prominent UWB antennas of the period was Nils Lindenblad's coaxial horn element [20, 21]. Lindenblad improved on the idea of a sleeve dipole element [shown in Figure 1.11(a)], adding a gradual impedance transformation to make it more broadbanded. The Radio Corporation of America (RCA) chose Lindenblad's element [seen in cross-section in Figure 1.11(b)] for experimental use in television transmission. RCA's concept of television operations imagined all television broadcasts under centralized control with multiple channels broadcast from a single location. Thus, a wideband antenna capable of transmitting multiple video signals simultaneously was essential to RCA's plans. For several years during the 1930s, a turnstile array of Lindenblad's coaxial horn elements graced the top of the Empire State Building in New York City, where RCA located its experimental television transmitter. Figure 1.11(c) displays a patent drawing of this array. The antennas at the top of the tower in Figure 1.11(c) (items 70–72) are folded dipoles used to transmit the audio portion of the television signal.

In fact, Lindenblad's coaxial element came to symbolize the entire television research effort. This UWB antenna has the distinction of being perhaps the only antenna to have been featured prominently on the cover of a mainstream periodical (see Figure 1.12) [22].

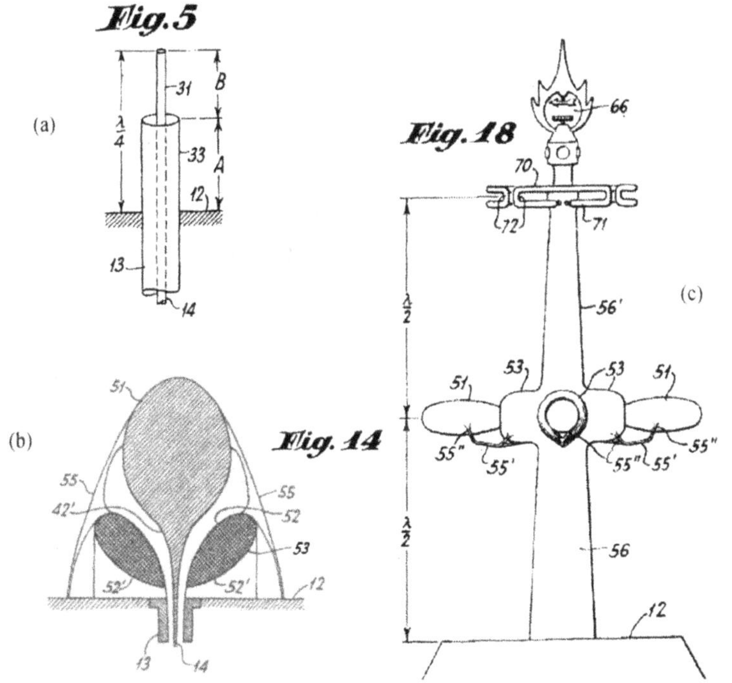

Figure 1.11   (a) Lindenblad's sleeve dipole starting point, (b) Lindenblad's element in cross-section, and (c) a turnstile array of Lindenblad elements for television transmission [20].

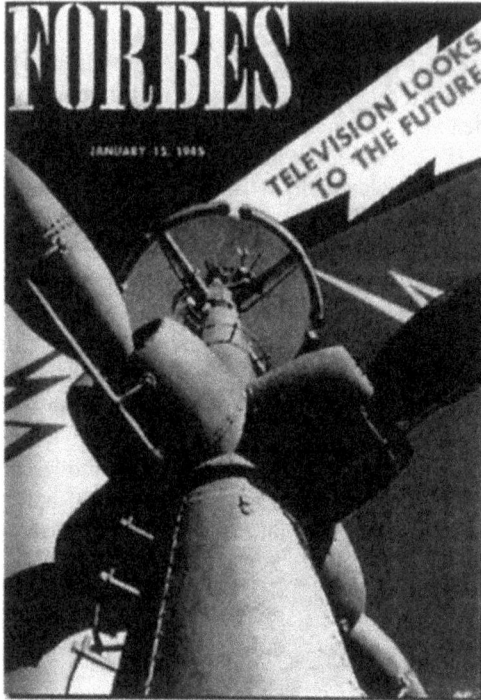

Figure 1.12 Lindenblad's antenna graces the cover of *Forbes* magazine, January 15, 1945 [22]; [©1945, Forbes and Globe Photo].

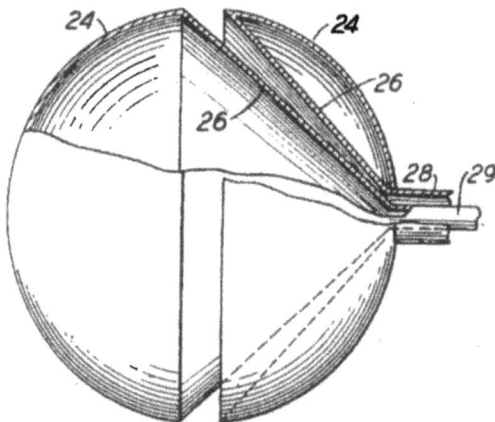

Figure 1.13 Schelkunoff's spherical dipole antenna [23].

Lindenblad was a pioneer in using bulbous, or "fat," antenna elements, but others were not far behind. Sergei Schelkunoff investigated spherical antenna designs [23, 24]. These bulbous antenna shapes distribute antenna current over a much larger area than a conventional thin-wire antenna. As a consequence, fat, or bulbous, antennas have lower reactive energy and more broadband behavior than an equivalent thin-wire antenna. Figure 1.13 shows Schelkunoff's 1941 spherical dipole design.

The fundamental UWB antenna principle, "fatter is better," was well understood by the 1940s. For instance, R. W. P. King observed that "...if an antenna is to be operated over a fairly wide band of frequencies, as in television, it has a smaller change in input reactance if it is thick than if it is thin" [25]. With Harald Friis, Schelkunoff was a pioneer in considering a variety of thick antennas, particularly biconical antennas and related shapes. Figure 1.14 presents a 1952 figure from Schelkunoff and Friis's classic work on antennas [26]. Note the assumption that monopole impedance is necessarily half the impedance of a corresponding dipole. Chapter 3 will address this misconception in detail.

Other bulbous shapes were suggested in the 1940s and 1950s. Figure 1.15(a) shows the teardrop biconical of Schelkunoff and Friis [27]. This modification to the biconical helps reduce reflections due to the discontinuity in taper at the end of a traditional biconical. Figure 1.15(b) shows John Kraus's "volcano smoke" antenna [28]. Kraus's bulbous monopole antenna yields a 5:1 impedance bandwidth and an omni pattern.

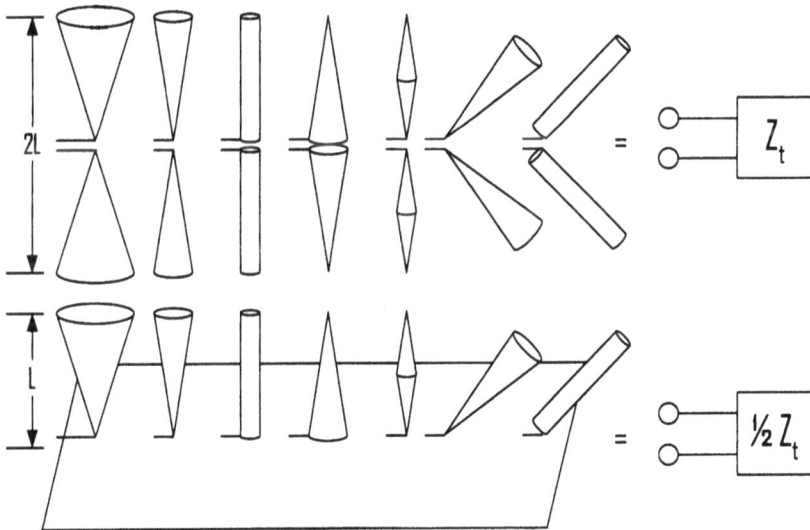

Figure 1.14   Schelkunoff and Friis's thick dipole and monopole antennas (After [26]). The commonly accepted principle that monopoles have half the impedance of the corresponding dipole does not apply to UWB antennas.

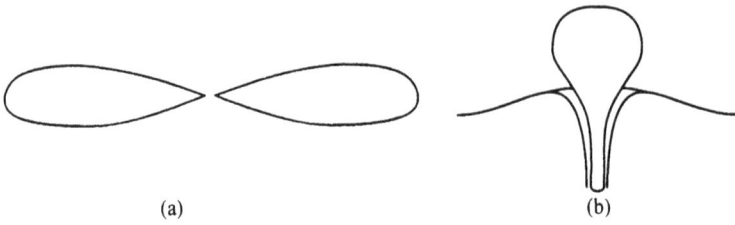

Figure 1.15 (a) Schelkunoff's and Friis's teardrop biconical (After [27]), and (b) Kraus's volcano smoke antenna (After [28]).

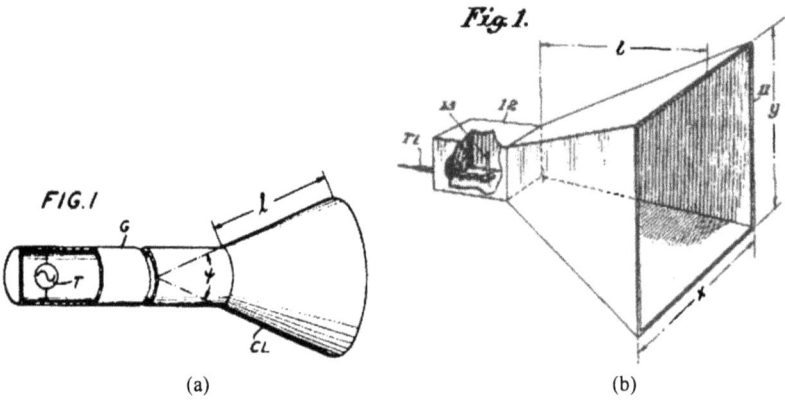

Figure 1.16 (a) King's conical horn [30]. (b) Katzin's pyramidal horn [31].

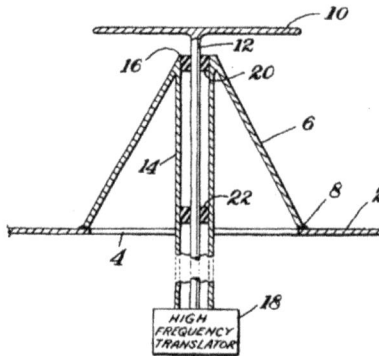

Figure 1.17 Kandoian's discone antenna [33].

Other researchers also pursued the idea of constructing antennas from coaxial transitions, yielding elaborate, elegant surface-of-revolution and solid antenna designs. Leon Brillouin introduced coaxial horns, both omnidirectional and directional (see Figure 3.9) [29]. These bulbous-shaped antennas offer outstanding performance. The disadvantage of shapes like these is the difficulty in manufacturing solid surface-of-revolution antennas with complicated curved shapes.

Designers of the 1940s also revisited the horn designs pioneered by Bose. Figure 1.16(a) shows a conical horn patented by Archie King [30], and Figure 1.16(b) depicts a pyramidal horn invented by Martin Katzin [31]. As microwave technology began to catch up with Bose's pioneering millimeter wave work, his "collecting funnel" antennas were independently rediscovered.

Armig Kandoian devised an improvement to the biconical antenna by replacing one of the conical elements with a disk [32, 33]. Kandoian's "discone" antenna offers a multiple-octave impedance match and a dipolelike pattern over at least a 2:1 span in frequency. Figure 1.17 shows Kandoian's design. Chapter 6 considers the discone antenna in greater detail.

The shortwave and television era led to tremendous progress in broadband antenna design. Advances in technology drove demand for larger bandwidths and pushed practical radio frequencies high enough and wavelengths small enough for complicated quarter-wavelength-scale structures to be practical. The shortwave and television era of the 1930s and 1940s also saw the development of tapered feed antennas by Carter, Lindenblad, and others. The principal that fatter is better became well understood. Bose's horn designs were rediscovered to meet the need for pencil beam microwave antennas.

1.1.3   More-Recent Advances

The line between the spark-gap pioneers and the shortwave and television era is clear and distinctive. There was a long hiatus as progress in radio technology slowly caught up with the pioneering designs of Lodge, Bose, and others. The line dividing the shortwave and television era from more recent advances is not nearly as clearcut.

The distinction between these periods is the increasing emphasis on the practical, readily manufactured antenna designs essential for wide-scale commercialization of broadband antennas. There is considerable overlap, however, between the shortwave and television era and more modern developments. In the early 1950s, Lodge's bow tie antenna was revived and examined by George Brown and O. M. Woodward [34]. Similarly in the 1940s, Robert W. Masters proposed an inverted triangular or "diamond" dipole for use with an ultrahigh frequency (UHF) television receiver [see Figure 1.18 (a)] [35]. Chapter 6 discusses the diamond dipole in greater detail.

B. J. Lamberty revived Marconi's square plate elements for microwave applications [36]. Figure 1.18(b) shows Lamberty's square plate monopole—a simple-to-build, compact, inexpensive UWB monopole element. Chapter 6 discusses these and similar antennas in more detail.

Also during the 1950s, Victor Rumsey identified the principle at the heart of a large family of broadband antennas [37]. The impedance and pattern properties of an antenna will be frequency-independent if the antenna shape is specified only in terms of angles. These frequency-independent antennas have bandwidth limited only by the range over which a repetitive geometry is scaled. Examples include the spiral antennas of Figure 1.19(a), the log periodic antenna of Figure 1.19(b), and the conical spiral antenna of Figure 1.19(c).

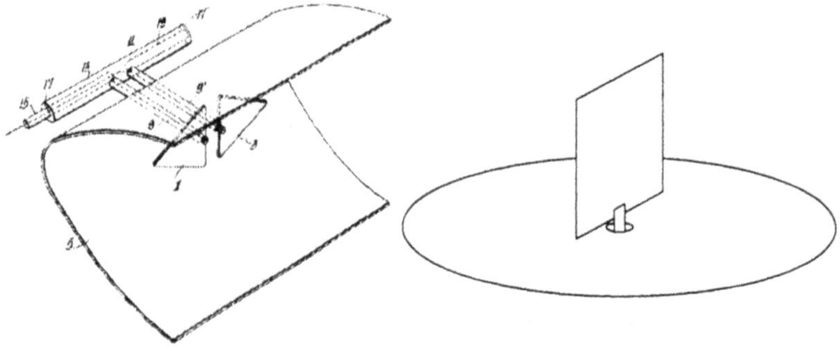

(a)   (b)

Figure 1.18   (a) Masters's diamond dipole [35], and (b) Lamberty's square plate monopole (After 36]).

(a)

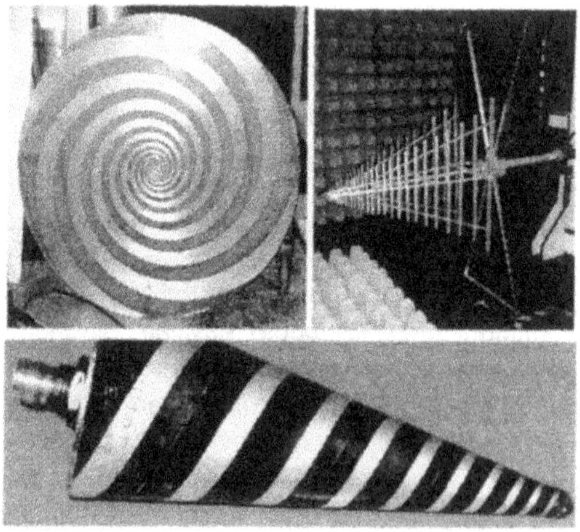

(b)

(c)

Figure 1.19   (a) A spiral antenna (courtesy of the ElectroScience Lab, Columbus, Ohio), (b) a log periodic antenna (courtesy of the Centre for Wireless Communications, Oulu, Finland), and, (c) a conical spiral antenna (courtesy of Next-RF, Inc., Huntsville, Alabama).

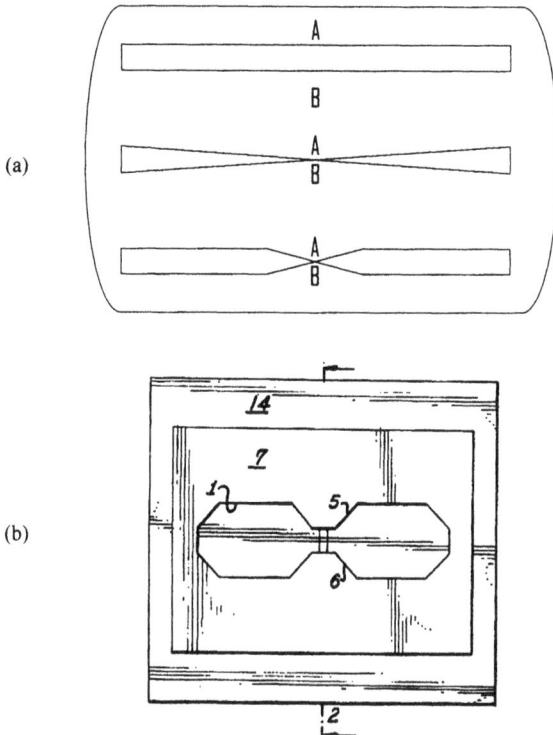

Figure 1.20 (a) Three slot antennas fed across the A-B gap (After [26, p. 557]), and (b) Marié's tapered slot antennas [40].

In each of these frequency-independent antennas, a small-scale portion radiates high frequencies, and a large-scale portion radiates low frequencies. The effective origin, or phase center, of the antenna moves with frequency. Thus, as will be seen in Chapter 2, these antennas are prone to dispersion and not suitable for many UWB applications.

Magnetic slot antennas were also developed in the late 1940s to early 1950s [38, 39, 40]. The tapered-feed concept was soon applied to extend the bandwidth of these antennas. Figure 1.20(a) shows a standard slot antenna and two tapered slot antennas presented by Schelkunoff and Friis. Figure 1.20(b) shows a tapered slot antenna invented by Georges Robert-Pierre Marié [41].

Henning Harmuth's Large Current Radiator (LCR) antenna is a more recent development in magnetic antennas [42]. Ideally, this magnetic antenna looks like a current sheet. Because the sheet will radiate from both sides, designers typically employ a lossy ground plane to limit undesired resonances and reflections. This tends to limit the efficiency and performance of large current radiators. Figure 1.21 shows an LCR antenna.

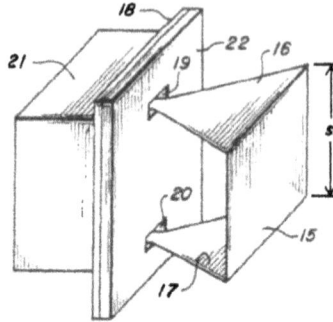

Figure 1.21 Harmuth's LCR antenna [42].

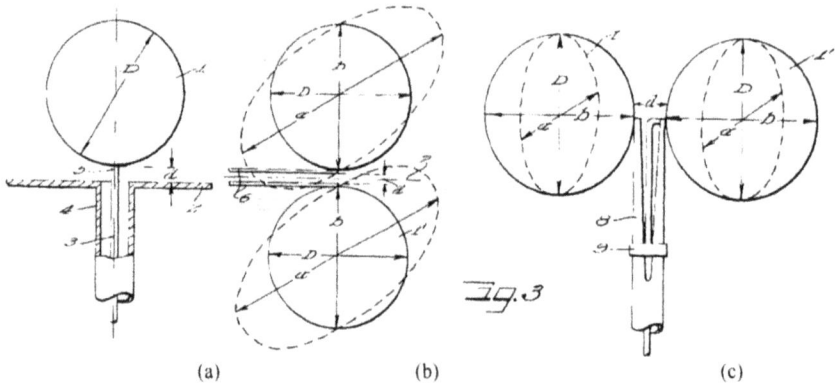

(a)   (b)   (c)

Figure 1.22 (a) Stöhr's spherical monopole, (b) Stöhr's spherical or ellipsoidal dipole, and (c) an alternate embodiment of Stöhr's spherical dipole [43].

Walter Stöhr discovered that the elaborately tapered antenna elements of Lindenblad, Kraus, and others are not essential for UWB antenna performance [43]. Stöhr's relatively simple spherical and ellipsoidal elements work well as either dipole or monopole elements. Figure 1.22(a–c) shows three of Stöhr's designs.

For most practical applications, however, planar antennas are superior to solid antennas. Planar antennas have the advantage of easy, low-cost manufacture on printed circuit boards. In general, a cross-section of a well-performing, solid UWB antenna yields a well-performing planar UWB antenna. Lodge's bow tie antenna is effectively a cross-section of a biconical antenna, for instance. Planar circular elements are thus a natural evolution of Stöhr's spherical elements.

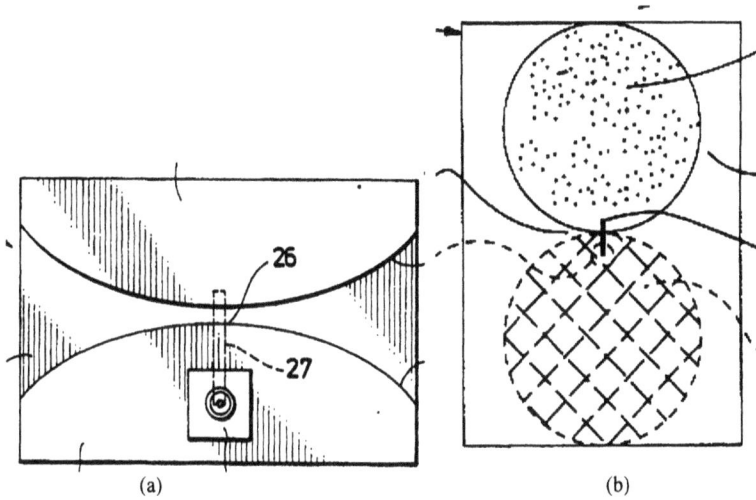

Figure 1.23 (a) The planar notch antenna of Lalezari et al. [44], and (b) the planar circle dipole of Thomas et al. [45].

Farzin Lalezari, Charles Gilbert, and John Rogers introduced the notch, or half-circle, planar antenna [44]. For relatively low frequencies, this antenna is effectively a dipole with semicircular elements. As frequency increases, this antenna transitions to a dual-notch horn mode. The discontinuity at the end of the continuous taper is the most significant limiting factor in this antenna's performance. The reflections from this discontinuity can make a good match more difficult to obtain. The advantage, though, is a lower frequency response than might be obtained by a more continuously tapered element shape. Thus, the design of Lalezari et al. is well suited for applications where an electrically small antenna is desirable. Figure 1.23(a) shows their planar notch antenna.

Mike Thomas and Ronald Wolfson introduced an element shape that is continuously tapered: a circular dipole element [45]. This planar circle dipole antenna is a well-behaved antenna with an excellent match. Figure 1.23(b) shows Thomas's planar circle dipole antenna. Chapter 6 discusses both Lalezari's planar notch antenna and Thomas's circle dipole antenna in more detail.

This section surveyed a variety of UWB antennas from a historical perspective. Hertz's foundation led to Lodge's biconical and bow tie antennas as well as Bose's horn antennas. Marconi applied ideas principally from Hertz and Lodge to create the first commercial radio systems. These early systems were narrowband by design but broadband in practice. As radio electronics became increasingly capable of meeting narrowband design goals, early UWB antennas were largely forgotten in favor of easier-to-build, narrowband designs like thin-wire monopoles and thin-wire loop antennas.

1860s　*Maxwell's Equations formulated*

1865: Maxwell's equations

1880s　*Maxwell's theory proven and refined*

1885: Poynting-Heaviside theory
1887-1888: Hertz discovers radio waves;
uses parabolic reflectors

Practical applications of radio waves

1890s　1892: Crookes speculates about radio
communications

1895: Marconi demonstrates radio system with
square plate antennas

1896: Lodge proposes "syntonic" radio with
biconicals and bow ties.

1897: Bose demonstrates mm-wave
apparatus, including horn antennas

1900s
1901: Marconi claims trans-Atlantic signals

1910s

1912: Titanic signals for help using wireless

1920s

Commercial AM broadcasting begins:
BW ~10 kHz

*Higher frequencies, television and broadband
signals drive renewed interest in UWB antennas*

1930s　1937: Carter rediscovers biconicals, invents
tapered feeds;
"Alpine Horn."

METAL
HORNS

1939-1941: Lindenblad develops TV antenna;
Schelkunoff invents spherical antenna

1940s　1942: King rediscovers
conical horn;
Kraus invents corner
reflector
1945: Kraus invents volcano
smoke antenna

1946: Katzin rediscovers pyramidal horn;
Kandoian invents discone

1948: Brillouin designs coaxial horns

1949: Masters invents
diamond dipole

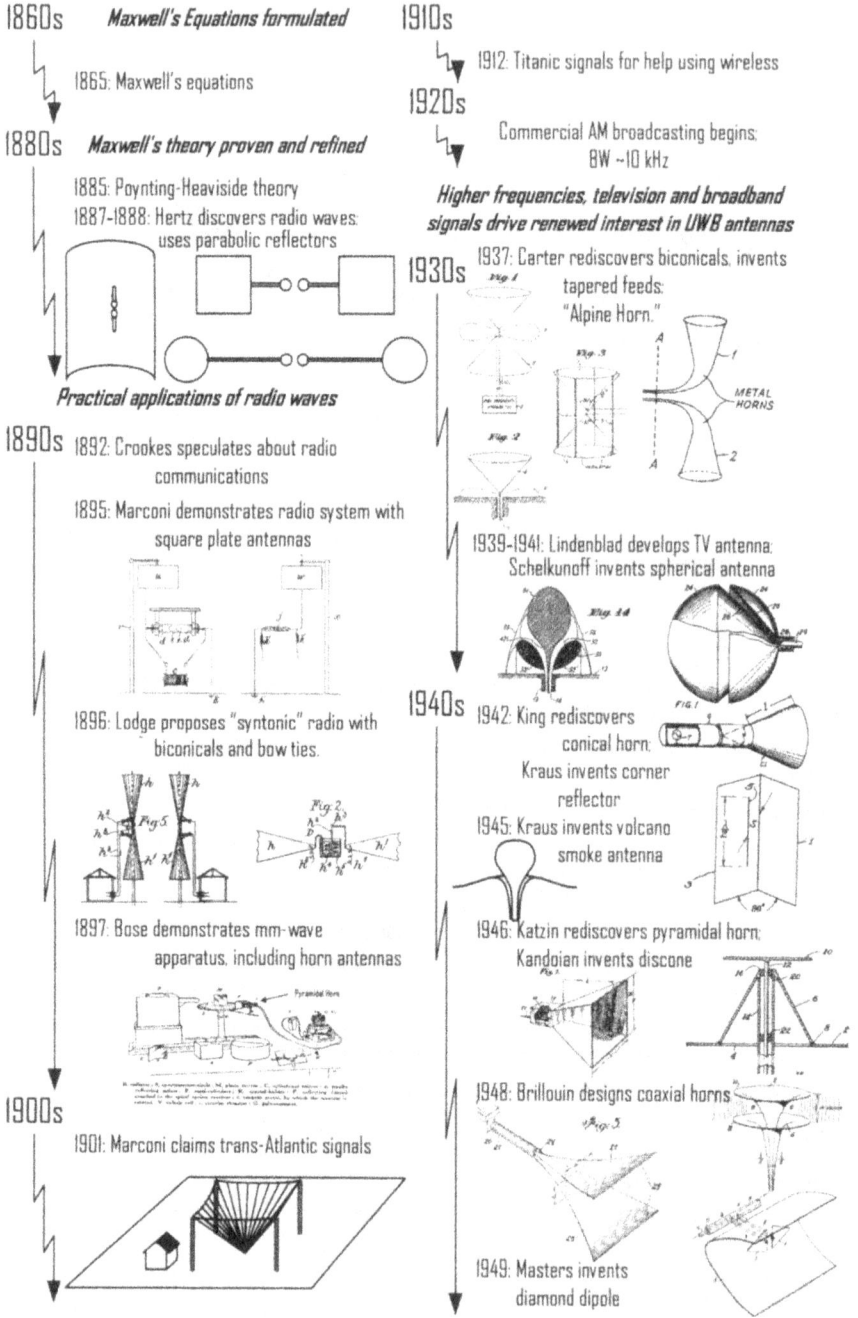

Figure 1.24　Time line of UWB antenna development.

1950s

More interest in practical, manufacturable designs

1952: Schelkunoff and Friis present teardrop biconical, tapered slots, and various conical elements

1957: Rumsey introduces frequency-independent antennas

1958: Lamberty describes square plate monopole

1960s

Ground Plane

Feed

1961: Turner and Turner design the scimitar antenna

1962: Marié invents a stepped slot antenna

1968: Stohr develops ellipsoidal antennas

1970s

1974: Lee proposes ellipsoidal antennas

1980s

1982: Burnside+ Chuang's rolled edge horn debuts

1985: Harmuth introduces the LCR antenna; Nester invents a microstrip notch horn

1989: Lalezari et al. invent a dual-notch antenna/semicircular dipole

1990s

Dawn of "Modern" UWB

1992: Honda et al. describe a circular monopole; Snyder and Peisley invent a rolled edge discone

1994: Thomas et al. invent circle dipoles

Too many modern designs to mention!

2000s

2002: FCC approves 3.1-10.6GHz UWB

Explosion of interest in UWB technology

Gap

Connector

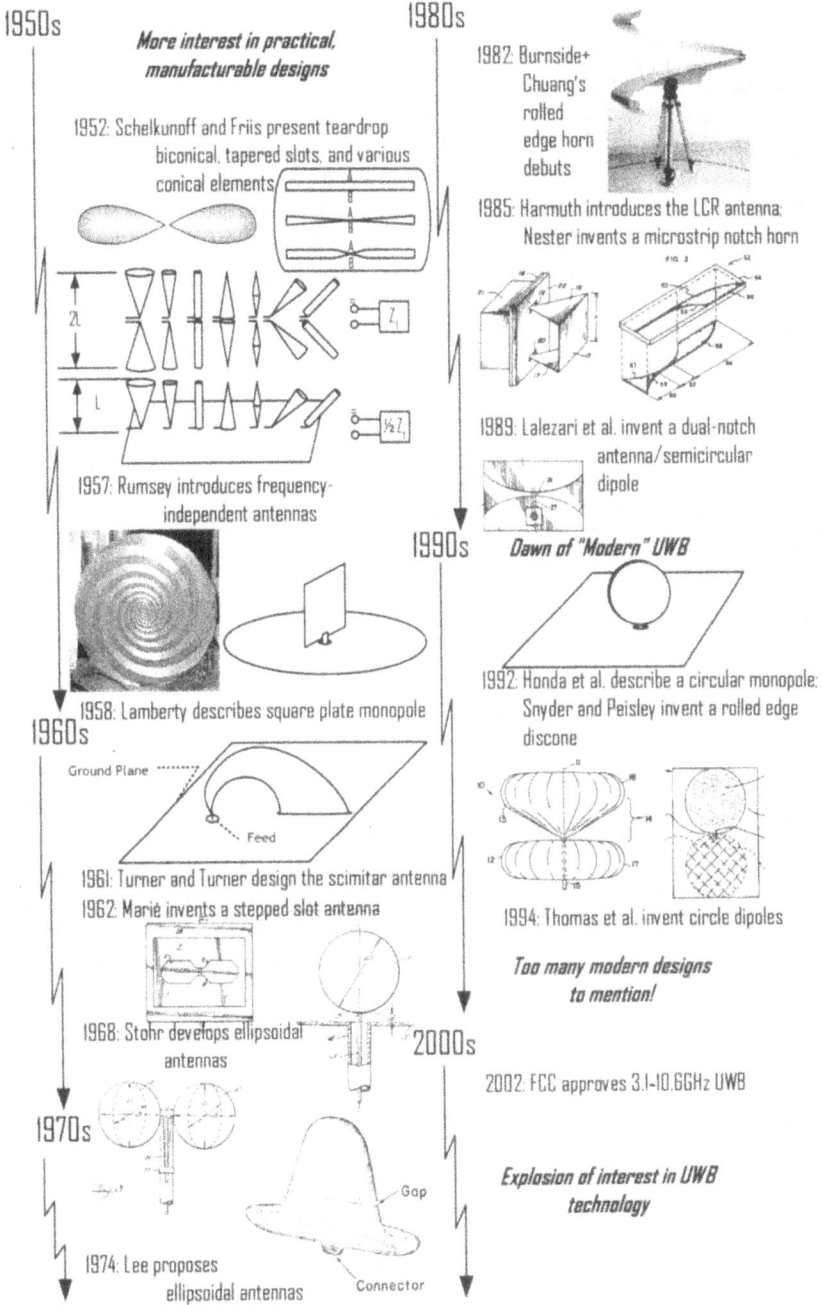

Figure 1.24 Time line of UWB antenna development (continued).

Improvements in radio electronics led to higher and higher frequencies making complicated quarter-wavelength-scale structures more practical. Increasingly, broadband signals, like those under consideration for television, fueled demand for broadband antennas. A new generation of engineers rediscovered bow ties, biconicals, and horn antennas. Moving further in response to advances in radio technology, they developed a new generation of UWB antennas with continuously tapered feeds for improved broadband matches and with thicker, fatter, more bulbous elements for improved broadband performance. In more recent decades, the early solid and surface-of-revolution antennas gave way to planar designs that could be implemented in easy-to-manufacture and inexpensive printed-circuit-board (PCB) layouts. Tapered microstrip feeds and rolled edges were among the later advances.

Figure 1.24 offers a brief graphical overview of the historical development of UWB antennas. The antennas depicted in this overview are either described in this chapter or will be discussed in later chapters.

This brief historical introduction has attempted to introduce a few of the more important and influential designs at the root of modern UWB antenna practice. Although some of these designs represent historical dead ends, others are widely used today. Chapter 6 presents a taxonomy that will revisit UWB antennas in a more rigorous fashion.

1.2   WHAT IS AN ANTENNA?

Section 1.1 presents an extensive gallery of antennas. Having strolled the aisles and gazed at a variety of exotic shapes and structures, we are now in a better position to step back and consider some of the theoretical methods one might apply to understand these antennas: how they work and how to use them. This section provides a top-level introduction to some ways of thinking about antennas in general and UWB antennas in particular. The first and most obvious question deserving of an answer in a book on antennas is, of course, What is an antenna?

An antenna makes an RF wireless device possible. A transmit antenna takes signals from a transmission line, converts them into electromagnetic waves, and broadcasts them into free space, just as a planter scatters or broadcasts seeds [46]. On the other end of the link, a receive antenna collects the incident electromagnetic waves and converts them back into signals. Definitions and thinking about antennas necessarily depend on context. Different ideas and different ways of thinking may be useful in different situations.

Various antenna perspectives are helpful to UWB technologists, RF systems engineers, amateur radio operators, and antenna designers. This section briefly summarizes four of these perspectives. The next four chapters examine these four perspectives at length. Although not necessarily an exhaustive list, these four antenna perspectives are among the most useful when working with UWB antennas.

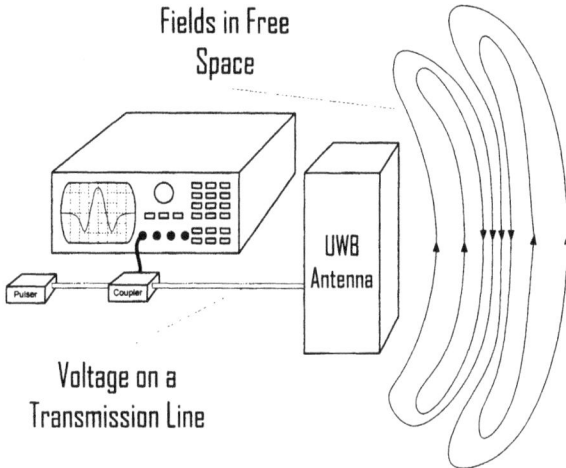

Figure 1.25 An antenna may be thought of as a transducer that couples between signals on a transmission line and electromagnetic radiation in free space.

1.2.1   Antennas as Transducers

The first perspective treats antennas as transducers. For many applications, an antenna may be treated as a black box whose performance is parameterized by such quantities as gain, pattern, polarization, bandwidth, dispersion, and match. An RF systems engineer, for instance, may not need to know how to design an antenna, only how to make good use of an antenna in an RF system.

Chapter 2 presents antennas as transducers coupling between electromagnetic waves in free space and signals on a transmission line. It also surveys basic antenna parameters and examines how typical narrowband antenna parameters like pattern, gain, and matching must be extended to apply to UWB antennas. Figure 1.25 illustrates how an antenna may be treated as a transducer.

1.2.2   Antennas as Transformers

The second perspective treats antennas as transformers. Unlike narrowband antennas, the impedance of a UWB antenna is subject to a significant degree of control through appropriate design. The tapered antenna designs of Lindenblad and Kraus are examples of how a continuously tapered impedance profile can lead to an excellent broadband match. Thomas's circle dipole elements, for instance, cooperate to form a gradually tapered impedance.

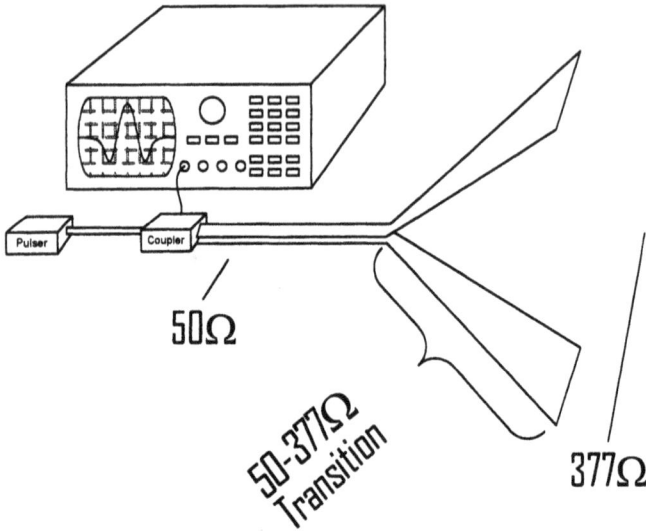

50Ω

50-377Ω Transition

377Ω

Figure 1.26 An antenna may be thought of as a transformer that couples between line impedance and free space impedance.

Chapter 3 examines antennas as transformers coupling between a line impedance and the impedance of free space. The concepts in Chapter 3 will appeal not only to UWB antenna designers but also to RF and systems engineers interested in broadband-matching an antenna to an RF front end. Figure 1.26 depicts how an antenna may be considered as a transformer.

1.2.3   Antennas as Radiators

The third perspective treats antennas as radiators. This classical view of antennas interprets them as current-bearing structures that radiate in a pattern governed by the current geometry. Voltages induce and accelerate charge distributions giving rise to currents. These time varying currents generate radiation fields. The connection between charges and currents and their associated fields is given by Maxwell's equations. Maxwell's equations and this electromagnetic field viewpoint are the subjects of Chapter 4.

Chapter 4 presents some simple rules to estimate and predict time domain radiation patterns from small element UWB antennas. Chapter 4 also includes a more rigorous exploration of time domain electromagnetics as applied to antennas, and discusses the time domain fields of point dipoles. Figure 1.27 shows how an antenna may be treated as a radiator of fields.

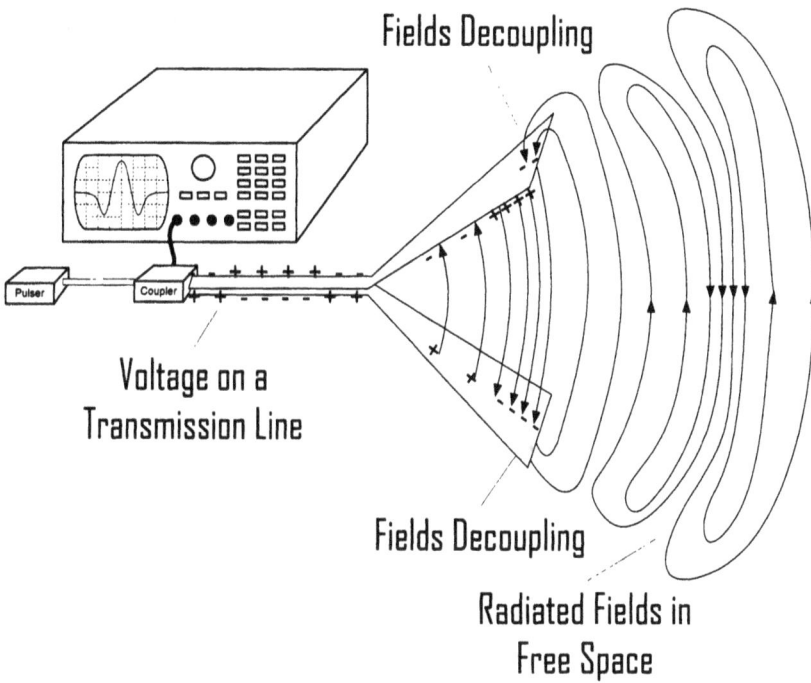

Figure 1.27 An antenna may be thought of as a radiator of electromagnetic fields.

1.2.4 Antennas as Energy Converters

The fourth perspective treats antennas as energy converters. Traditional analysis techniques described in Chapter 4 start from a known antenna geometry and predict such properties as pattern. But the usual electromagnetic analysis is descriptive, not normative. It can describe how a known geometry behaves but does not guide a designer to find new geometries. Understanding electromagnetic energy flow, however, opens the possibility of designing antennas that conform to a desired energy-flow pattern. From this point of view, an antenna may be thought of as a device that converts guided energy into radiation energy with a minimum of reactive energy as a by-product.

Chapter 5 discusses the electromagnetic energy flow around antennas and explains how energy analysis can be used to select antenna elements and to set fundamental limits on the performance of UWB antennas. Figure 1.28 depicts how an antenna may be treated as an energy-conversion device.

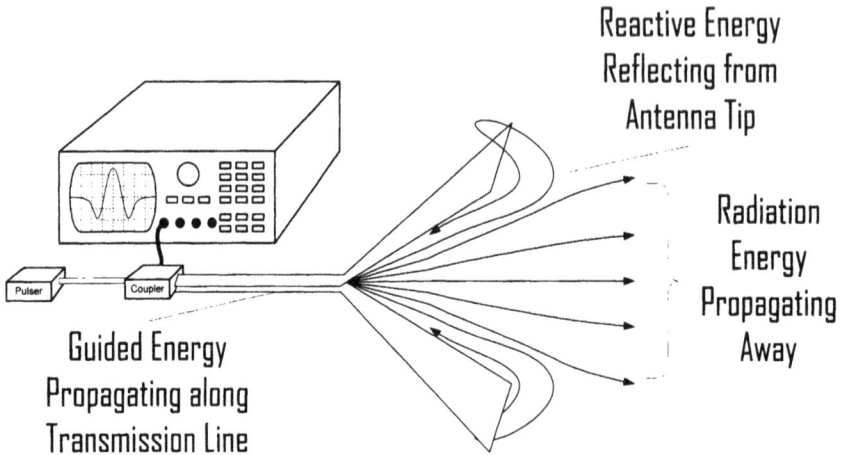

Figure 1.28   An antenna may be thought of as an energy conversion device, converting guided energy to radiation energy with a minimum of reactive energy.

The list of antenna perspectives introduced in this chapter and discussed at length in this book is not meant to be exhaustive and all encompassing. These are perspectives useful for considering antennas in general and particularly UWB antennas. Other perspectives would be valuable for considering other kinds of antennas. For instance, an examination of narrowband antennas might benefit from Schelkunoff's analysis of antennas as leaky resonators. In any event, the antenna perspectives surveyed in this section lay a theoretical framework within which UWB antennas can be understood and discussed.

1.3   UWB ANTENNAS

This chapter presents a historical survey of UWB antennas as well as an outline of some useful perspectives for analyzing antennas. With this foundation, we are in a better position to begin combining the empirical data from the historical survey with the more theoretical and abstract thinking about antennas and how they work from Section 1.2. The present section identifies what distinguishes UWB antennas from antennas in general. Additionally, this section identifies three classes of UWB antennas distinguished by their intended application. Then, this section presents an alternate way to classify UWB antennas according to form and function. Finally, this section reviews some of the system considerations significant to those who wish to design and use UWB antennas.

1.3.1 What Distinguishes a UWB Antenna?

It should come as no surprise that large bandwidth is what distinguishes a UWB antenna from antennas in general. There are two criteria available for identifying when an antenna may be considered ultrawideband. One definition (ascribed to a 1990 Defense Advanced Research Projects Agency (DARPA) report [47]) requires a UWB antenna to have a fractional bandwidth greater than 0.25. An alternate and more recent definition (due to the Federal Communications Commission (FCC) [48]) places the limit at 0.20. Using the fractional bandwidth defined in Chapter 2,

$$bw = 2\frac{f_H - f_L}{f_H + f_L} \geq \begin{cases} 0.25 & DARPA \\ 0.20 & FCC \end{cases} \tag{1.3}$$

where f_H is the upper, or high, end of the antenna's operational band and f_L is the bottom, or low, end of the antenna's operational band. Additionally, the FCC provides an alternate definition whereby a UWB antenna is any antenna with a bandwidth greater than 500 MHz. According to the FCC, the upper and lower ends of the operational band are defined by the points where the radiated power is down 10 dB from its peak level. This FCC definition does not, strictly speaking, define antenna bandwidth because the radiated power also depends on the spectral response of the transmitted power. Chapter 2 discusses the subtleties involved in defining the frequencies that bound a bandwidth.

Historically, there have been at least three different classes of UWB antennas. These classes are based on applications. First is the "DC-to-daylight" class. These antennas are designed to have maximal bandwidth. Typical applications include ground penetrating radars (GPRs), field measurement or electromagnetic compatibility (EMC), electromagnetic weapons, impulse radars, and covert communications systems. The design goal of these antennas is to grab as much spectrum as possible. Second is the "multinarrowband" class. These are antennas designed as scanner or signal-intelligence antennas for receiving or detecting relatively narrowband signals across broad swaths of frequencies. The design goal of multinarrowband antennas is similarly to grab as much spectrum as possible but to only use small sub-bands at any given time. Third is what might be called "modern" UWB antennas. These are antennas designed for use in conjunction with the approximately 3:1 bandwidth, 3.1–10.6-GHz UWB systems as authorized by the FCC (Part 15, Subpart F). The bandwidth requirements for a modern UWB antenna are much narrower than for DC-to-daylight antennas.

These modern UWB antennas have certain implications that distinguish them from the other more traditional classes of UWB antennas. First, instead of trying to grab maximal bandwidth, these modern UWB antennas must live within a certain spectral mask. In this context, excessive bandwidth degrades system response and is counterproductive. Second, unlike a multinarrowband antenna, a modern UWB antenna potentially uses much, if not all, of its bandwidth at the same time. Thus, a modern UWB antenna must be well behaved and consistent across the antenna's operational band. Its properties include pattern, gain, matching, and a requirement for low or no dispersion. A wide variety of antennas meet the demands of modern UWB systems. Section 1.3.2 offers a brief overview.

1.3.2   A Taxonomy of UWB Antennas

UWB antennas may be categorized according to form and function into four different classes:

- **Frequency-independent antennas**: These antennas rely on a variation in geometry from a smaller-scale portion to a larger-scale portion. The smaller-scale portion contributes higher frequencies while the larger-scale portion contributes lower frequencies. Because the effective source of the radiated fields varies with frequency, these antennas tend to be dispersive. Examples of frequency-independent antennas include spiral, log periodic, and conical spiral antennas.

- **Small-element antennas**: These antennas tend to be small, omnidirectional antennas well suited for commercial applications. Examples of small-element antennas include Lodge's biconical and bow tie antennas, Masters' diamond dipole, Stohr's spherical and ellipsoidal antennas, and Thomas's circle dipole.

- **Horn antennas**: A horn antenna is an electromagnetic funnel concentrating energy in a particular direction. Horn antennas tend to have high gain and relatively narrow beams. Horn antennas also tend to be larger and bulkier than small-element antennas. These antennas are well suited for point-to-point links or other applications where a narrow field of view is desired. Examples include Bose's original horn antennas, and the coaxially tapered horns of Brillouin.

- **Reflector antennas**: A reflector antenna also concentrates energy in a particular direction. Like horn antennas, reflector antennas tend to have high gain and are relatively large. Reflector antennas tend to be structurally simpler than horn antennas and are easier to modify and adjust by manipulating the antenna feed. Hertz's original parabolic cylinder reflector is an example.

Many of these antenna types are well known from conventional antenna practice. Others are more obscure. This book will focus on those specific antenna designs best suited for modern UWB systems. Chapter 6 presents a more detailed overview of UWB antennas, organized according to this taxonomy.

1.3.3   UWB Device and Systems Considerations

Ultimately, a UWB antenna is worthless unless it can be successfully integrated with a UWB device. Traditional antenna practice takes an antenna's impedance as a given and designs a matching network to couple the antenna to the front end of an RF system. Matching over an ultrawide band of frequencies is no trivial task, so UWB device designers no longer have this luxury.

Instead, a UWB antenna needs to be designed so as to meet specific matching and response goals as a function of frequency. The performance of a UWB antenna needs to be uniform and consistent over the entire operational band. As a result, UWB device design is a holistic process in which an antenna designer and an RF engineer must cooperate to ensure that the overall device (antenna + RF front end) meets a desired spectral mask and other performance goals.

Further, a UWB device will make the cash register ring only if it is part of an overall system that meets the needs of potential customers. UWB devices must interact with each other in a network. UWB devices must also coexist with other narrowband devices and services occupying the same bandwidth. Modification of UWB antennas to introduce frequency notches or other filtering behavior is also an important consideration. In addition, the directionality of UWB devices can play a significant role in meeting overall system goals.

1.4   CONCLUSION

The aim of this chapter was to provide an introduction to UWB antennas and an overview of the book. UWB antennas live in a fascinating realm of RF engineering in which science and art combine to create results that might be confused with magic.

UWB antennas trace their roots back to the earliest days of radio, yet remain a field of active interest and exciting developments. Analysis and application of UWB antennas benefit from a variety of perspectives. System designers may think of UWB antennas as transducers coupling between signals and fields. Understanding a few simple rules of the road allows UWB antennas to be taken out for a spin without ever having to look under the hood. RF engineers and antenna designers alike need to be able to match antennas to an RF front end. Thinking of UWB antennas as impedance transformers is helpful in this context. The classical approach to antennas, using electromagnetic analysis to understand how antenna geometries and currents give rise to fields, is essential to the hard core antenna designer. A newer approach—looking at electromagnetic energy flow around antennas—helps designers select optimal element shapes as well as understand fundamental antenna limits. A wide variety of UWB antennas are available to meet device and system goals. But antennas are just one small piece of an overall system design. A system architect must integrate antennas to devices to create a viable product. A system architect must further combine devices within a system in order to create a viable solution to real-world problems. Perhaps the most significant advice this book can offer is that UWB antennas necessitate a holistic approach to the overall design process. All these topics and more are addressed in the following chapters.

Endnotes

[1]   Clarke, Arthur C., *Profiles of the Future: An Inquiry Into the Limits of the Possible*, London: Indigo, 1999. This dictum is often referred to as "Clarke's Third Law."

[2]   Heaviside, Oliver, *Electromagnetic Theory*, Vol. 1, London: "The Electrician" Printing and Publishing Company, 1893, p. 10.

[3]   Randall, John Herman, Jr., *Aristotle*, New York: Columbia University Press, 1960, p. 274.

[4]   Hertz, Heinrich, *Electric Waves*, London: Macmillan and Co., 1893. Hertz is justly acclaimed for the accomplishments documented in *Electric Waves*. This work demonstrates Hertz's painstaking attention to detail and his mastery of both Maxwell's theory as well as the experimental techniques necessary to create, detect, and manipulate radio waves. Hertz is one of very few physicists who were simultaneously brilliant theoreticians and accomplished experimentalists.

[5]   Crookes, William, "Some Possibilities of Electricity," Fortnightly Review, February 1, 1892, pp. 174–176. This article is available online at: earlyradiohistory.us/1892fort.htm at the time of this writing.

[6]   Aitken, Hugh G. J., *Syntony and Spark: The Origins of Radio*, Princeton: Princeton University Press, 1985. This excellent book, now regrettably out of print, provides a comprehensive history of the earliest days of radio. Aitken explains technical history and early radio engineering in a clear and understandable fashion.

[7]   Aitken, Hugh G. J., *The Continuous Wave: Technology and American Radio, 1900–1932*, Princeton: Princeton University Press, 1985. Picking up where *Syntony and Spark* leaves off, *The Continuous Wave* documents the evolution of radio from the early spark-gap technology to more modern, narrowband techniques. This book is also, unfortunately, no longer in print.

[8]   Lewis, Tom, *Empire of the Air: The Men Who Made Radio*, New York, HarperCollins, 1991. This book focuses on the lives and careers of the three individuals most responsible for taking Marconi's point-to-point radio and making the concept of broadcasting a commercial success: Lee DeForest (who invented the vacuum tube), David Sarnoff (who led RCA in its commercialization of radio), and Edwin Howard Armstrong (the brilliant electrical engineer who invented the heterodyne receiver as well as practical FM). Lewis's book was made into an outstanding documentary by Ken Burns, now available on DVD.

[9]   Schantz, Hans, "A brief history of UWB antennas," IEEE Aerospace and Electronic Systems Magazine, Vol. 19, No. 4, April 2004, pp. 22-26. The present chapter includes some material originally presented in this article. This article reproduces material originally presented in a conference paper: Schantz, Hans, "A brief history of UWB antennas," 2003 IEEE Conference on Ultra Wideband Systems and Technologies, Nov. 16–19, 2003, pp. 209–213.

[10] Aitken, Hugh G. J., *Syntony and Spark*, Op. Cit., p. 95.

[11] Hertz, Heinrich, *Electric Waves*, Op. Cit., pp. 172–173, 183, 185.

[12] Lodge, Oliver, "Electric telegraphy," U.S. Patent 609,154, August 16, 1898.

[13] Emerson, Darrel T., "The work of Jagadis Chandra Bose: 100 years of millimeter wave research," IEEE Transactions on Microwave Theory and Techniques, Vol. 45, No. 12, December 1997, pp. 2267-2273. See also www.tuc.nrao.edu/~demerson/bose/bose.html. The author is indebted to Dr. Emerson for his insightful presentation and for permission to reproduce Figures 1.1(c) and 1.6(a).

[14] Masini, Giancarlo, *Marconi*, New York: Marsilio Publishers, 1976, p. 77. This is an excellent biography, translated from Italian, and the author's primary source of Marconi information.

[15] Marconi, Guglielmo, "Transmitting electrical signals," U.S. Patent 586,193, July 13, 1897.

[16] Belrose, John S., "A radioscientist's reaction to Marconi's first transatlantic wireless experiment—revisited," 2001 IEEE Antennas and Propagation Society International Symposium. Boston, Massachusetts. July 8-13, 2001, Vol. 1, pp. 22-25. In fact, recent investigation casts doubt on Marconi's ability to have received a trans-Atlantic Morse code "S" signal on December 12, 1901, as he claimed.

[17] Ghirardi, Alfred A., *Radio Physics Course*, New York: Farrar & Rinehart, Inc., 1930, p. 330.

[18] Carter, Philip S., "Shortwave antenna," U.S. Patent 2,175,252, October 10, 1939.

[19] Carter, Philip S., "Wide band, shortwave antenna and transmission line system," U.S. Patent 2,181,870, December 5, 1939.

[20] Lindenblad, Nils E., "Wide band antenna," U.S. Patent 2,239,724, April 29, 1941.

[21] Lindenblad, Nils E., et al., RCA Review, April 1939.

[22] Forbes Magazine, "Television Looks to the Future," January 15, 1945. The cover of this issue prominently displays Lindenblad's television antenna.

[23] Schelkunoff, Sergei A., "Ultra shortwave radio system," U.S. Patent 2,235,506, 1941.

[24] Schelkunoff, Sergei A., *Advanced Antenna Theory*, New York: John Wiley and Sons, 1952, p. 160.

[25] King, R. W. P., et al., *Transmission Lines, Antennas, and Waveguides*, New York: McGraw Hill, 1945, p. 107.

[26] Schelkunoff, Sergei A., and Harald Friis, *Antennas: Theory and Practice*, New York: John Wiley and Sons, 1952, p. 424.

[27] Schelkunoff, Sergei A., and Harald Friis, Op. Cit., pp. 314-319.

[28] Paulsen, Lee, et al., "Recent investigations on the volcano smoke antenna," 2003 IEEE Antennas and Propagation Society International Symposium. Columbus, Ohio. June 22-27, 2003, Vol. 3, pp. 845-848.

[29] Brillouin, Leon N., "Broadband antenna," U.S. Patent 2,454,766, November 30, 1948.

[30] King, Archie P., "Transmission, radiation, and reception of electromagnetic waves," U.S. Patent 2,283,935, May 26, 1942.

[31] Katzin, Martin, "Electromagnetic horn radiator," U.S. Patent 2,398,095, April 9, 1946.

[32] Kandoian, Armig G., "Three New Antenna Types and Their Applications," Proceedings of the IRE, February 1946, pp. 70W-75W.

[33] Kandoian, Armig G., "Broadband antenna," U.S. Patent 2,368,663, 1945.

[34] Brown, George H. and O.M. Woodward, "Experimentally Determined Radiation Characteristics of Conical and Triangular Antennas," RCA Review, Vol. 13, December 1952, pp. 425-452.

[35] Masters, Robert W., "Antenna," U.S. Patent 2,430,353, November 4, 1947.

[36] Lamberty, B. J., "A Class of Low Gain Broadband Antennas," *1958 IRE Wescon Convention Record*, August 1958, pp. 251-259.

[37] Rumsey, Victor H., "Frequency-Independent Antennas," IRE National Convention Record, pt. I, 1957, pp. 114-118.

[38] Booker, H. G., "Slot Aerials and their Relation to Complementary Wire Aerials (Babinet's Principle)," IEE Journal (London), Vol. 93, Part IIIA, No. 4, 1946, pp. 620-626.

[39] Riblet, H. J., "Microwave Omnidirectional Antennas," Proceedings of the IRE, Vol. 35, May 1947, pp. 474-478.

[40] Lindenblad, Nils E., "Slot Antennas," Proceedings of the IRE, Vol. 35, December 1947, pp. 1472-1479.

[41] Marié, Georges Robert-Pierre, "Wide band slot antenna," U.S. Patent 3,031,665, April 24, 1962.

[42] Harmuth, Henning, "Frequency-independent shielded loop antenna," U.S. Patent 4,506,267, March 19, 1985.

[43] Stöhr, Walter, "Broadband ellipsoidal dipole antenna," U.S. Patent 3,364,491, January 16, 1968.

[44] Lalezari, Farzin, Charles E.Gilbert, and John M. Rogers, "Broadband notch antenna," U.S. Patent 4,843,403, June 27, 1989.

[45] Thomas, Mike, and Ronald I.Wolfson, "Wideband arrayable planar radiator," U.S. Patent 5,319,377, June 7, 1994.

[46] In fact, the term "broadcast" originally referred to the spreading of seed, before it was applied to and preempted by description of radio transmissions.

[47] Foster, Charles, et al., "Assessment of Ultra-Wideband (UWB) Technology," IEEE Aerospace and Electronic Systems Magazine, November 1990, pp. 45-49. This is the classical citation for the DARPA $bw = 0.25$ standard, but I found no such reference therein.

[48] U.S. 47 C.F.R. Part 15 Subpart F §15.503d Ultra-Wideband Operation (October 1, 2003 edition), p. 767. "...a fractional bandwidth equal to or greater than 0.20 or ... a UWB bandwidth equal to or greater than 500 MHz, regardless of the fractional bandwidth." See endnote 3 of Chapter 2 for additional detail.

Chapter 2

Antennas as Transducers

An understanding of the mechanism by which energy is radiated from a circuit and the derivation of equations for expressing this radiation involve conceptions which are unfamiliar to the ordinary engineer.

Frederick Emmons Terman, 1932

Much has changed since Frederick Emmons Terman's day, when the process of electromagnetic radiation was an esoteric subject not of interest to the "ordinary engineer" [1]. Even today, however, one need not delve into the technical details of how antennas work in order to make good use of them. In many applications, an antenna may be thought of as a transducer that couples between signals on a transmission line and radio waves in space. From this point of view, antennas are black boxes whose properties are defined by such parameters as bandwidth, pattern, gain, and matching, among others. The transducer viewpoint is valuable for anyone who wants to know how to use an antenna and what an antenna does but is not interested in the details of how an antenna works or how to design one.

Traditionally, engineers study antennas from the "frequency domain" point of view. The frequency domain point of view breaks down physical processes into sine waves of infinite duration—with no particular beginning or end. The frequency domain further assumes "steady-state," or time average, behavior. Thus, initiates in frequency domain analysis generally speak of antennas as radiating "power." The frequency domain is best suited for studying single-frequency or relatively narrowband systems operating in a steady-state mode.

An alternate way of looking at antennas is from the "time domain" point of view. The time domain point of view treats physical processes as transient excitations of finite duration—with a definite beginning and a definite end. The time domain assumes transient, or time-varying, behavior. Time domain devotees think of an antenna in terms of radiating "energy." The time domain is best suited for studying broadband, UWB, or other systems that operate in a transient fashion.

Mathematically, a frequency domain analysis and a time domain analysis are exactly equivalent, and (properly used and interpreted) both kinds of analyses will yield the same answer. When looking at UWB systems however, time domain analysis generally yields a far cleaner, easier-to-understand and easier-to-interpret answer than does a corresponding frequency domain analysis.

A UWB engineer needs to be conversant in both the time domain and frequency domain, able to switch seamlessly from one realm to the other as the nature of the problem demands. For instance, a UWB engineer may think of an antenna as radiating energy, but must also implicitly recognize that the radiated energy is the time integral of the radiated power. F. Scott Fitzgerald claimed that "the test of a first-rate intelligence is the ability to hold two opposed ideas in the mind at the same time, and still retain the ability to function" [2]. UWB requires a talent much more rare and far more useful: the ability to hold two complementary ideas in the mind at the same time all the while differentiating between them and properly applying them as the situation requires. Chapter 4 will delve deeper into the similarities, differences, and connections between time domain and frequency domain analysis.

This chapter will treat a UWB antenna as a black box, discussing basic antenna properties and how they apply in the UWB limit. These properties include the bandwidth of an antenna, waveform dispersion, antenna pattern, directivity, gain, aperture, field of view, polarization, and matching. These basic antenna properties are traditionally understood and explained from a frequency domain point of view. Thus, this chapter will introduce antenna properties from a perspective consistent with a narrowband, frequency domain point-of-view, but it will also explain how these concepts apply to UWB systems.

2.1 BANDWIDTH

A UWB antenna is distinguished by its large bandwidth. Thus, considering bandwidth is an appropriate place to begin an examination of a UWB antenna's properties. Bandwidth (*BW*) is simply the difference between the upper and lower operating frequencies (f_H and f_L, respectively)

$$BW = f_H - f_L \qquad (2.1)$$

There are various ways to express bandwidth, including fractional or percent bandwidth. These measures of relative bandwidth require a calculation of center frequency—either the arithmetic or geometric average of the upper and lower frequencies. There are also various ways to define the upper and lower frequencies of a UWB antenna or of a UWB system. Spectrum managers often use a variety of different bandwidth definitions, and a careful engineer needs to be alert to abuses. This section will discuss how to calculate bandwidth, center frequency, and relative bandwidth. Then it will consider a few of the many ways to define antenna bandwidth. Finally it looks at the FCC's definition of radiated bandwidth.

2.1.1 Calculating Bandwidth

The large bandwidths of UWB antennas may be described in a variety of ways. One may consider the ratio of the upper frequency to the lower frequency, $f_H : f_L$. Thus, a UWB system operating from f_L = 200MHz to f_H = 1GHz may be called a 5:1 bandwidth system. Certain ratios have special names. A 2:1 band is an "octave," and a 10:1 band is a "decade."

In addition, the bandwidth of a system is often described relative to the center frequency (f_C). Often, the center frequency is defined as the arithmetic average of the upper and lower frequencies

$$f_C = \tfrac{1}{2}(f_H + f_L) \qquad (2.2)$$

An arithmetic average yields the center frequency when frequency is considered on a linear scale. An alternate definition of center frequency involves the geometric average

$$f_C = \sqrt{f_L f_H} \qquad (2.3)$$

A geometric average yields the center frequency when frequency is considered on a logarithmic scale. For instance, the center frequency of an f_L = 1MHz to f_H = 100MHz antenna is f_C = 50.5MHz if treated arithmetically and f_C = 10MHz if treated geometrically. The geometric average is less commonly used to determine center frequency, so the arithmetic average should be assumed unless otherwise specified. Although tradition leans toward the arithmetic average, the geometric average is the more rigorous and meaningful measure according to fundamental antenna physics (as explained in Section 5.5.3).

The fractional bandwidth (bw) of a system is the ratio of the bandwidth to the center frequency

$$bw = \frac{BW}{f_C} \qquad (2.4)$$

Alternatively, fractional bandwidth may be defined on a percentage basis

$$bw\% = \frac{BW}{f_C} 100\% \qquad (2.5)$$

Using the arithmetic average definition of center frequency, a UWB system operating from f_L = 200MHz to f_H = 1GHz has a fractional bandwidth

$$bw = \frac{BW}{f_C} = \frac{f_H - f_L}{f_C} = 2\frac{f_H - f_L}{f_H + f_L} = 1.33 \qquad (2.6)$$

and a percent fractional bandwidth of $bw\% = 133$. Using the geometric average definition of center frequency, a UWB system operating from $f_L = 200\text{MHz}$ to $f_H = 1\text{GHz}$ has a fractional bandwidth

$$bw = \frac{BW}{f_C} = \frac{f_H - f_L}{\sqrt{f_L f_H}} = \frac{800}{447.2} = 1.79 \qquad (2.7)$$

and a percent fractional bandwidth of $bw\% = 179$. Since the geometric definition of center frequency always yields a smaller center frequency than the arithmetic average, fractional bandwidths calculated using the geometric definition are always larger than their arithmetic counterparts.

2.1.2 Determining Antenna Bandwidth

If the upper and lower operating frequencies are known, then characterizing the bandwidth is a trivial task. The subtlety and difficulty lie in how those frequencies are defined. The ways in which bandwidth can be defined include impedance, pattern, gain, and radiated bandwidths. A holistic approach that takes into account all of the properties of an antenna important to a particular application is the preferred way to describe antenna bandwidth.

One approach uses the impedance bandwidth of an antenna. Consider a particular impedance goal such as a -10dB S_{11}. The upper and lower operating frequencies may be defined as the endpoints of the frequency range across which the antenna meets or exceeds the impedance goal. The upper operating frequency of many UWB antennas is limited only by the frequency dependence of the feed mechanism. Thus, an antenna may still accept and radiate energy long after pattern variations or changes in other properties have rendered the antenna's performance no longer acceptable. Further, an impedance bandwidth may be deceiving. Just because an antenna accepts energy, that does not necessarily mean energy radiates away. The subject of matching and impedance will be revisited in Section 2.5. From the point of view of fundamental antenna physics, the bandwidth of a UWB antenna with negligible resistive losses is usually taken as the -3dB or half power points (see Section 5.5.3).

An alternate method for defining antenna bandwidth utilizes the "gain bandwidth" of an antenna. For a constant gain antenna, the upper and lower operating frequencies may be taken as those points where the gain falls -3dB or -10dB below the in-band constant gain. This approach also suffers from some problems. The gain of many antennas increases as a function of frequency. For instance, the gain of a constant aperture antenna varies as f^2. Thus, gain increases 6dB per octave. A two-octave antenna thus experiences a 12dB increase in gain across its operating band. An arbitrary fixed gain cutoff becomes difficult to justify for such an antenna.

Still another method employs pattern bandwidth. With this approach, the upper and lower frequencies are those between which an antenna's pattern meets a particular specification. For instance, one may desire an omnidirectional antenna in the azimuthal plane. The ends of the antenna's operating bandwidth may be taken as those frequencies where gain varies by more than 3dB in the azimuthal plane. Here again, this may not accurately reflect changes in other parameters, such as matching, that may render the antenna unsuitable for use.

A holistic approach considers whichever antenna properties are significant in a particular application. In practice, a UWB system will place demands on an antenna for matching, gain, pattern, spectral response, and other properties. Therefore, the range of frequencies over which the antenna meets desired specifications defines its bandwidth.

2.1.3 Radiated Bandwidth

In the context of radiated power, the definition of bandwidth is more clear cut. Recently, the FCC issued a report and order that defines a UWB system in terms of the −10dB power bandwidth. Thus, to the FCC, the upper and lower frequencies are those where the radiated spectral power density is −10dB down from the center frequency (see Figure 2.1). The FCC uses the arithmetic definition of bandwidth. For a UWB system, the FCC requires a −10dB bandwidth greater than 500 MHz, or a fractional bandwidth greater than 0.2 where fractional bandwidth is defined as in (2.6) [3]. In the example of Figure 2.1, f_L = 3.05GHz, f_C = 4.00GHz, f_H = 4.95GHz, BW = 1.90GHz, and bw = 0.475.

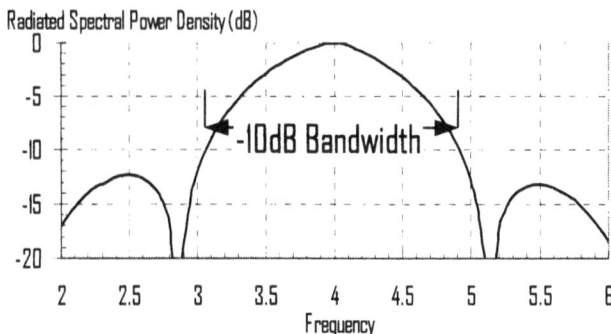

Figure 2.1 The FCC defines radiated bandwidth in terms of the −10dB points of the received spectral density.

2.2 DISPERSION

The large bandwidth of a UWB antenna introduces another complication: signal dispersion. Dispersion is the stretching out of a UWB signal waveform into a longer, more distorted waveform. Alternatively, dispersion is a variation in a UWB signal waveform as a function of look angle. The effective origin of signals from an antenna is sometimes referred to as the "phase center." If the phase center moves as a function of frequency, the resulting radiated waveform will be dispersive [4].

Where this dispersion occurs in a controlled and predictable fashion, it can be possible to compensate for the dispersion [5]. In a point-to-point link, for instance, the dispersion remains the same. Thus, it is possible to construct a filter to "undo" the dispersive effects of an antenna. Unfortunately, dispersion will tend to be different at different look angles, making compensation a challenge. Although it is possible to quantitatively describe dispersion [6], this section will look at dispersion from a more qualitative viewpoint.

2.2.1 Example of a Dispersive Antenna

Classical frequency-independent antennas rely on variations in geometry to obtain their broadband behavior. A smaller-scale portion of a frequency-independent antenna radiates high frequency components and a larger-scale portion radiates lower frequency components of a signal. Because the phase center moves as a function of frequency, frequency-independent antennas radiate dispersed signals.

For instance, consider a log spiral antenna. Figure 2.2 shows a 1–11GHz log spiral antenna. This antenna is fed from the tip of the cone where the spiral has a smaller scale. Lower-frequency components must propagate to the larger-scale structure at the base of the antenna before they can be radiated. Figure 2.3 shows the result of this physical structure on a radiated signal. The transmit antenna accepts a transmitted impulse voltage signal (left) at its terminals. A receive antenna then yields a received impulse voltage signal (right) at its terminals. The dispersion of the log spiral antennas used for both transmit and receive results in a dispersed receive signal. This dispersion is clearly manifest in two respects. First, the received signal has a temporal extent over twice as long as the transmit signal. Second, the received signal shows a distinct "chirp." The earlier portion of this signal exhibits relatively high-frequency content with a shorter time duration between zero crossings. The later portion of the received signal exhibits relatively low-frequency content with a longer time duration between zero crossings.

A third disadvantage of dispersion is not immediately evident in Figure 2.3's boresight response. Because the phase center moves as a function of frequency, the temporal extent and "chirpiness" of the received signal will vary as a function of look angle. A different response will be obtained for different look angles relative to the antenna.

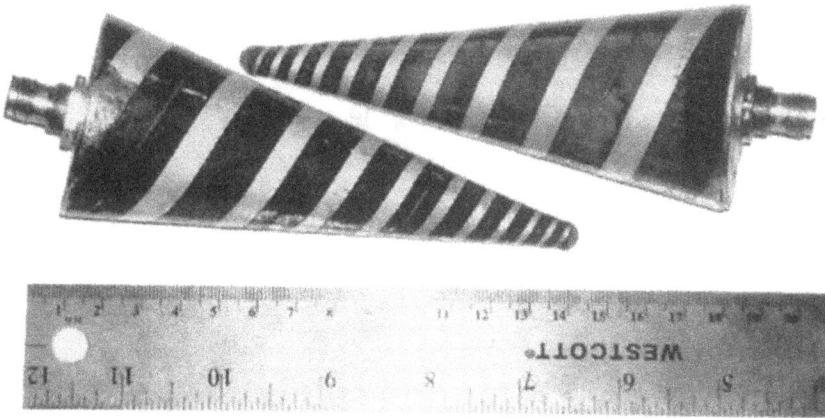

Figure 2.2 A pair of 1–11GHz log spiral antennas.

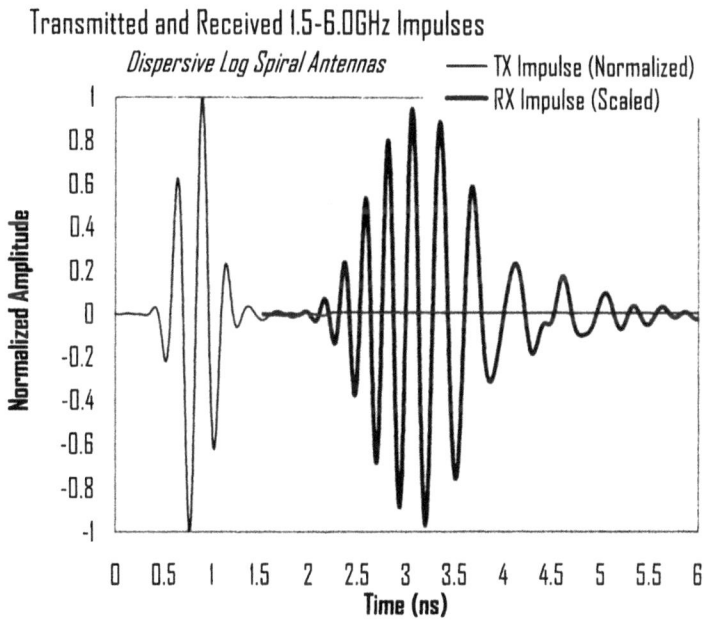

Figure 2.3 Time domain response of a pair of log spiral antennas.

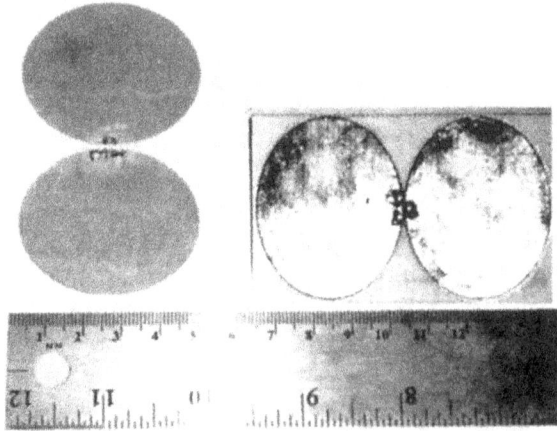

Figure 2.4 A pair of 1.5–6.0GHz planar elliptical dipole antennas.

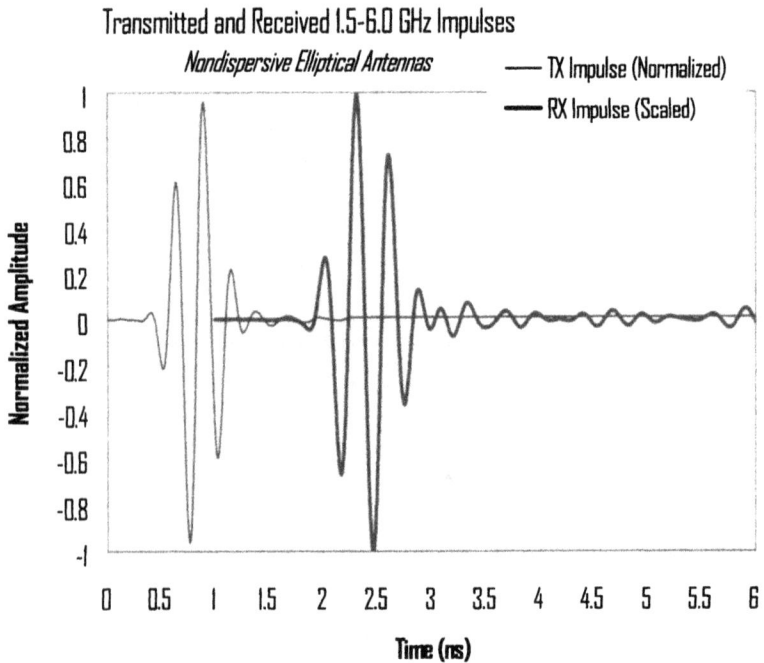

Figure 2.5 Time domain response of a pair of log spiral antennas.

Certainly, it may be possible to compensate for dispersion [3]. Nevertheless, good system performance is easier to achieve using UWB antenna elements that are as nondispersive as possible. Chapter 1 discusses the discovery of frequency-independent antennas from a historical perspective, and Chapter 6 provides a brief discussion of both frequency-independent and fractal antennas.

2.2.2 Example of a Nondispersive Antenna

Figure 2.4 shows nondispersive planar elliptical dipole antennas [7]. This particular antenna operates from 1.5GHz to 6.0GHz. These antennas have a 3.8cm minor axis and a 4.8cm major axis (a 1.25:1 axial ratio). A network analyzer transmits a 1.5GHz–6.0GHz impulse from one planar elliptical dipole antenna. The other planar elliptical dipole antenna receives this impulse. Figure 2.5 depicts the radiated and received impulses. Again, the received signal is normalized to be as large as the original transmitted signal for ease of comparison. Very little waveform distortion occurs between these antennas. In fact, the received signal is almost an exact copy of the original transmitted signal, only inverted.

Chapter 4 explains how the signal radiated from a UWB dipole is the second time derivative of the signal incident at the antenna terminals. In the frequency domain, a time derivative corresponds to a multiplication by $j\omega$. If the bandwidth is only an octave or two (typical of most modern UWB systems), the variation in angular frequency ($\omega = 2\pi f$) is negligible and multiplication by j is equivalent to a 90° rotation. A double time derivative thus implies $j^2 = -1$, or in other words a 180° rotation or a phase inversion. Of course, inverting one of the antennas in a link will undo the inversion from the double derivative and yield a received waveform (to a reasonable approximation) identical to the signal incident at the antenna terminals. This approximation is reasonable, however, only when the antennas in question are well-designed, nondispersive antennas with uniform and consistent behavior across the operational band.

A small, compact UWB antenna element tends to be nondispersive. Such antennas have a fixed phase center and radiate a compact, mostly nondispersive waveform. A nondispersive antenna is preferred for UWB applications. Typical propagation environments, particularly ones heavy in multipath with large delay spreads, tend to distort waveforms and degrade signals. To resolve signals in these environments, nondispersive antennas are useful.

2.2.3 Angular Dependence of Dispersion

Even a relatively small and compact antenna may exhibit variations in waveform as a function of angle. For instance, consider the behavior of a planar loop antenna excited by an impulse signal [8]. A first signal radiates from the near side of the antenna. A second signal radiates from the far side of the antenna. The far signal is delayed relative to the near signal by the additional amount of time required for the signal to traverse the antenna's diameter. Figure 2.6(a) illustrates this configuration.

If this were a fixed delay, then the waveforms radiated in any given direction would be more or less identical. Delay also depends upon the distance on the antenna along which the signal must propagate from the feed before it radiates. For small angles where the direction of radiation is close to the feed, the near signal has very little distance to propagate to the effective source of the radiated signal. The large signal, however, must traverse a longer path on the antenna. For large angles where the direction of radiation is far from the feed, the situation is reversed: the near-side signal has to travel much further than the close side. The result is a superposition between two waveforms with a relative delay that depends upon the angle of radiation.

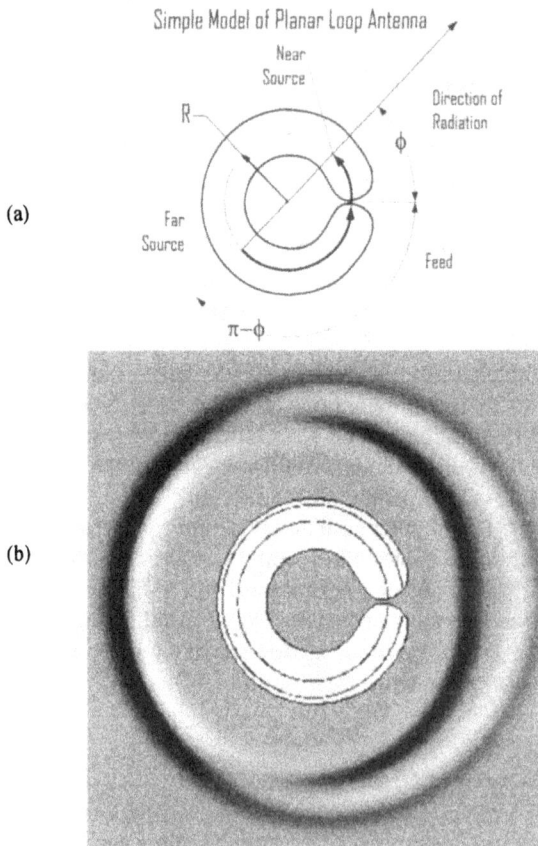

Figure 2.6   (a) A simple model of a planar loop antenna, and (b) result of a simple model of the radiation from a planar loop antenna.

Figure 2.6(b) shows the variation in waveform with angle as the near and far components sweep past each other with a relative delay that depends on angle. This simple model does an excellent job of capturing the essential physics and predicting the waveform shape and dispersion from a small-element antenna. Further details and examples of this signal tracing around an antenna are presented in Section 4.3.

The results of Figure 2.6(b) demonstrate that even a compact antenna element can have a dispersive radiation pattern. To avoid dispersion requires a symmetric pattern of currents contributing radiated signals with substantially fixed relative delays as a function of angle.

Dispersion is a problem for multiband systems as well as for those where a waveform occupies the entire band. If an antenna behaves differently at different look angles or at different frequencies, then a UWB system must compensate for these differences. This may be a difficult and resource-intensive process. Further, there is no guarantee that the performance of a dispersive antenna may be made equivalent to that of a nondispersive antenna. In summary, the general rule for a UWB system is the less dispersion, the better.

2.3 WHERE ENERGY GOES

One of the most fundamental antenna-design trade-offs is that between the field of view or pattern of an antenna and the antenna's sensitivity or gain. When an antenna radiates a signal, it directs the signal energy according to its pattern into a particular field of view. The smaller the field of view, the more concentrated the energy radiated by the antenna. This energy concentration is referred to as the antenna's gain.

With conventional narrowband antennas, properties like pattern and gain are essentially constant across the relatively narrow operating band of the antenna. UWB antennas introduce added complexities. The bandwidth of a UWB antenna is often large enough that there are significant variations in pattern and gain within the operational band of the antenna. This section will begin by reviewing basic antenna properties such as pattern, directivity, and gain and will end by considering how these parameters may be applied to a UWB antenna.

2.3.1 Antenna Pattern

An antenna pattern is a measure of the field of view of an antenna. It describes the spatial distribution of energy radiated or received by an antenna. The widest possible field of view is that of an "isotropic" antenna, one that radiates uniformly in all directions. More practical and achievable is an omnidirectional, or "omni," antenna, one that provides coverage at all angles in a particular plane. For some applications, a directional "pencil beam," or high gain, antenna is preferred. These three kinds of patterns will be considered in turn.

2.3.1.1 Isotropic Antennas

An isotropic antenna radiates and receives energy uniformly in all directions (see Figure 2.7). A true isotropic antenna is a topological impossibility because electromagnetic waves are transverse [9, 10]. Consider a radiation field pattern projected on a spherical shell surrounding an antenna. The field lines must be continuous. For instance, suppose we begin with a vertical polarization pattern with field lines arranged north-south like the longitude lines on a globe. The "source" at one pole and the "sink" at the other pole represent nulls in the antenna pattern, so this "longitude" field pattern is not isotropic. Consider instead a horizontal polarization pattern with field lines arranged east-west like the circles of latitude on a globe. Here again, the poles are nulls in the antenna pattern, so this "latitude" field pattern is not isotropic either. Similarly, it is not mathematically possible to define an antenna pattern without at least a single null.

Of course, there are ways to work around the topological difficulties and achieve a desired isotropic coverage. For instance, one can alternate the use of a pair of antennas arranged so that each provides coverage in the null of the other. A turnstile antenna, for instance, can achieve an average isotropic coverage by rotating a nonisotropic pattern. Nevertheless, a true isotropic antenna cannot be built and serves only as a convenient theoretical reference point against which one may compare other antennas.

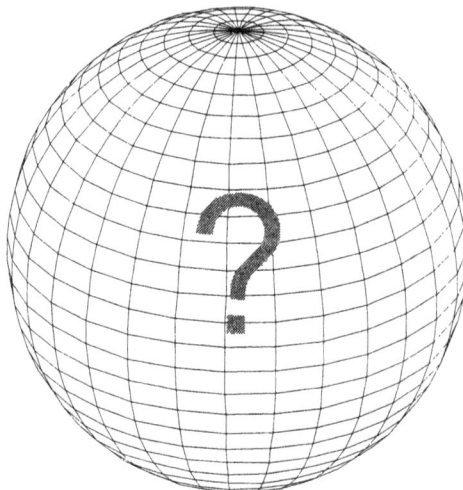

Figure 2.7 An isotropic antenna radiates and receives uniformly in all directions ([11]; © 2003 IEEE).

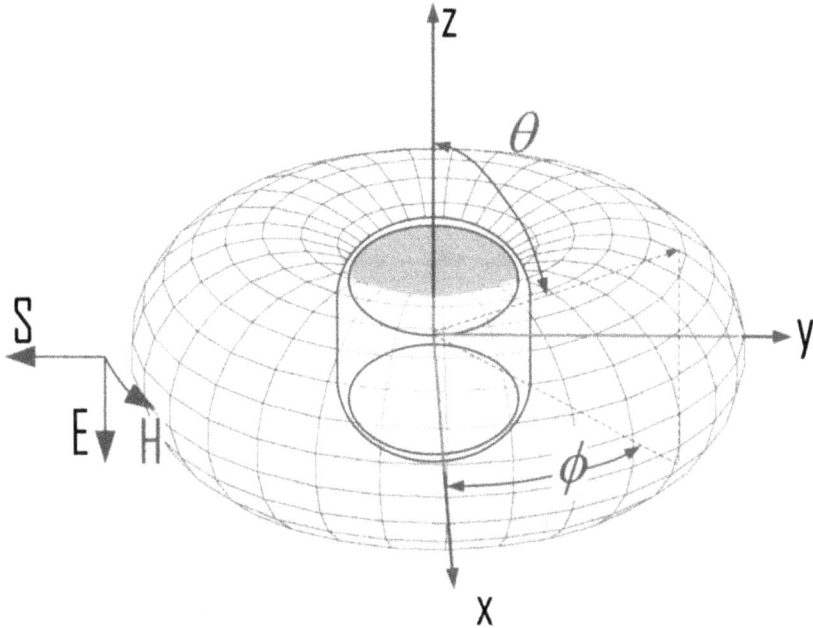

Figure 2.8 An omnidirectional antenna radiates in all directions in a plane ([11]; © 2003 IEEE).

2.3.1.2 Omni Antennas

An omni antenna radiates and receives energy uniformly in all azimuthal directions (i.e., in a particular plane) (see Figure 2.8). An omni pattern may be readily achieved. In fact, dipole and small loop antennas are examples of omni antennas. This omni pattern is referred to as a "doughnut" pattern since the nulls of the pattern resemble the holes in a doughnut. Typically, the elevation angle θ is defined from the axis of the dipole (the z axis in Figure 2.8), and the azimuthal angle ϕ is defined in the plane perpendicular to the dipole axis (the x-y plane in Figure 2.8). The electric field (**E**) is oriented in the elevation, or θ, direction and the magnetic field (**H**) is oriented in the azimuthal, or ϕ, direction. The radiated energy flux (**S**) propagates in a direction perpendicular to the electric and magnetic fields.

Consumer electronic devices (such as cell phones) typically employ an omni-pattern. These devices are generally in a vertical or near-vertical orientation but at an arbitrary alignment with respect to the other end of the link (such as a cell phone tower). Thus, an omni pattern allows the device to remain connected with other devices lying anywhere (more or less) in the same general plane.

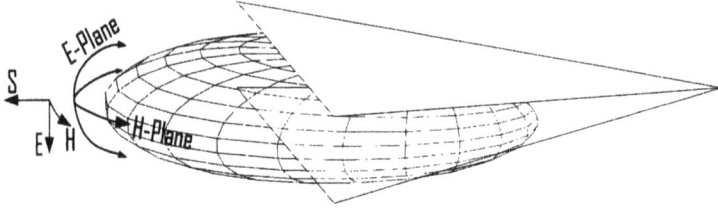

Figure 2.9 A directional antenna focuses energy in a beam ([11]; © 2003 IEEE).

2.3.1.3 Directional Antennas

Figure 2.9 depicts a directional, or gain, antenna. Directional antennas focus energy in a particular narrow field of view. Thus, a directional antenna is ideal for a point-to-point link or any other application where an antenna can be aligned to aim directly at the other end of the link. Typical gain antennas include horns and reflectors. As electromagnetic radiation launches from a horn antenna, the electric fields must be perpendicular to the antenna elements. Therefore, an examination of the orientation and arrangement of the antenna elements generally reveals the orientation of the electric fields.

2.3.1.4 Pattern, Field of View, and Beamwidth

An antenna pattern function, $P(\theta, \phi)$, describes the angular distribution of energy radiated or received by an antenna. The "half-power," or -3dB, points define an antenna's field of view . For an isotropic antenna, $P(\theta, \phi) = 1$, and the pattern is uniform in all directions. An infinitesimal dipole antenna has a pattern $P(\theta, \phi) = \sin^2\theta$. This antenna is omnidirectional since the pattern is independent of ϕ. The half-power points are found by solving $P(\theta, \phi) = \sin^2\theta = 0.5$ to yield $\theta = \pm45°$. The beamwidth of an antenna is defined as the angular measure between the -3dB points. Thus, a small dipole has a 90° elevation beamwidth and an omni pattern in the azimuthal plane. Table 2.1 shows pattern functions for a few typical antennas.

2.3.2 Antenna Directivity, Gain, and Bandwidth

The pattern of an antenna describes how well an antenna directs, or focuses, energy in a particular direction. An antenna's "directivity" quantifies its ability to direct energy. As an antenna directs more and more energy into a smaller and smaller field of view, the signals it receives or radiates (within the mainlobe of its pattern) become larger and larger. This signal enhancement relative to an isotropic antenna signal is referred to as "gain." Gain is directly proportional to directivity.

The proportionality constant is the antenna efficiency. Finally, the "aperture" of an antenna describes the antenna's effective area–its ability to intercept a portion of an incident electromagnetic wave.

2.3.2.1 Directivity

The directivity (D) of an antenna is the ratio of how much energy it directs in a particular direction divided by the average energy

$$D = \frac{peak\ energy}{average\ energy} = \frac{\left|P(\theta,\phi)\right|_{max}}{\frac{1}{4\pi}\oint_S \left|P(\theta,\phi)\right| d\Omega}$$

$$= \frac{1}{\frac{1}{4\pi}\int_{-\pi}^{+\pi}\int_{0}^{+\pi} \left|P(\theta,\phi)\right|\sin\theta\ d\theta\ d\phi}$$

$$(2.8)$$

The directivity of an isotropic antenna is $D = 1$. For a point dipole, directivity is $D = 1.5$. Table 2.1 also provides the directivity for a few typical antennas.

A relationship also exists between the directivity and the size of an antenna. A low gain or omni antenna is typically about a quarter wavelength in size or smaller. A directional or high gain antenna will tend to be much larger–typically several wavelengths in dimension or more. The fundamental relationships between antenna size, directivity, and bandwidth are the subject of a later chapter.

Table 2.1

Pattern, Directivity, Aperture, and Field of View for a Few Typical Antennas

Antenna	Pattern $P(\theta, \phi)$:	Directivity D	Aperture A (λ^2)	Field of View
Isotropic	1	1.00 (0.00dBi)	0.0796	Isotropic
Point dipole	$\sin^2\theta$	1.50 (1.76dBi)	0.119	θ: 45°–135° ϕ: omni
$\lambda/2$ dipole	$\dfrac{\cos^2\left(\frac{\pi}{2}\cos\theta\right)}{\sin^2\theta}$	1.64 (2.15dBi)	0.131	θ: 51°–129° ϕ: omni
Dipole with plane reflector	$\sin^2\theta\cos^2\frac{\phi}{2}$	3.00 (4.77dBi)	0.239	θ: 45°–135° ϕ: –90°–+90°
Sector or pencil beam antenna	$\begin{cases}1 & \theta_1 \le \theta \le \theta_2; \\ & \phi_1 \le \phi \le \phi_2 \\ 0 & else\end{cases}$	$\sim\dfrac{32,400}{(\theta_2-\theta_1)(\phi_2-\phi_1)}$	$\sim\dfrac{2,580}{(\theta_2-\theta_1)(\phi_2-\phi_1)}$	θ: θ_1–θ_2 ϕ: ϕ_1–ϕ_2

2.3.2.2 Gain

Directivity and gain are related by "efficiency" (η). Efficiency is also the ratio of the radiated power (P_{rad}) to the input power (P_{in})

$$\eta = \frac{P_{rad}}{P_{in}} = \frac{G}{D} \tag{2.9}$$

For an ideal or highly efficient antenna ($\eta \cong 1$), gain is identical to directivity ($G = D$). Gain and directivity are intimately connected. The more directive an antenna–the more it concentrates energy to or from a particular direction–the higher the gain. The connection between gain and directivity is merely the application of conservation of energy to antenna behavior. Section 7.2 discusses efficiency in greater detail.

Unless otherwise specified, gain is typically interpreted as the peak, or maximum, gain of an antenna. Sometimes this is also called the "boresight gain." Gain is usually expressed relative to the gain of an isotropic antenna in units of decibels

$$G_{dB} = 10\log G \tag{2.10}$$

Thus, an ideal ($\eta = 1$) isotropic antenna has a gain $G_{dB} = 0$dBi. An ideal small dipole has a maximum gain $G = D = 1.5$ and $G_{dB} = 1.76$dBi. A 50% (−3.00dB) efficient small dipole will have a maximum gain of about 1.76dBi − 3.00dB = −1.24dBi.

For an isotropic antenna with gain $G = 1$, the transmitted power (P_{TX}) is uniformly distributed over the area of a spherical shell at a particular range (r) from the antenna. Thus, the radiated power density (S) is given by

$$S = \frac{P_{TX}}{4\pi r^2} \tag{2.11}$$

For a nonisotropic antenna with gain $G_{TX} \neq 1$, the radiated power density is

$$S = \frac{P_{TX}G_{TX}}{4\pi r^2} \tag{2.12}$$

Finally, the effective isotropic radiated power (*EIRP*) is defined as

$$EIRP = P_{TX}\,G_{TX} \tag{2.13}$$

where G_{TX} is the transmit antenna's maximum, or peak, gain.

2.3.2.3 Aperture

The effective aperture of an antenna is its effective capture area: the area of an incident wave front that the antenna can intercept. The effective aperture (A) of an antenna may be found from the gain and either the wavelength or frequency

$$A = \frac{\lambda^2 G}{4\pi} = \frac{c^2 G}{4\pi f^2} \tag{2.14}$$

Normally, effective aperture is defined in terms of an antenna's maximum gain. An isotropic antenna has a gain $G = 1$, so an isotropic antenna has an aperture $A = \lambda^2/(4\pi)$. The aperture of an isotropic antenna is equivalent to a disk of radius $r = \lambda/(2\pi) = c/(2\omega)$. The sphere defined by this radial distance is sometimes called the "radiansphere" [12]. This spherical shell defines the boundary around a small antenna on which the near, or inductive, fields are comparable in magnitude to the far, or radiation, fields. Thus, it serves as the boundary between the near and the far zones around an antenna. This radian distance, $r = \lambda/(2\pi)$, is also the distance across which fields can couple. For instance, parallel conducting wires spaced $\lambda/(2\pi)$ apart will block about half of the copolarized incident RF energy.

Effective aperture describes the area an antenna can intercept from an incident radio wave. The apertures of a few typical antennas are presented in Table 2.1. Received power (P_{TX}) is simply the product of the effective aperture area and the incident radiation power density

$$P_{RX} = AS \tag{2.15}$$

Suppose a receive antenna with aperture A and gain G_{RX} receives a signal from a transmit antenna with gain G_{TX} at a particular range (r). Combining the results of (2.14) with (2.11) and (2.13) yields

$$P_{RX} = AS = \frac{P_{TX} G_{TX}}{4\pi r^2} \frac{\lambda^2 G_{RX}}{4\pi}$$
$$= P_{TX} \frac{G_{TX} G_{RX} \lambda^2}{(4\pi r)^2} = P_{TX} \frac{G_{TX} G_{RX} c^2}{(4\pi r)^2 f^2} \tag{2.16}$$

Equation 2.16 presents a very important relationship known as Friis's Law [13]. Friis's Law defines a basic link between two antennas in free space and forms the starting point for understanding RF propagation.

One interesting feature of Friis's Law causes much confusion. For a pair of constant gain antennas, Friis's Law indicates that the received power decreases as $1/(4\pi r^2)$. This makes sense because at a range r, the transmit power is distributed over the surface of a spherical shell of area $4\pi r^2$. Friis's Law also indicates that the received power decreases as $1/f^2$ with increasing frequency. These two effects are sometimes lumped together under the term "path loss," creating the impression that somehow propagation through free space creates a frequency-dependent attenuation. Of course, this is not the case.

The real reason lies in the definitions of gain and aperture. A constant gain antenna has a constant pattern with respect to changes in frequency. For instance, an antenna with the pattern of a half-wave dipole scales in electrical size as frequency changes. If frequency doubles, the pattern remains the same, but the size of the antenna is cut in half. The antenna has only one-fourth the original effective area and only captures one-fourth the energy possible with the larger, lower-frequency version of the antenna. This is why Friis's Law includes a $1/f^2$ frequency dependence. Section 2.3.3 further explains the behavior of constant gain and constant aperture behavior.

2.3.2.4 Antenna Factors and Antenna Height

One can also directly relate a received voltage signal to an incident field strength. An "antenna factor" (AF) is the ratio of incident electric field strength $|E_{inc}|$ to received voltage (V_{rec})

$$AF = \frac{|E_{inc}|}{V_{rec}} \qquad (2.17)$$

Antenna factors have units of inverse length and are typically provided for calibrated laboratory-grade antennas designed for use in precision electric field measurements, such as for assessing electromagnetic compatibility or unintentional emissions. Often, antenna factors are expressed in decibels. Since antenna factors are not a dimensionless ratio, it is essential to know the units used in deriving them. A good table of antenna factors will always explain what units are assumed. A frequent choice of units defines incident electric field intensity in decibel microvolts per meter ($dB_{\mu V/m}$) and received voltage in decibel microvolts ($dB_{\mu V}$). For these units, a calculation of incident field strength as a function of received voltage follows

$$|E_{inc}|[dB_{\mu V/m}] = AF[dB] + V_{rec}[dB_{\mu V}] + C[dB] \qquad (2.18)$$

where C is the antenna cable loss in decibels. Often antenna factors and cable loss may be entered into a spectrum analyzer to allow the analyzer to read out received field intensity directly.

Antenna factors may be related back to other antenna parameters. The relationship between antenna factors and antenna aperture (A) is

$$AF = \sqrt{\frac{Z_S}{Z_{load} A}} \qquad (2.19)$$

where $Z_S = 376.7\Omega$ is the impedance of free space, and Z_{load} is the load impedance. Assuming $Z_{load} = 50.0\Omega$, applying (2.14) to (2.19) yields

$$AF = \frac{9.73}{\lambda[m]\sqrt{G}} = \frac{f[MHz]}{30.8\sqrt{G}} \qquad (2.20)$$

or

$$AF[dB] = 20\log_{10}\left(\frac{9.73}{\lambda\sqrt{G}}\right) \qquad (2.21)$$

The units of the antenna factor will be the inverse of the units used for wavelength. Antenna factors are the inverse of another quantity referred to as "effective antenna height" (h_{eff})

$$|E_{inc}| = h_{eff}V_{rec} \qquad (2.22)$$

2.3.3 Pattern, Gain, and UWB Antennas

Much of this chapter's discussion has focused on a review of antenna basics that apply to narrowband and UWB antennas alike. This chapter has considered many antenna parameters that are functions of frequency and wavelength. For a narrowband antenna, the relatively small change in frequency across the operational band of the antenna means that these antenna parameters (like gain or aperture) are essentially constant. These antenna parameters may be treated as simple algebraic quantities when considering narrowband antennas.

For UWB antennas, this is no longer the case. Since gain, power, and aperture are functions of frequency, instead of algebraic quantities these antenna parameters become density functions. For instance, three very important relationships are

$$EIRP(f) = P_{TX}(f)\, G_{TX}(f) \qquad (2.23)$$

$$A_{RX}(f) = \frac{c^2 G_{RX}(f)}{4\pi f^2} \qquad (2.24)$$

and Friis's Law for UWB antennas

$$P_{RX}(f) = \frac{P_{TX}(f)G_{TX}(f)G_{RX}(f)c^2}{(4\pi r)^2 f^2} = \frac{EIRP(f)}{4\pi r^2} A_{RX}(f) \qquad (2.25)$$

This section will consider some of the difficulties that appear when trying to extend traditional narrowband antenna concepts to UWB antennas. To illustrate and explain these difficulties, this section will consider two archetypical UWB antennas: constant gain antennas and constant aperture antennas.

2.3.3.1 Constant Gain Antennas

For an efficient antenna, gain equals directivity. If an antenna has constant gain and constant efficiency as functions of frequency, then it will also have constant directivity. This means that an antenna's pattern will not vary across the bandwidth for which its gain remains constant. A well-designed small-element dipole will usually have a constant pattern across a significant frequency band, as shown in Figure 2.10. Typically (but not always), a constant gain antenna exhibits low gain and a wide, or omni, field of view.

A constant pattern as a function of frequency is usually desirable for a UWB antenna. This means that the UWB antenna's field of view remains fixed with respect to (w.r.t.) frequency. Thus, the high-frequency and low-frequency components of a signal are uniformly transmitted, resulting in minimal dispersion of a time domain signal.

Examining (2.25), if gain is constant with respect to frequency, then a constant transmit power density yields a constant *EIRP*. Since typical regulatory limits, such as those imposed by the FCC, include a constant *EIRP* across a particular operating band, a constant gain antenna requires a constant transmit power density. Examining (2.24), constant gain implies that antenna aperture must vary inversely with the square of the frequency ($A \sim 1/f^2$). Since effective antenna aperture decreases with increasing frequency, a constant gain antenna receives less energy as frequency increases. Figure 2.11 illustrates these relationships.

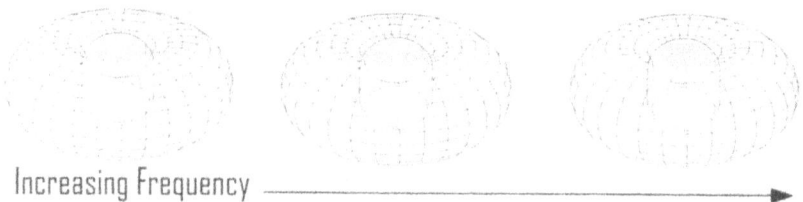

Increasing Frequency ⟶

Figure 2.10 A constant gain antenna has a consistent pattern across the operating frequency band ([11]; © 2003 IEEE).

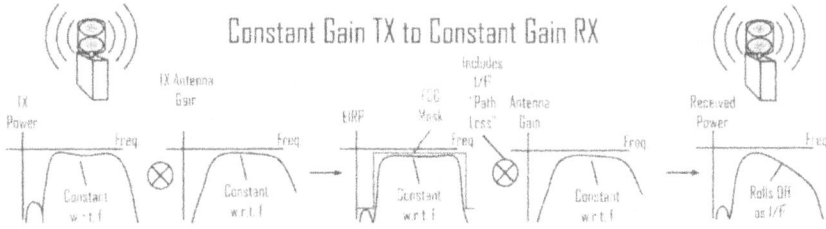

Figure 2.11 A constant transmit power density applied to a constant gain antenna yields a constant *EIRP*. If a constant gain antenna receives a constant *EIRP* signal, the receive power density rolls off as $1/f^2$ in band ([11]; © 2003 IEEE).

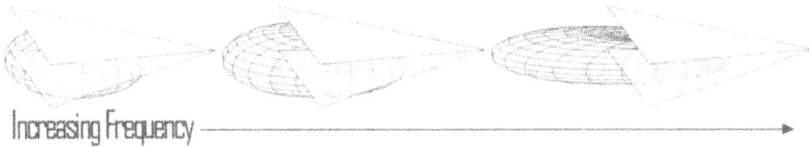

Figure 2.12 A constant aperture antenna has a narrowing pattern and increasing gain as frequency increases ([11]; © 2003 IEEE).

In summary, three general properties characterize a constant gain antenna. First, a constant gain antenna will have a fixed pattern and minimal dispersion. Second, a constant gain antenna will require a constant power-density excitation to radiate a constant *EIRP* signal. Finally, if a constant gain antenna receives a constant *EIRP* signal, the resulting received power density rolls off as $1/f^2$ in band.

2.3.3.2 Constant Aperture Antennas

The effective aperture of a constant aperture antenna remains fixed across its operating band. Thus, according to (2.24), the gain of a constant aperture antenna is proportional to f^2. The gain increases 6dB for every octave increase in frequency. If gain increases, then the field of view and the pattern must narrow. Thus, as frequency increases, constant aperture antennas are characterized by increasing gain and narrowing field of view. Radiated and received waveforms will experience dispersion at the edge of the field of view due to the changing gain of lower-frequency components relative to higher ones. Horn and reflector antennas sometimes (but not always) exhibit constant apertures. Constant aperture antennas tend to be directional, high-gain, and relatively large. Figure 2.12 illustrates the pattern behavior of a constant aperture antenna.

Because the gain of a constant aperture antenna increases as f^2, it cancels out the $1/f^2$ roll off in Friis's Law [2.25]. Thus, if a constant aperture antenna is used on the receive end of a link, a constant *EIRP* signal yields a constant received power density. Since the gain of a constant aperture antenna tends to be relatively high, the received power will be higher than expected from a lower-gain omni antenna. If an application can tolerate a limited field of view that narrows further with increasing frequency, then a constant aperture antenna offers good performance for a receive antenna. Figure 2.13 shows the details of how a link using a constant aperture receive antenna works.

The advantages of a constant aperture antenna are less significant when it is used at a transmitter. The higher gain of a constant aperture antenna must be compensated for by a lower transmit power to meet a fixed *EIRP* regulatory limit. Here again, a constant aperture receive antenna yields a constant received-power spectral density. The relatively high gain of a constant aperture receive antenna yields a higher receive power than an equivalent omni receive antenna. The narrow field of view of a typical constant aperture antenna may be advantageous to direct radiated energy only where required for an application, thus avoiding interference. Otherwise, there is no advantage to using a constant aperture transmit antenna relative to an omni transmit antenna. Figure 2.14 illustrates how a constant aperture antenna works on both sides of a link.

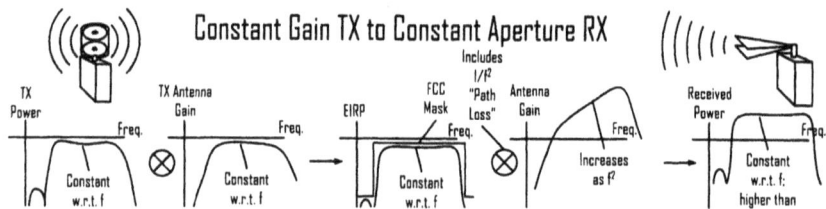

Figure 2.13   A constant transmit power density applied to a constant gain antenna yields a constant *EIRP*. If a constant aperture antenna receives a constant *EIRP* signal, the receive power density is constant in band ([11]; © 2003 IEEE).

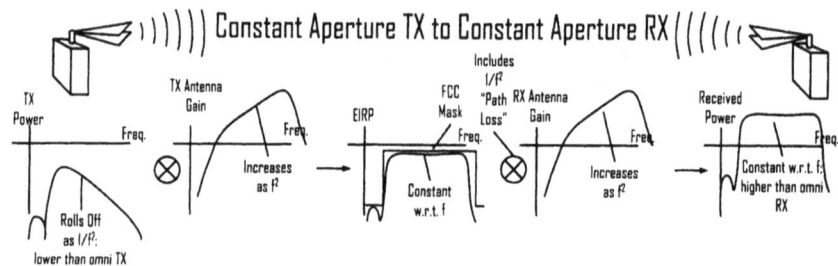

Figure 2.14   A constant aperture antenna requires a transmit power density that rolls off as $1/f^2$ to yield a constant *EIRP*. If a constant aperture antenna receives a constant *EIRP* signal, the receive power density is constant in band ([11]; © 2003 IEEE).

In summary, three general properties characterize a constant aperture antenna. First, a constant aperture antenna will have a narrowing field of view and exhibit waveform dispersion at the edge of its field of view. Second, a constant aperture antenna will require a power density that rolls off as $1/f^2$ in band so as to radiate a constant *EIRP* signal. Finally, if a constant aperture antenna receives a constant *EIRP* signal, the resulting received power density is constant in band.

2.3.3.3 Other UWB Antennas

Constant gain and constant aperture antennas are archetypes that illustrate important UWB antenna behavior. The typical constant gain antenna is a small-element antenna with dipolelike pattern and behavior that remains consistent across the antenna's operational band. The typical constant aperture antenna is a horn antenna with a relatively narrow beam that narrows further across the antenna's operational band. Important as these archetypes are, they do not exhaustively categorize UWB antennas. This section will discuss a few of the more common deviations from constant gain and constant aperture behavior.

A constant pattern is a very desirable characteristic for a UWB antenna since it avoids the waveform dispersion and distortion associated with a pattern that varies with frequency. A constant pattern usually implies constant gain. However, it is possible to combine a constant pattern with increasing gain by introducing inefficiencies that suppress gain at the low end of the operational band. An electrically small antenna will exhibit this kind of behavior, becoming increasingly efficient at higher wavelengths. The disadvantage of such an antenna is that performance is necessarily worse than that of a corresponding constant gain antenna. Figure 2.15 illustrates this behavior, and Chapter 6 provides additional examples.

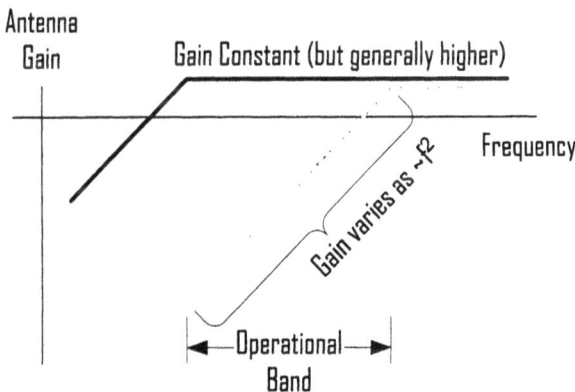

Figure 2.15 An electrically small antenna may have a constant pattern but an increasing gain (   )
generally less than that of a larger constant gain antenna.

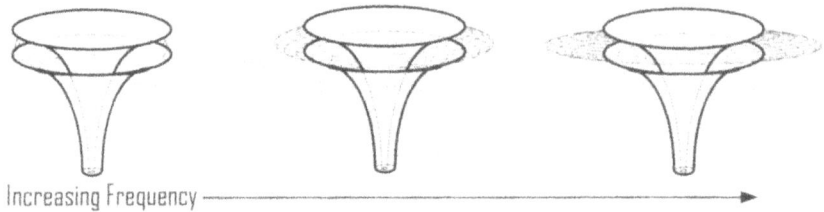

Increasing Frequency ─────────────────────────────────►

Figure 2.16 An omnidirectional antenna may have increasing gain with frequency if the elevation pattern narrows with frequency.

It is also possible to construct an efficient omnidirectional antenna with increasing gain by narrowing the elevation pattern with increasing frequency as shown in Figure 2.16. If the desired field of view lies close enough to the azimuthal plane, then the dispersion due to this pattern narrowing may be neglected. Leon Brilluoin's coaxial horn antennas (discussed in Section 3.3.2) can potentially exhibit an omni pattern with increasing gain due to a narrowing of elevation pattern [14].

Finally, constant gain usually implies low directivity. It is possible, however, to construct a high-gain horn with a constant gain. In short, although many UWB antennas may be characterized as constant gain or constant aperture, there are also many antennas that do not fall into either classification.

2.4 POLARIZATION

Electromagnetic waves are transverse to the direction of propagation. The orientation of the electric field may lie anywhere in the plane normal to the direction of propagation. Polarization is defined in terms of the electric field orientation. The electric field of a vertically polarized wave is vertically oriented. The electric field of a horizontally polarized wave is horizontally oriented. As the polarization axes of a transmit antenna and a receive antenna are moved out of alignment by an angle α, the received signal power will decrease as $\cos^2 \alpha$. This relationship was first observed in the context of optics by the French army engineer Étienne Louis Malus (1775–1812) [15]. Figure 2.17 depicts this geometry. If the antenna axes are orthogonal to each other ($\alpha = 90°$), no signal will be received.

The polarization of a wave does not have to be fixed in any particular direction. Figure 2.18 displays a "chiral" polarization signal. A chiral polarization signal has a rotating electric field vector that traces out a corkscrew path as a signal propagates. The corkscrew may be either clockwise (left hand) or counter-clockwise (right hand) relative to the direction of propagation. The former is "left hand" chiral polarization and the latter is "right hand" chiral polarization. These are the UWB analogs of circular polarization for narrowband signals. Efficient reception of a chiral polarization signal requires a chiral polarization antenna of the same sense. A linear polarization antenna will only be able to accept half of the energy in a chiral polarization signal. A chiral polarization antenna of the opposite sense will not receive the signal. Mark Barnes and Larry Fullerton have investigated using chiral UWB signals [16].

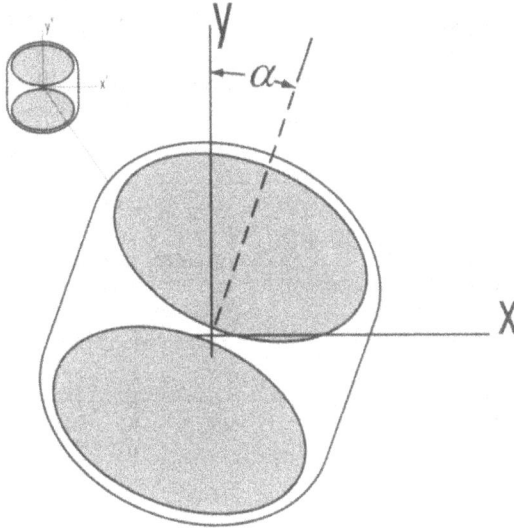

Figure 2.17   The polarization axes of two antennas are offset by an angle α.

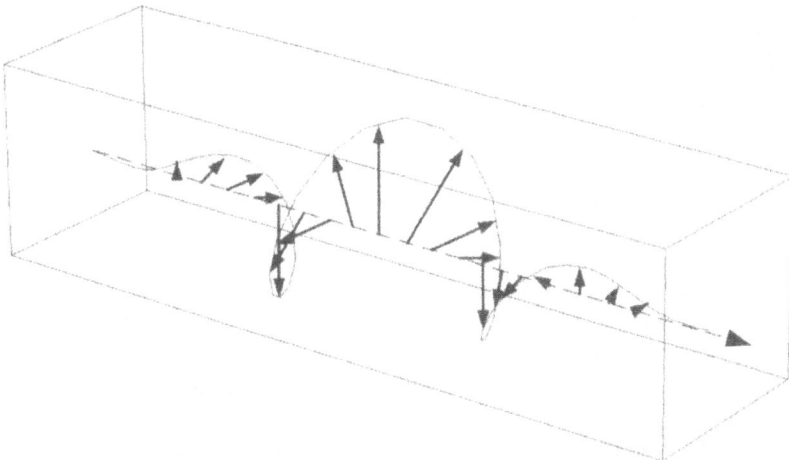

Figure 2.18   A "chiral" polarization signal has a rotating electric field vector that traces out a corkscrew path.

2.5 ANTENNA MATCHING

A final important antenna parameter is impedance matching. A good impedance match maximizes the power transfer and efficiency of an antenna. A poor impedance match yields undesirable reflections and reverberation in a UWB system. Additionally, a poor match reflects energy that might otherwise be transmitted, yielding a power loss. The quality of an antenna match is usually measured in terms of either voltage standing wave ratio (VSWR) or a scattering parameter (S_{11}). Chapter 3 presents the mathematical details of these parameters. Table 2.2 shows typical values of VSWR, LogMag S_{11} (in decibels), and power loss (in %) for a range of antenna matches.

Table 2.2
Matching in Terms of VSWR, LogMag S_{11}, and Power Loss

Match	VSWR	LogMag S_{11} (dB)	Power Loss (%); (dB)
Marginal	3.00:1	−6.0	25.0; −1.25
Good	2.00:1	−9.5	11.1; −0.511
Good	1.92:1	−10.0	10.0; −0.458
Excellent	1.50:1	−14.0	4.0; −0.177
Superb	1.22:1	−20.0	1.0; −0.043

Characterizing a particular match as acceptable depends upon the application. The labels in Table 2.2 are intended to provide general guidance valid in many but not all cases. For some applications, a 3:1 VSWR might be acceptable. With this "marginal" match, 25% of incident power is reflected away and lost. A "good" match with a VSWR around 2:1 and $S_{11} \sim -10$dB loses about 10% of incident power. An "excellent" match with a VSWR around 1.5:1 and $S_{11} \sim -14$dB loses about 4% of incident power. A "superb" match with a VSWR around 1.22:1 and $S_{11} \sim -20$dB loses about 1% of incident power.

Some applications may be more or less tolerant than others of mismatch reflection. For instance, a radar antenna that reflects copies of transmitted signals will radiate a replica of the intended transmit pulse delayed by the two-way reflection time of the antenna transmission line. Such a reflection is likely to yield erroneous "ghost" images. Thus, radar applications tend to be less tolerant of antenna mismatch. On the other hand, a low-pulse-rate positioning system that relies on capturing the leading edge of a received waveform may be very tolerant of mismatch. Any late-arriving echoes are unlikely to confound a measurement of leading-edge arrival.

Evaluating an antenna solely on the basis of matching can be highly deceptive. A 50Ω termination, for instance, provides a superb match to a 50Ω transmission line, but neither radiates nor receives any energy. Resistive losses can make an antenna appear well matched yet render it inefficient and ineffectual. A comprehensive evaluation of an antenna requires an assessment of its ability to radiate and receive energy, such as a measurement of gain.

Adding loss to an antenna to improve match or extend impedance bandwidth is a tempting way to make a bad antenna seem better. It is preferable, however, to design a good match into an antenna from the beginning rather than to rectify a bad match by wasting valuable signal energy in resistive loading. These topics and more are the subject of Chapter 3.

2.6 ANTENNAS AS TRANSDUCERS

This chapter has introduced UWB antennas by treating an antenna as a transducer that couples energy between signals on a transmission line and radio waves propagating through space. For many applications, an understanding of basic properties like the bandwidth of an antenna, waveform dispersion, antenna pattern, directivity, gain, aperture, field of view, polarization, and matching are all that are needed to make good use of a UWB antenna.

To improve the performance of a UWB antenna or to design one from scratch, an understanding of how UWB antennas work is essential. Chapter 3 will begin an in-depth review of fundamental UWB antenna science by looking at transmission lines, impedance, and antennas. In Chapter 3, UWB antennas will be looked at as transformers that couple the impedance of a transmission line to the impedance of free space.

Endnotes

[1] Terman, Frederick Emmons, *Radio Engineering*, New York, McGraw Hill, 1932, p. 494. Dr. Terman was a professor at Stanford University and a widely respected authority on radio technology. He is generally recognized as one of the grandfathers of Silicon Valley.

[2] Fitzgerald, F. Scott, *The Crack-Up*, New York: New Directions Publishing Corporation, 1993, originally published in 1936.

[3] U.S. 47 C.F.R. Part 15 Subpart F §15.503(a-d) Ultra-Wideband Operation (October 1, 2003 edition), p. 767. The relevant regulations read as follows:

"(a) *UWB bandwidth.* For the purpose of this subpart, the UWB bandwidth is the frequency band bounded by the points that are 10 dB below the highest radiated emission, as based on the complete transmission system including the antenna. The upper boundary is denoted f_H and the lower boundary is denoted f_L. The frequency at which the highest radiated emission occurs is designated f_M.

"(b) *Center frequency.* The center frequency f_C equals $(f_H + f_L)/2$.

"(c) *Fractional bandwidth.* The fractional bandwidth equals $2(f_H - f_L)/(f_H + f_L)$.

"(d) *Ultra-wideband (UWB) transmitter.* An intentional radiator that, at any point in time, has a fractional bandwidth equal to or greater than 0.20 or has a UWB bandwidth equal to or greater than 500 MHz, regardless of the fractional bandwidth."

[4] Schantz, Hans, "Dispersion and UWB antennas," 2004 International Workshop on Ultra Wideband Systems. Joint with Conference on Ultrawideband Systems and Technologies. Kyoto, Japan. May 18-21, 2004, pp. 161-165. Some of the material in Section 2.2 was originally presented in this paper.

[5] Hertel, Thorsten, and Glenn Smith, "On the dispersive properties of the conical spiral antenna and its use for pulse radiation," IEEE Transactions on Antennas and Propagation, Vol. 51, No. 7, July 2003, pp. 1426-1433.

[6] Lamensdorf, Dennis, and Leon Susman, "Baseband-pulse-antenna techniques," IEEE Antennas and Propagation Magazine, Vol. 36, No. 1, February 1994, pp. 20-30.

[7] Schantz, Hans, "Planar elliptical element ultra-wideband dipole antennas," 2002 IEEE Antennas and Propagation Society International Symposium, San Antonio, Texas, June 16-21, 2002, Vol. 3, pp. 44-47.

[8] Schantz, Hans, "Planar loop antenna," U.S. Patent 6,593,886, July 15, 2003.

[9] Mathis, H. F., "A Short Proof That an Isotropic Antenna Is Impossible," Proceedings of the IRE, Vol. 39, 1951, p. 970.

[10] Mathis, H. F., "On Isotropic Antennas," Proceedings of the IRE, Vol. 42, 1954, p. 1810.

[11] Schantz, Hans, "Introduction to UWB antennas," 2003 IEEE Conference on Ultra Wideband Systems and Technologies. Reston, Virginia. November 16-19, 2003, pp. 1-9.

[12] Wheeler, Harold H., "The Radiansphere around a Small Antenna," Proceedings of the IRE, Vol. 47, August 1959, pp. 1325-1331.

[13] Friis, Harald T., "A Note on a Simple Transmission Formula," Proceedings of the IRE, Vol. 34, No. 5, 1946, pp. 245-256.

[14] Brillouin, Leon.N., "Broad band antenna," U.S. Patent 2,454,766, November 30, 1948.

[15] Jenkins, Francis A. and Harvey E. White, *Fundamentals of Optics,* 4th ed., New York: McGraw-Hill, 1976, pp. 503-504.

[16] Barnes, Mark A., and Larry W. Fullerton, "Chiral and dual polarization techniques for an ultra-wide band communications system," U.S. Patent 5,764,696, June 9, 1998.

Chapter 3

Antennas as Transformers

From this point of view, the arms of an antenna form the banks of a channel in which the waves excited by the source are confined before they emerge into unlimited space. In this sense antennas are similar to waveguides.

Sergei Schelkunoff, 1952

The discovery of the relationship between voltage (V) and current (I) by Georg Simon Ohm (1789–1854) was a giant leap forward in understanding electrical physics. Ohm placed the concept of electrical resistance (R) on a mathematical footing by identifying it as the ratio of voltage to current:

$$R = V/I \tag{3.1}$$

Heaviside generalized Ohm's concept of electrical resistance to the case of alternating current and dubbed it "impedance." Later, Schelkunoff extended Heaviside's transmission-line impedance concept to include the impedance of fields in free space [1]. Schelkunoff identified the field impedance (Z) as the ratio of the electric (E) to the magnetic (H) field intensities:

$$Z = E/H \tag{3.2}$$

The concept of impedance bridges the gap between fields guided by transmission lines on the one hand and fields propagating in free space on the other. Both the RF system engineer and the antenna designer need to understand impedance to achieve a good, high-efficiency match between a UWB antenna and an RF front end.

This chapter will begin by introducing the idea of antenna impedance, by discussing some common transmission line geometries and by providing examples of how to couple transmission lines into antenna elements. Then, this chapter will examine matching and some techniques to transform a UWB antenna impedance

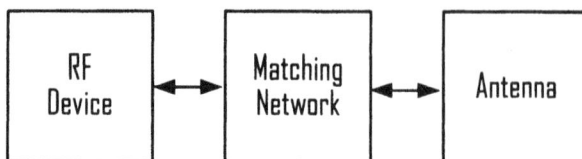

Figure 3.1   Traditional narrowband systems bridge the gap between an RF device and an antenna by using a matching network.

to a line impedance. Finally, this chapter will survey some "balun" transformer geometries that allow efficient coupling from balanced lines to unbalanced antennas and from unbalanced lines to balanced antennas. The aim of this chapter is to explain the concepts of impedance and matching and to provide a unified picture of how an antenna may be thought of as an impedance transformer that couples between a transmission-line impedance and the impedance of free space.

3.1 ANTENNA IMPEDANCE

A half-wave thin-wire dipole has an impedance of about 73Ω. A thin wire quarter-wave monopole, or whip, has an impedance of half that, or about 36Ω. Figure 3.1 illustrates how traditional narrowband systems couple between an RF front end and an antenna by using a matching network. A matching network allows compensation across even a relatively large impedance difference, at least for a narrow bandwidth. As bandwidth increases, however, so also does the difficulty in achieving a good match. This difficulty is aptly summarized by "Munk's Law:" *To match to an ultrawideband antenna begin with a well-matched antenna* [2]. There are a variety of excellent texts that address impedance matching [3, 4]. The goal of this chapter is to describe how to design a well-matched antenna, not how to design a broadband impedance-matching network for a poorly matched antenna.

3.1.1 UWB versus Narrowband Antenna Impedance

UWB impedance differs from that of a narrowband antenna in one crucial respect. To a large extent, the impedance of a UWB antenna is a design choice, not an intrinsic property of the antenna. The designer of a UWB antenna can exercise significant control over antenna impedance, selecting a desired impedance from a wide range of possibilities.

In the narrowband realm where antenna impedance is an intrinsic property of an antenna, a monopole has half the impedance of a dipole. This is not necessarily the case for UWB antennas. For instance, elliptical elements yield an excellent

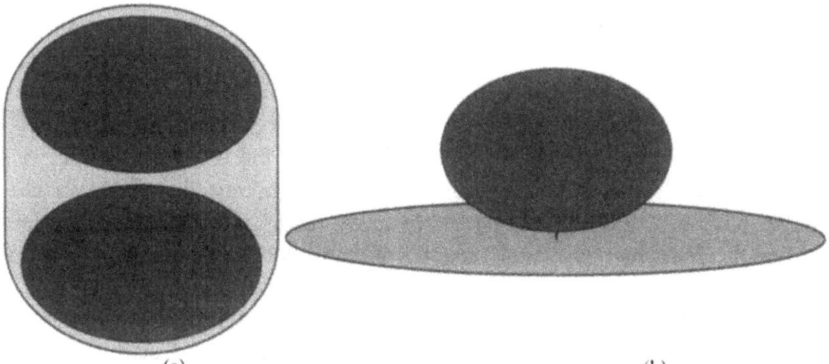

(a)   (b)

Figure 3.2   (a) A UWB dipole and (b) a UWB monopole may have the same impedance. This is
contrary to expectations from the narrowband case where a monopole has half the
impedance of a corresponding dipole.

match to 50Ω [5]. Extrapolating from narrowband experience, one would expect
an elliptical monopole to have an impedance of 25Ω. In fact, elliptical monopoles
also offer a good match to 50Ω [6]. Figure 3.2(a) shows a UWB dipole, and
Figure 3.2(b) shows a UWB monopole with the same 50Ω impedance. Both these
antennas rely on a controlled impedance variation from the feed point to free
space. Thus, impedance of these antennas can be adjusted by varying the gap
between the dipole elements and by varying the gap between a monopole element
and the ground plane. In each case, a proper design yields a 50Ω match, despite
the 2:1 ratio of impedance one might expect from narrowband experience.

Of course, nothing is particularly magical about 50Ω. The standard
characteristic impedance of 50Ω was chosen as a compromise between minimum
attenuation and maximum power-handling capability for an air-filled coaxial line
[7]. Since 50Ω is the standard input impedance for most RF test equipment, 50Ω
is a convenient (but not required) design goal for an antenna's input impedance.

3.1.2 Controlling Antenna Impedance

There are two general methods for achieving a desired antenna impedance:
resistive loading and geometry control. Resistive loading, the introduction of lossy
material to an antenna structure, is a quick and easy way to obtain a good match.
This lossy material has a significant impact on antenna performance. A preferred
method to create a desired antenna impedance is geometry control. By controlling
antenna geometry, a significant degree of impedance control is possible. First
though, consider resistive loading.

A designer may achieve an arbitrarily good match by using resistive loading. In fact, the most compact, best-matched, UWB antenna type currently available on the market uses resistive loading to achieve a –30dB S_{11} match across decades of bandwidth. This antenna type is marketed by a variety of companies under the designator, "50Ω load." Despite excellent matching, the radiation characteristics of this "antenna" leave much to be desired, illustrating the disadvantage of resistive loading. Resistive loading trades away radiation efficiency to achieve a good match. A typical resistively loaded antenna will have a gain at least 3dB lower than a comparable unloaded antenna.

There are some cases in which resistive loading is justified. For instance, an antenna with stringent requirements for pattern might employ resistive loading to control side lobes and back lobes. An electrically small antenna may have to incorporate some loss mechanism to maintain a reasonable impedance across a wide bandwidth. Using a diplexing filter, one may introduce loss selectively, for instance, to out-of-band frequency components. And of course, a system designer always has the option of increasing transmit power to compensate for loss in a transmit antenna.

A lossy receive antenna, however, irrevocably throws away signal, resulting in a lower signal-to-noise ratio (SNR) and lower overall system performance. Thus, a receive antenna should use resistive loading only as a last resort, not as first aid in meeting matching goals.

Unlike resistive loading, geometry control yields a good match without the need to sacrifice antenna performance in other areas. Intelligent control of antenna geometry is by far the preferred method for obtaining a desired match. Using antenna geometry to control impedance is simple in principle but may be challenging in practice. Obtaining a good match requires a smooth and uniform transition of characteristic impedance from the antenna's connector through to free space. Thus, this section considers how an antenna may be broken down into distinct zones. Figure 3.3 shows how an antenna is composed of three parts:

- A *feed line* to connect an antenna to an RF front end;
- A *feed region* to transition between a feed line and one or more radiating elements;
- *Radiating elements* that serve to couple between radiation fields and the guided fields in the antenna's feed line.

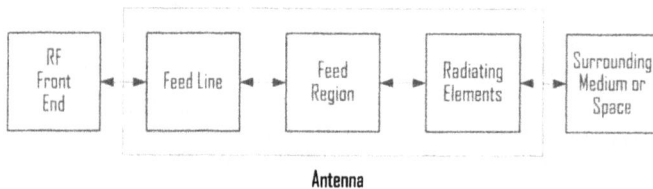

Figure 3.3 An antenna comprises (1) a feed line, (2) a feed region, and (3) radiating elements.

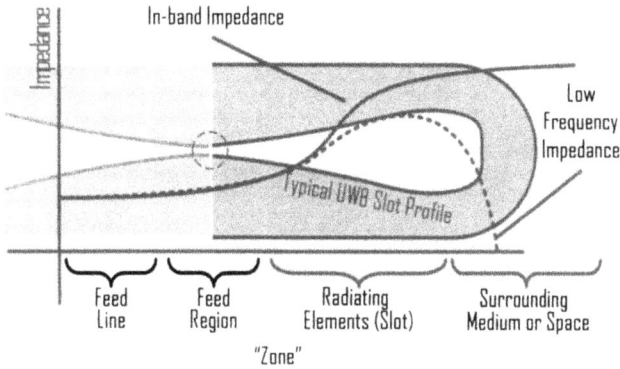

Figure 3.4 Qualitative figure of antenna impedance along the length of a slot antenna. Ideally, a smoothly varying and continuous impedance with length characterizes a feed line (not shown), a feed region, radiating elements, and a surrounding medium.

A well-matched UWB antenna will have a smooth and continuous impedance transformation designed into the feed line, the feed region, and the radiating elements. Properly done, this impedance transition makes the antenna an organic whole that radiates efficiently and provides an excellent match to an RF front end.

Consider a UWB slot antenna, such as the antenna shown in Figure 3.4. A designer adjusts the slot width to yield a 100Ω impedance at the feed region. The two directions of the slot present two 100Ω transmission lines in parallel to the feed region. This yields a good match to a 50Ω source impedance. By increasing the slot width, the slot impedance transitions from 100Ω to the much higher impedance of free space. The end of the slot is a short. Thus, at low frequencies, the impedance goes to zero. Nevertheless, in-band, a properly tapered slot presents a resistive load due to the radiation resistance of the fields decoupling from the antenna. Figure 3.4 illustrates this behavior and shows the effective impedance versus length for in-band frequency components and for out-of-band components.

In this sense, a UWB antenna may be thought of as an extension of a transmission line. Thus, to understand how to build a well-matched UWB antenna, one must first understand transmission lines.

3.2 TRANSMISSION LINES

Figure 3.5(a) depicts a twin-lead transmission line, and Figure 3.5(b) presents a coaxial transmission line. Twin lead is used for lower frequencies, typically very high frequency (VHF) band or lower. This is because twin lead is prone to radiate at higher frequencies. At any given instant, at any location along the transmission line, equal and opposite currents (+ and −) flow along the paired symmetric conductors. This configuration is thus a "balanced" line. If the conductor diameter is D and the distance between conductors is d, then the characteristic impedance of the twin-lead transmission line is

$$Z = \frac{Z_0}{\pi} \cosh^{-1} \frac{d}{D} \qquad (3.3)$$

where $Z_0 = 376.7\Omega$ is the impedance of free space.

Coaxial line is better at higher frequencies because it is formed from an inner conductor and a ground shield. At any given instant, at any location along the line, a current (+) flows on the center conductor, and return currents flow on the inside surface of the ground sheath (0). This configuration is thus an "unbalanced" line. A coaxial line is usually filled with a dielectric to provide mechanical stability. Additionally, the dielectric acts to slow down the signals and decrease the wavelength. If the dielectric constant of the medium is ε_r, and if the speed of light is c, then the velocity of signal propagation is

$$v = \frac{c}{\sqrt{\varepsilon_r}} \qquad (3.4)$$

and the wavelength is

$$\lambda = \frac{c}{f\sqrt{\varepsilon_r}} \qquad (3.5)$$

If the dielectric constant of the material in the line is ε_r, the inner conductor radius is r_i, and the outer ground shield radius is r_0, then the characteristic impedance of the line is

$$Z = \frac{Z_0}{2\pi\sqrt{\varepsilon_r}} \ln \frac{r_0}{r_i} \qquad (3.6)$$

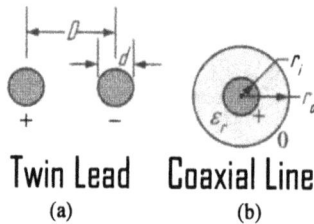

Twin Lead   Coaxial Line
   (a)   (b)

Figure 3.5 (a) A twin lead, and (b) a coaxial line.

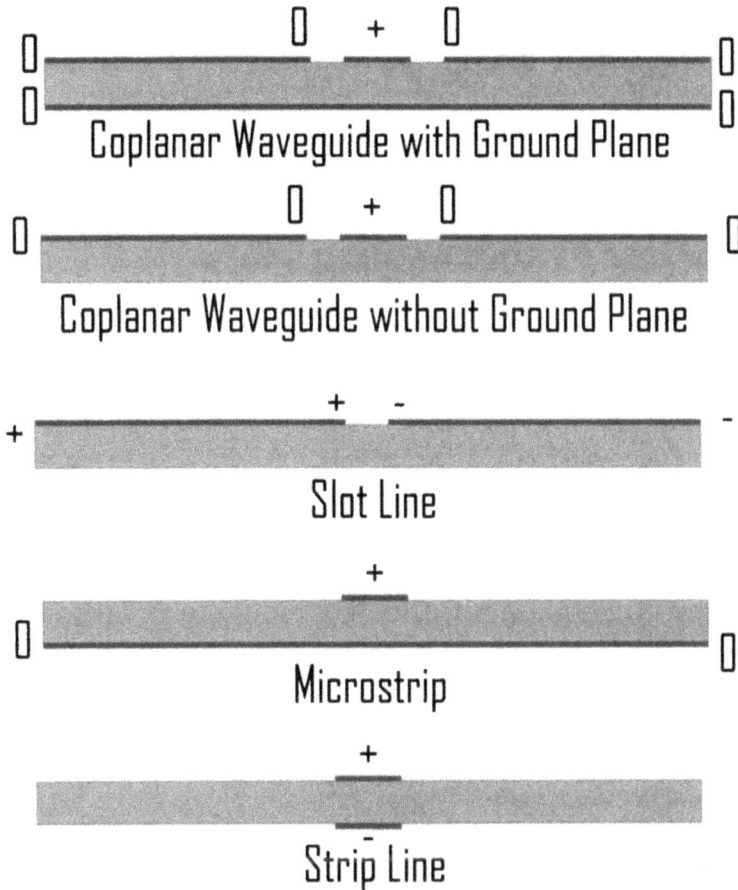

Figure 3.6 A wide variety of planar printed circuit board transmission lines are possible.

Low cost and ease of manufacture make printed circuit board (PCB) transmission lines popular. Figure 3.6 depicts a few of the most popular PCB transmission-line geometries. Excellent references are available to explain the relationship between cross-section geometry and line impedance [8, 9].

These models do not necessarily apply directly to antenna behavior. Transmission-line models, by their nature, assume bound fields and slowly changing field geometries. For instance, a typical model of a slot line assumes that the edges of a slot are orthogonal to each other. When the width of a slot varies rapidly with length, this assumption is no longer valid. Nevertheless, typical transmission-line models provide a valuable starting point for a UWB antenna designer [10].

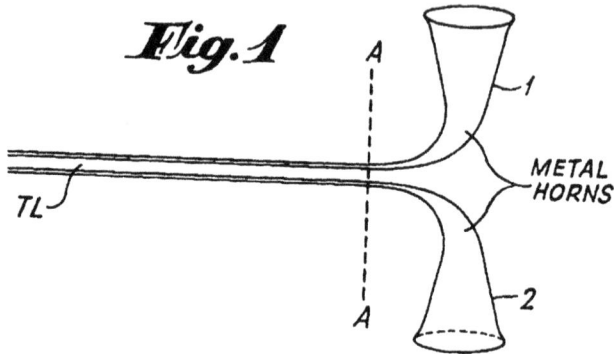

Figure 3.7   Carter's alpine horn antenna presents a tapered transition from a twin-lead transmission line to biconical elements [11].

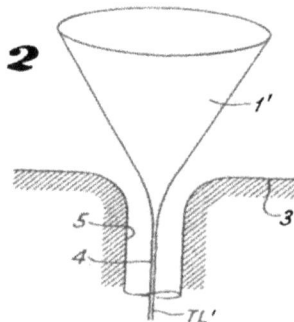

Figure 3.8   Carter's conical monopole element presents a tapered transition from a coaxial line to the conical monopole [11].

3.3 TRANSITION FROM FEED LINE TO FREE SPACE

The transition between the impedance of a particular transmission line and free space impedance is limited only by properties of the geometry in question and the designer's imagination. A wide variety of geometry choices confront a designer. This section will present several examples from engineering practice to illustrate how a designer can achieve an appropriate taper and match.

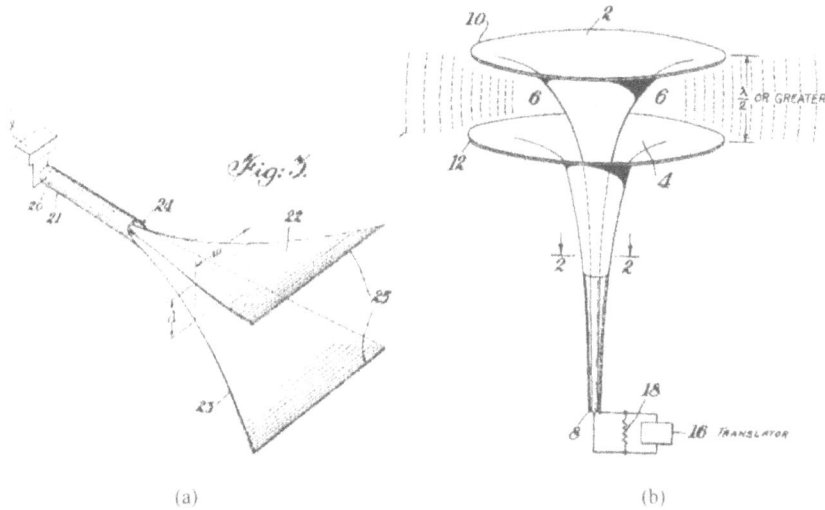

Figure 3.9 (a) Brillouin's coaxial line to parallel plate horn antenna, and (b) Brillouin's coaxial line to omnidirectional horn antenna [12].

3.3.1 Twin-Lead Transition

Carter introduced one of the earliest tapered antenna feeds, the alpine horn antenna [11]. Figure 3.7 is an illustration of this antenna from Carter's patent. Carter's alpine horn antenna was the first major improvement in biconical antenna design since Lodge invented the biconical antenna back in the 1890s. The Alpine Horn antenna presents a tapered transition from a twin-lead transmission line to biconical elements. Carter found that "a continuous and gradual change" in the transmission line geometry led to a "substantially constant surge impedance."

3.3.2 Coaxial Transitions

In the same patent, Carter also showed how a coaxial feed line can be tapered to feed a conical monopole element. Figure 3.8 shows this structure.

Within the next few years, Brillouin devised two other broadband antennas that rely on tapered coaxial feeds [12]. One antenna had a transition from a coaxial line to parallel-plate-horn type radiating elements. Another of Brillouin's designs had a transition from a coaxial line to an omnidirectional horn antenna. Figure 3.9(a, b) shows Brillouin's two designs. Lindenblad's antenna element (featured in Chapter 1) is also an example of a coaxial transition to UWB antenna elements.

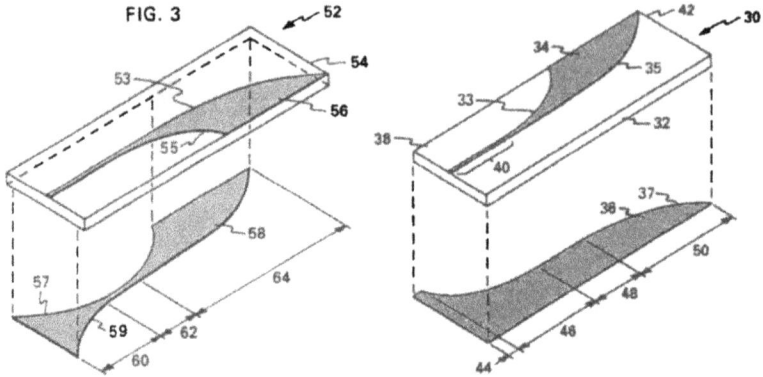

Figure 3.10 Two variations of Nester's microstrip notch antenna ([13]; gray added).

3.3.3 Planar Transmission Line Transitions

The elegantly tapered solid antenna structures of Carter, Brillouin, Lindenblad, and others offer superb UWB performance but are relatively difficult and expensive to manufacture. In more recent years, planar PCB antennas have emerged as a low-cost yet high-performance alternative to classical UWB antenna designs. Many of these antennas rely on tapered PCB transmission lines to achieve broadband matches. This section will review a few examples of antennas involving planar transmission-line transitions.

Planar horn antennas offer high gain in a simple, low-cost structure. One of the earliest planar horn designs was William Nester's microstrip notch antenna [13]. Nester's design involves a gradual transition from a microstrip transmission line to a slot transmission line that then flares out to form horn-type radiating elements. Figure 3.10 shows two variations of Nester's microstrip notch antenna.

Transitions to a slot are of great importance in UWB antenna design. A single slot line may flare out to form a horn or may close up to form a UWB slot antenna. Also, the gap between UWB antenna radiating elements may be thought of as a slot line. The appropriately tapered slot line thus defines the preferred shape for these radiating elements.

I have invented a number of antennas that involve transitions from planar PCB transmission lines to slots [14]. One antenna design takes a CPW line and gradually tapers to the slot between two planar elliptical antenna elements [see Figure 3.11(a)]. Such CPW transitions may be implemented with or without ground planes. Another design takes a microstrip transmission line and gradually tapers to a dual stripline to feed elliptical dipole elements [see Figure 3.11(b, c)]. These designs are examined in greater detail in Chapter 6.

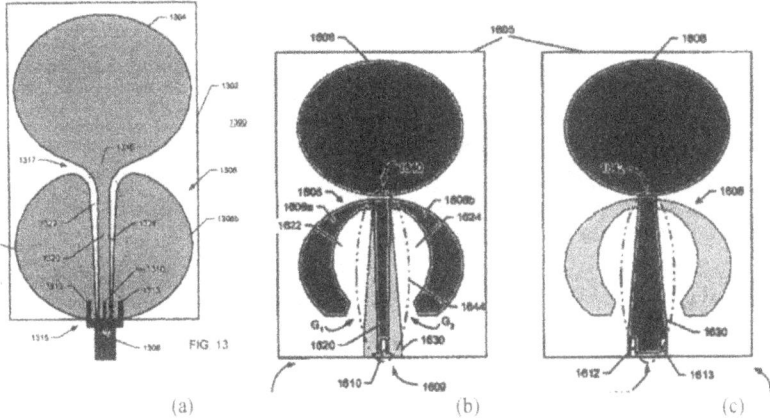

Figure 3.11 (a) Bottom-fed elliptical dipole antenna with a coplanar waveguide to dual notch transition; (b) front side of an alternate bottom-fed elliptical dipole antenna by the author that features a microstrip-to-dual-notch transition; and (c) back-side view of the antenna shown in (b) ([14]; gray added).

This section's examples have thus far exhibited gradual changes in antenna geometry used to achieve good broadband matches. What really matters, however, is obtaining a gradual impedance transition. Thus, even a profound discontinuity in transmission-line geometry is possible if one takes care to maintain a smooth and gradual impedance variation through the transmission-line geometry discontinuity.

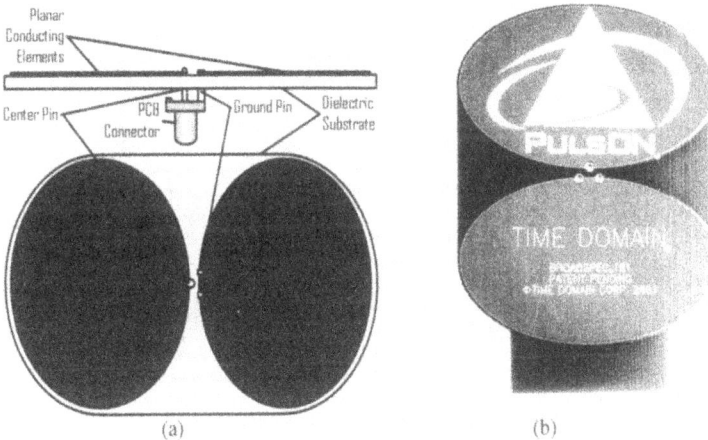

Figure 3.12 (a) A planar elliptical dipole through hole PCB connector that can be used to connect a coaxial feed line directly to a slot line (After [14]), and (b) typical planar elliptical dipole (courtesy of the Time Domain Corporation, Huntsville, Alabama).

For instance, Figure 3.12(a, b) presents an elliptical dipole antenna with a through-hole PCB connector that can be used to connect a coaxial feed line directly to a slot line. One can obtain a good match by carefully spacing the through-hole connector relative to the PCB.

The literature describes a wide variety of transitions [8]. Figure 3.13(a) shows one well-known transition that couples a microstrip transmission line to a slot. One side of the slot flares out to form a notch horn antenna, while the other side is terminated in an open. The same transition may be used to feed the slot gap between two radiating elements. Figure 3.13(b) shows a similar transition, where a quarter-wave open and a quarter-wave short terminate the microstrip and the slot line, respectively [15].

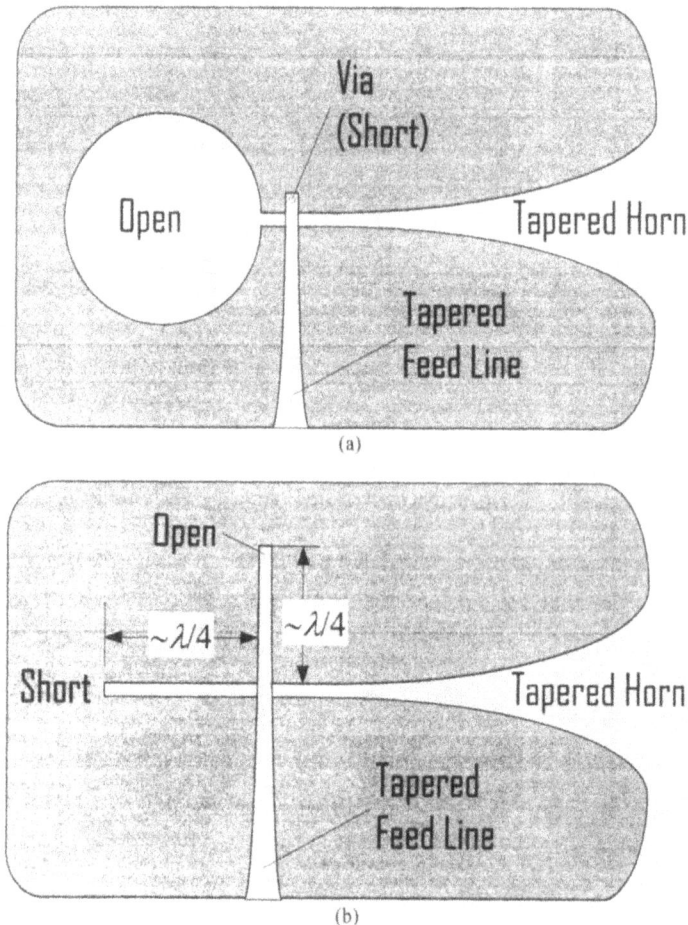

Figure 3.13 (a) Transition from a microstrip line to a slot with a shorted stub, and (b) transition from a microstrip line to a slot with an open stub (After [15]).

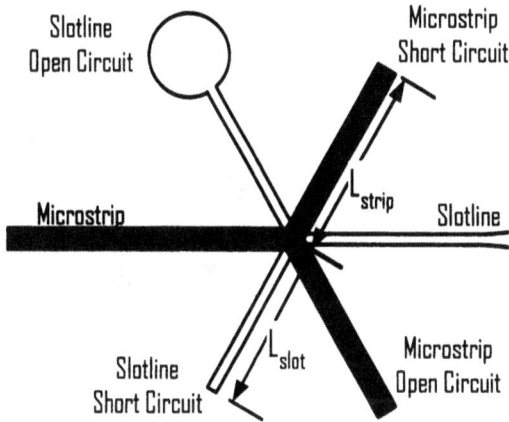

Figure 3.14 A Y-Y balun offers an excellent microstrip-to-slot line transition that operates over a decade of bandwidth or more (After [8]).

Another very broadband transition is the Y-Y balun of Figure 3.14 [16]. This transition couples from a microstrip transmission line to a slot line and operates over a decade bandwidth or more. A microstrip line comes to a junction point and branches out into two equal-length microstrip lines. One microstrip line branch terminates in a short, and the other line terminates in an open. Thus, equal and opposite reflected signals return to the junction point and cancel. Similarly, a slot line comes to the same junction point and branches out into two other slot lines. One slot line branch terminates in an open, and the other terminates in a short. Here again, equal and opposite reflected signals return to the junction point and cancel. This geometry forces signals on the microstrip transmission line to couple into the slot line and vice versa. A wide variety of Y-Y baluns are possible [17, 18]. If board space permits, the Y-Y balun is an excellent choice.

3.4 IMPEDANCE TRANSFORMATION AND MATCHING

Section 3.3 provides a few examples of potential broadband transitions from one transmission-line structure to another. A smooth and continuous transition in a transmission line is helpful to obtain a good match but not sufficient to optimize an antenna. The answer to which geometric profile is optimal rests on which impedance profile is optimal.

This section will examine the connection between impedance and matching and consider how to taper an impedance profile to optimize matching. Then, the challenge facing an antenna designer is to vary the antenna geometry as a function of distance from the feed so as to meet the desired impedance profile.

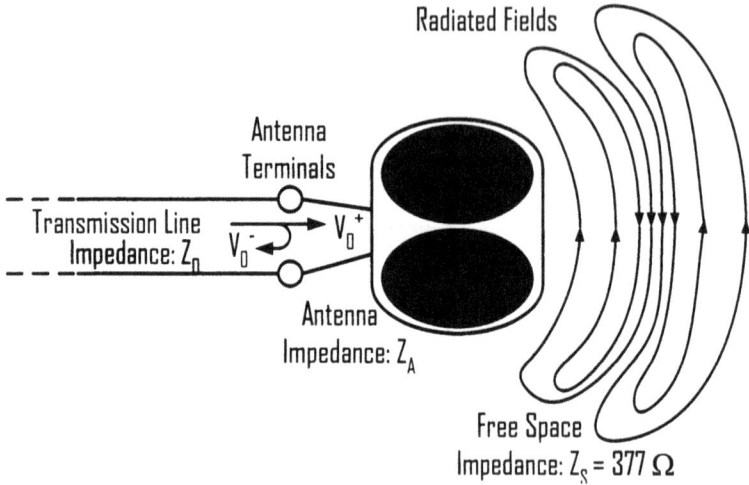

Figure 3.15 Reflection and mismatch from an antenna depend upon transmission line and antenna impedances.

3.4.1 Matching

The impedance of a transmission line (Z_0) is the ratio of the electric to magnetic fields of an electromagnetic signal propagating along the line. The impedance of an antenna is the ratio (Z_A) of the electric to magnetic fields at the antenna's terminals. Note that these impedances are often complex valued quantities since the electric and magnetic fields are not necessarily in phase.

If the transmission-line impedance (Z_0) and the antenna impedance (Z_A) are not identical, then there will be a mismatch at the antenna terminals, and some of the incident signal will be reflected back to the source. This reflection is characterized by a reflection coefficient (Γ), which is the ratio of the reflected voltage (V_0^-) to the transmitted voltage (V_0^+):

$$\Gamma = \frac{V_0^-}{V_0^+} = \frac{Z_A - Z_0}{Z_A + Z_0} \tag{3.7}$$

This reflection coefficient is typically a complex valued quantity since the transmitted and reflected voltages are not necessarily in phase. The reflection coefficient is related to the transmission-line and antenna impedances as shown in (3.7). Figure 3.15 provides an illustration of the quantities involved in the reflection coefficient.

Chapter 2 supplies a superficial overview of matching. Looking deeper into the subject, parameters like return loss and VSWR follow from more fundamental properties like the reflection coefficient, transmission-line impedance, and antenna impedance. The reflection coefficient is the ratio of the transmitted to the reflected voltage, so the power fraction reflected, return loss, or S_{11} is proportional to the reflection coefficient squared. Expressed logarithmically (in decibels), the return loss is:

$$\text{Log Mag } S_{11} = 20\log|S_{11}| = 20\log|\Gamma| = 10\log|\Gamma^2| \qquad (3.8)$$

The other parameter frequently used to characterize matching is the VSWR. The VSWR is defined as the ratio of the peak voltage maximum to peak voltage minimum in the standing-wave pattern at an impedance discontinuity. One may also think of this as the variation in amplitude of the standing-wave envelope. If the transmission line is perfectly matched, then there is no reflected signal. The peak voltage envelope is uniform along the transmission line, so the VSWR is 1:1 for this case.

Suppose the transmission line is mismatched so that the reflected voltage signal is half the amplitude of the transmitted voltage ($|\Gamma| = 0.5$). As the reflected signal propagates back away from the mismatch, it will combine constructively and destructively with the forward-propagating signal. Where it combines constructively, the resulting standing wave will have an amplitude 1.5 times larger than the incident forward-traveling wave alone. Where the waves combine destructively, the resulting standing wave will have an amplitude 0.5 times the amplitude of the original forward-traveling wave. Thus, the ratio of maximum to minimum in the standing-wave envelope is 1.5:0.5 or 3:1. In general, the VSWR is related to the reflection coefficient:

$$\text{VSWR} = \frac{1+|\Gamma|}{1-|\Gamma|} \qquad (3.9)$$

An outstanding tool for visualizing and understanding reflection coefficients, matching, and VSWR is available online [19].

3.4.2 Impedance Transformation

Because signals reflect from impedance discontinuities, the art of impedance matching is the art of gradually varying impedance along a transmission line so as to minimize reflection. Impedance transformation is well covered in many books on microwave theory and techniques [4, 20]. This section reviews a few highlights important to UWB antenna design.

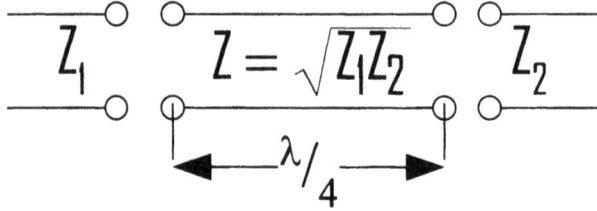

Figure 3.16 A quarter-wave matching section yields an optimal match between lines of different impedances when the matching section impedance (Z) is the geometric average of the impedances.

3.4.2.1 Quarter-wave Transformer

The optimum way to implement a quarter-wave match between two transmission lines with impedances Z_1 and Z_2 with a quarter-wave section of constant impedance line is to use a matching section whose impedance (Z) is the geometric average of the two line impedances:

$$Z = \sqrt{Z_1 Z_2} \tag{3.10}$$

Figure 3.16 shows this quarter-wave matching structure.

3.4.2.2 Exponential Taper Transformation

Of course, a designer is not limited to matching using discrete quarter-wave segments. One may obtain a better match by using a continuously tapered impedance. Figure 3.17 shows an infinitesimal unit length of transmission line with impedance $Z(x)$. Suppose the impedance of this infinitesimal unit length of transmission line is the geometric average of adjacent sections:

$$Z(x) = \sqrt{Z(x+\Delta x)Z(x-\Delta x)} \tag{3.11}$$

Expanding in a Taylor series to second order, the impedance as a function of length is

$$Z(x \pm \Delta x) = Z(x) \pm Z'(x)\Delta x + \tfrac{1}{2}Z''(x)(\Delta x)^2 + O[(\Delta x)^3] \tag{3.12}$$

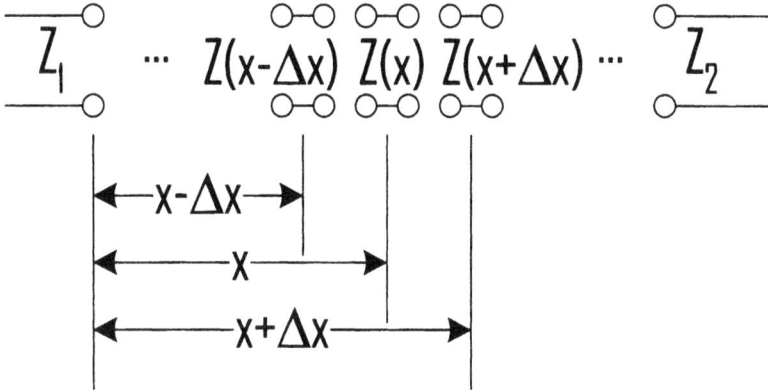

Figure 3.17 In the limit, as each infinitesimal transmission line segment is the geometric mean of adjacent segments, the resulting transmission line taper varies exponentially with length.

where $Z'(x) \equiv \frac{d}{dx} Z(x)$ and $Z''(x) \equiv \frac{d^2}{dx^2} Z(x)$. Ignoring terms of order $(\Delta x)^3$, the geometric average condition may be simplified as

$$Z(x) = \sqrt{Z(x + \Delta x) Z(x - \Delta x)}$$

$$\cong \sqrt{\left(Z(x) + Z'(x)\Delta x + \tfrac{1}{2} Z''(x)(\Delta x)^2\right)\left(Z(x) - Z'(x)\Delta x + \tfrac{1}{2} Z''(x)(\Delta x)^2\right)}$$

$$\cong \sqrt{Z^2(x) + Z(x)Z''(x)(\Delta x)^2 - Z'^2(x)(\Delta x)^2} \qquad (3.13)$$

$$Z^2(x) \cong Z^2(x) + Z(x)Z''(x)(\Delta x)^2 - Z'^2(x)(\Delta x)^2$$

$$Z'^2(x) \cong Z(x)Z''(x)$$

where again, terms of order $(\Delta x)^3$ and higher are ignored. Simplifying the resulting differential equation yields

$$Z'^2(x) \cong Z(x)Z''(x) \qquad\qquad Z'(x) = C_1 Z(x)$$

$$\int \frac{Z''(x)}{Z'(x)} dx \cong \int \frac{Z'(x)}{Z(x)} dx \qquad \int \frac{Z'(x)}{Z(x)} dx = \int C_1 dx \qquad (3.14)$$

$$\ln Z'(x) = \ln Z(x) + C_0 \qquad\qquad \ln Z(x) = C_1 x + C_2$$

$$Z(x) = C_3 \exp[C_1 x]$$

If the boundary conditions are $Z(x = 0) = Z_1$ and $Z(x = L) = Z_2$, the impedance varies exponentially with length according to the relation

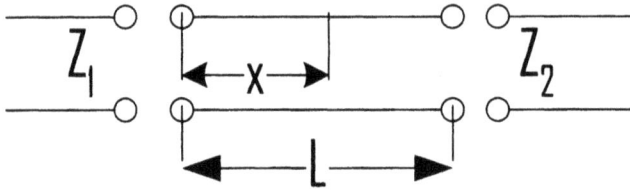

Figure 3.18 The geometry of a tapered impedance transformer.

$$Z(x) = Z_1 \exp\left[\frac{x}{L} \ln \frac{Z_2}{Z_1}\right]$$ (3.15)

An analysis of this impedance taper shows that the reflection coefficient is [20]:

$$\Gamma_{exp} = \Gamma_0 e^{-j\beta L} \frac{\sin \beta L}{\beta L}$$ (3.16)

Figure 3.18 shows the geometry of the exponential taper described by (3.15).

3.4.2.3 Klopfenstein Taper Transformation

An alternate impedance taper was first proposed by R. W. Klopfenstein [21]. The Klopfenstein taper yields a better in-band match than an exponential taper. Conversely, it can yield a comparable match for a shorter length of tapered line. Interestingly, a Klopfenstein taper is characterized by an impedance discontinuity at both ends of the taper, particularly in the limit of a larger passband reflection coefficient (Γ_m). A Klopfenstein taper is defined as follows:

$$\ln Z(x) = \frac{1}{2} \ln[Z_1 Z_2] + \frac{\Gamma_0}{\cosh A} A^2 \phi\left(\frac{2x}{L} - 1, A\right)$$ (3.17)

where the reflection coefficient at zero frequency is

$$\Gamma_0 = \frac{Z_2 - Z_1}{Z_2 + Z_1}$$ (3.18)

where the function $\phi(x,A)$ is defined

$$\phi(x, A) = -\phi(-x, A) = \int_0^x \frac{I_1\left[A\sqrt{1-y^2}\right]}{A\sqrt{1-y^2}} dy \qquad (3.19)$$

where the maximum ripple in the passband follows from the parameter A is

$$\Gamma_m = \frac{\Gamma_0}{\cosh A} \qquad (3.20)$$

and where $I_1[x]$ is the modified Bessel function. The reflection coefficient of a Klopfenstein taper transition is

$$\Gamma_{Klopf} = \frac{\Gamma_0 e^{-j\frac{\omega}{c}L}}{\cosh A}\begin{cases} \cos\sqrt{\left(\frac{\omega}{c}L\right)^2 - A^2} & \frac{\omega}{c}L \geq A \\ \cosh\sqrt{A^2 - \left(\frac{\omega}{c}L\right)^2} & \frac{\omega}{c}L < A \end{cases} \qquad (3.21)$$

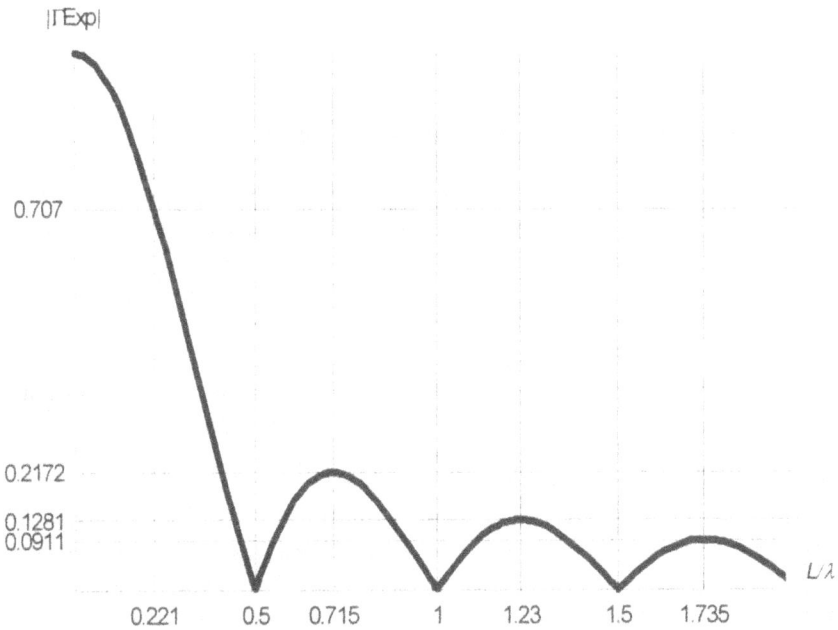

Figure 3.19 Reflection coefficient for an exponential taper in the large impedance difference limit.

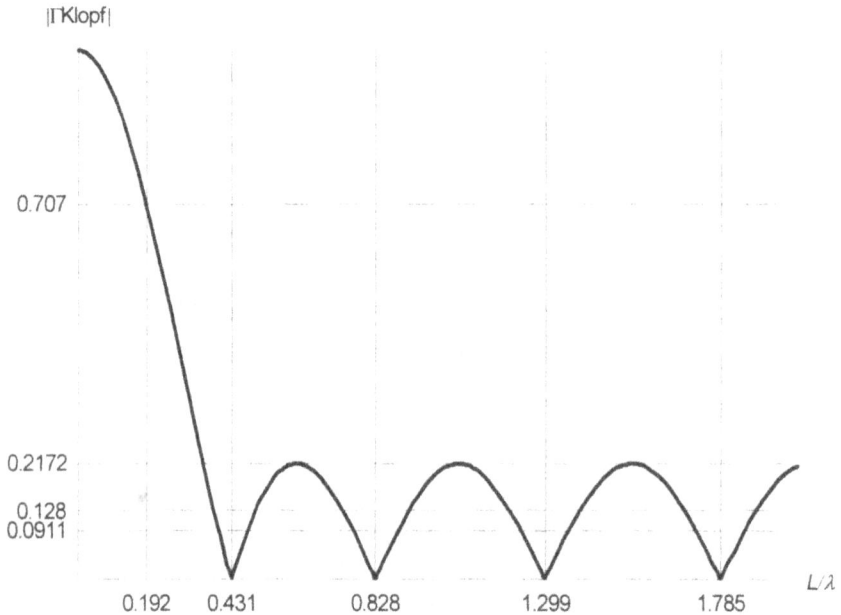

Figure 3.20 Reflection coefficient for a Klopfenstein taper in the large impedance difference limit.

3.4.2.4 Comparing Exponential and Klopfenstein Tapers

A few basic characteristics follow from an analysis of the exponential reflection coefficient of (3.16). Because of the $\sin(\beta L)/(\beta L)$ dependence, the in band reflection of the exponential taper can never be better than -13.3dB below the zero frequency reflection coefficient Γ_0. In the case of a large impedance difference, $\Gamma_0 \rightarrow 1$, and the reflection coefficient for the exponential taper will have a peak of magnitude $\Gamma_{exp} = 0.2172$. To compare the exponential and Klopfenstein tapers on a fair and balanced basis, set the in-band reflection coefficient to $\Gamma_m = 0.2172$. Figure 3.19 shows the exponential reflection coefficient plotted against the length of the taper in units of wavelength. Figure 3.20 shows the same plot for the Klopfenstein taper.

The -3dB limit for the exponential taper lies at $L = 0.221\lambda$ compared to $L = 0.192\lambda$ for the Klopfenstein taper. Thus, for the same in-band reflection coefficient, the Klopfenstein taper is about 13% shorter. The exponential taper does have the advantage that the reflection peaks decrease with increasing frequency. Figure 3.21 compares the return loss (S_{11}) of the exponential and Klopfenstein tapers.

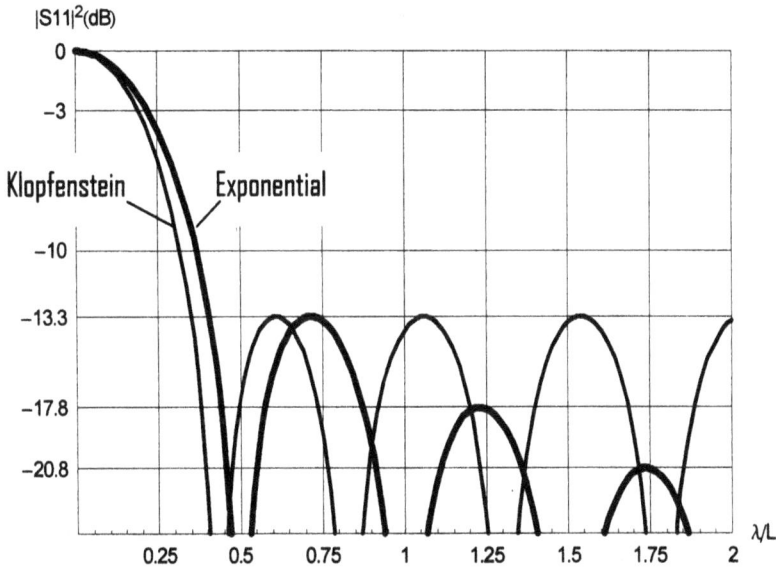

Figure 3.21 Return loss (S_{11}) for the exponential and Klopfenstein tapers.

The Klopfenstein taper has an additional degree of freedom relative to the exponential taper because the passband reflection coefficient (Γ_m) is a free parameter. The passband reflection coefficient may be increased to yield a shorter transition at the cost of a worse match. For instance, taking $\Gamma_m = 0.3333$ (VSWR = 2.00:1) yields a 0.173λ taper. Alternatively, a smaller passband reflection coefficient yields an improved match at the cost of a longer transition. By way of example, selecting $\Gamma_m = 0.1000$ (VSWR = 1.22:1) yields a 0.223λ taper, slightly longer than the 0.221λ taper of an exponential taper in the limit of a large impedance difference ($\Gamma_0 \to 1$).

3.5 COUPLING BALANCED AND UNBALANCED LINES

Yet another kind of transformation that is absolutely critical to the antenna designer and RF system engineer alike is the transformation between a balanced and an unbalanced line. Unbalanced lines (like a coaxial transmission line) are commonly used in conjunction with test equipment and with RF hardware. Even if impedances are perfectly matched, when an unbalanced line is connected to a balanced antenna (like a dipole), the discontinuity in impedance line structure can lead to currents on the outside of what should otherwise be a neutral, shielded, "cool" cable. See Figure 3.22 for an illustration of currents on the outside sheath of an unbalanced cable connected to a balanced antenna.

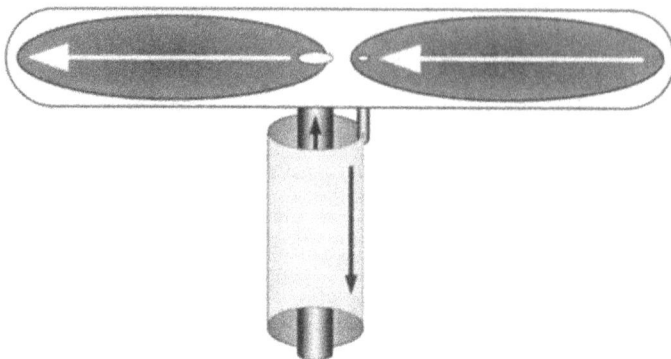

Figure 3.22 Connecting an unbalanced line (like a coax) to a balanced antenna (like a dipole) can yield sheath currents and lead to hot cables.

These sheath currents yield "hot" cables. Cables with exterior currents are hot in the RF sense: they will radiate or receive RF signals. Instead of the carefully crafted intended antenna pattern, the resulting system has an effective pattern due to the combined effect of the antenna, the cable, and the RF device. Hot cables are evident when physical contact with cables or attached RF devices yields a change in system performance. For instance, if touching a cable produces a visible change in the measured properties of an antenna attached to test equipment, then one should suspect hot cables.

Hot cables can cause a variety of problems. They can decrease the gain, particularly of a high-gain antenna. Hot cables are also liable to muddy or fill in antenna nulls, cause patterns to shift, or alter the desired polarization purity of the antenna. Hot cables may make a system more vulnerable to noise. Touching hot cables or associated RF devices can cause noticeable variations in system performance, yielding spurious or unexpected anomalies. Perhaps the most significant effect of a hot cable is to make the cable and the attached RF device part of the antenna, particularly at low frequencies. Below the operational band of the antenna, where the antenna is not an effective radiator or receiver of RF signals, hot cables connected to the conducting enclosure of a typical RF device may actually be a much better antenna system than the antenna itself. This makes the system as a whole vulnerable to out-of-band interferers and may make it difficult for the system to meet stringent out-of-band emissions limits.

There are three general solutions to the problem of hot cables. The first class of solutions involves using a choke to suppress sheath currents. The second type of solution is to use a broadband *bal*anced to *un*balanced (balun) transformer to couple between an unbalanced line and a balanced antenna (or vice versa). The final and best solution is to avoid the problem entirely by making the RF front end, the transmission line, and the antenna compatible by using appropriate feed and line geometries.

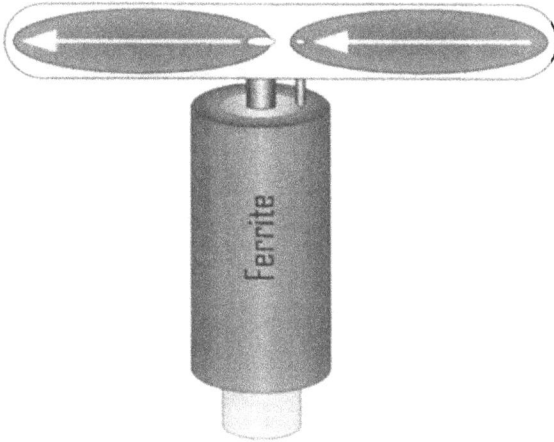

Figure 3.23 Ferrite loading absorbs sheath current energy, reducing the effect of hot cables.

3.5.1 Chokes

The first method for dealing with hot cables is to use a choke to suppress sheath currents and thus "cool" hot cables. Ferrite and other lossy cable coatings help dissipate sheath currents. "Bazooka baluns" help trap and reflect sheath currents. Finally, inductive loading helps increase the relative impedance of sheath current modes diverting currents back to the inside of cables where they belong.

The first type of choke is lossy loading. The intent behind lossy loading is to absorb and damp out sheath currents. Typical lossy loadings include ferrite beads. Steel wool wrapped around a cable may be a readily available and inexpensive substitute for ferrites. Lossy loading not only resistively dissipates sheath currents, it also tends to increase the impedance of sheath currents, diverting currents to the inside of the cable. In addition to the traditional ferrite clam shells and beads, lengths of ferrite-loaded flexible tubing are available [22]. A section of semirigid coaxial line may be slid through such tubing, significantly suppressing sheath currents. Figure 3.23 shows a ferrite choke on an antenna feed cable.

Figure 3.24 shows a bazooka balun. A bazooka balun is a conducting shield around a cable. The shield is approximately a quarter-wavelength long, open at the antenna end, and electrically connected to the sheath of the cable at the far end. A bazooka balun traps sheath currents inside the shield. Ideally, any sheath currents reflect from the short at the bottom of the bazooka balun and cancel out at the top. A bazooka balun tends to be relatively narrowband, however. Also, the bazooka balun looks like a capacitive stub in parallel with the antenna and may impact matching.

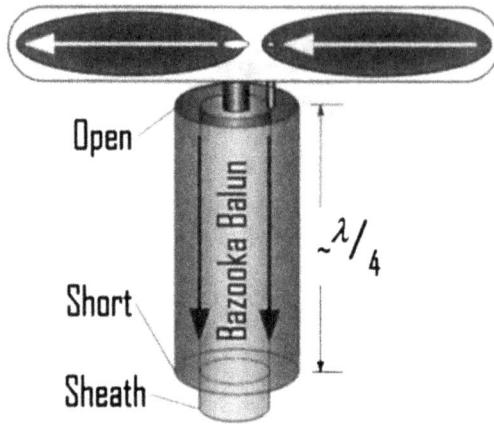

Figure 3.24 A bazooka balun traps sheath currents but tends to have a relatively narrowband response.

Yet another technique for controlling sheath currents is use of inductive loading. A simple inductive load may be implemented by merely coiling the feed cable into a solenoidal shape as illustrated in Figure 3.25. One must exercise care, though, not to overdo the coil. In addition to inductance, a coil like the one shown in Figure 3.25 will have a certain parasitic capacitance from one turn to the next. An unlucky designer may discover that his inductive coil forms a tank circuit resonating at an inconvenient frequency. Thus, using an inductive coil is a possible but not very promising solution.

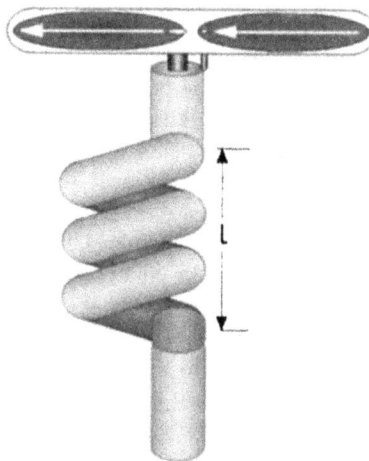

Figure 3.25 An inductive coil increases the inductance of sheath current modes, but is prone to resonate due to parasitic capacitance.

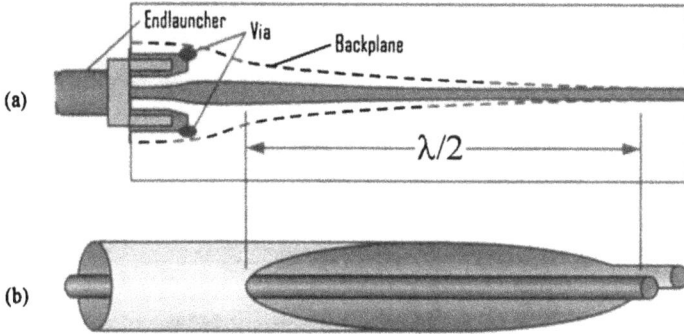

Figure 3.26 (a) A microstrip-to-stripline tapered-line balun transformer (After [25]), and (b) a coaxial-to-twin lead tapered-line balun transformer (After [28]).

3.5.2 Balun Transformers

Of the available choke techniques, ferrite loading is the most promising, particularly in the UWB context. But choke techniques merely alleviate the symptom, sheath currents, rather than address the underlying malady, incompatible transmission line geometries. There is, however, an alternate class of solutions: balun transformers.

A balun transformer couples a balanced line to an unbalanced line or vice versa. Ferrite balun transformers are well understood and offer excellent broadband performance [23]. There is also an outstandingly clear and accessible resource explaining ferrite transformers and how to build them [24]. Maybe someday ferrite structures that work at microwave frequencies will be available, but for the time being, alternate solutions are needed.

One particularly simple-to-implement balun transformer is a tapered line transformer. Figure 3.26(a) shows a microstrip to stripline tapered line balun transformer first introduced by Duncan and Minerva [25]. This microstrip tapered balun offers good broadband performance over about an octave of bandwidth with the limiting factor being even mode resonances [26]. More complicated balun structures can use N transmission lines to form broadband N^2:1 impedance-ratio baluns [27].

Figure 3.26(b) shows a coaxial-to-twin lead tapered-line balun transformer [28]. Ideally, the tapered section should be as long as space allows but at least approximately a half-wavelength at the low end of the operational band. For optimal matching, the impedance should be uniform along the line. If necessary, the taper may serve as an impedance transformer as well. These structures may be integrated into antennas also, as shown in Figures 3.10 and 3.11(a–c) for the microstrip-to-stripline taper and Figure 3.9 for the coaxial taper.

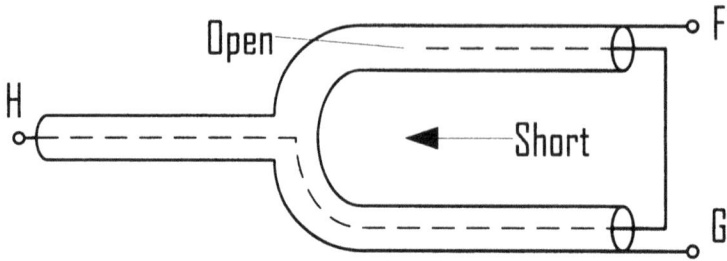

Figure 3.27 A Roberts balun. An unbalanced load connects at H, and a balanced load connects across
F and G. An open and short cancel each other out electrically to affect the unbalanced-to-
balanced transition (After [29]).

Willmar Roberts introduced a balun transformer in 1957 that has since seen
wide use [29]. The Roberts balun (shown in Figure 3.27) uses a combination of an
open and a short to cancel each other out electrically, forcing signals to make the
desired unbalanced-to-balanced transition. This balun can operate over more than
an octave in frequency with good performance and may even perform modest
impedance transformation, such as from a 50Ω feed line to a 70Ω antenna. The
Roberts balun may also be implemented in a PCB architecture, as shown in Figure
3.28. Similar PCB balun structures may also be used [30].

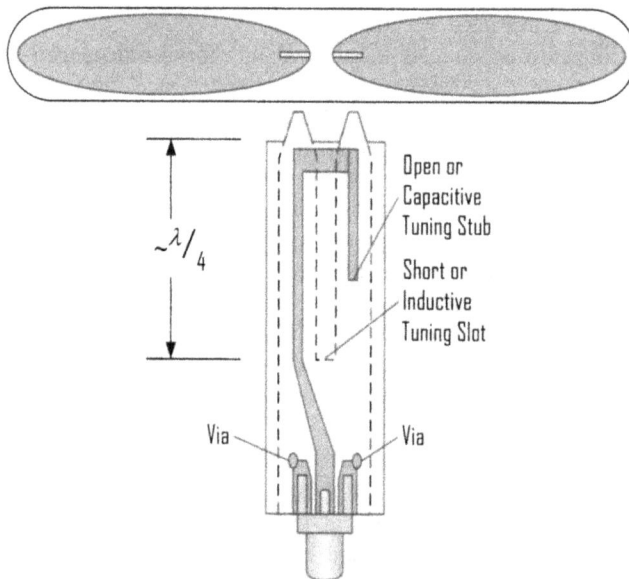

Figure 3.28 A PCB Roberts balun transformer converts an unbalanced line geometry to a balanced
line geometry (After [30]).

3.5.3 Compatibility

Both chokes and balun transformers act to remedy the problem of coupling balanced-to-unbalanced devices. Ultimately, however, the best solution is to avoid the problem entirely with a more integrated design that combines an RF front end with a compatible transmission line and antenna. Most test equipment is single ended and accepts coaxial transmission lines, so for development stage hardware, unbalanced hardware is attractive. If an application calls for a balanced, omnidirectional dipole, however, a differential RF front end and a balanced transmission line make sense if production quantities justify the additional difficulty in development, testing, and evaluation.

3.6 ANTENNAS AS TRANSFORMERS

This chapter discussed UWB antennas by considering them as impedance transformers. From this point of view, an antenna is a transformer coupling between a line impedance and free space impedance. Unlike the case of narrowband antennas, the impedance of a UWB antenna is a matter of design, not an intrinsic fact of nature. To design a UWB antenna with a desired impedance and match requires thinking of a UWB antenna as an extension of a transmission line in which impedance varies smoothly and continuously from the feed line through the feed region and radiating elements to free space. Exponential and Klopfenstein tapers provide a good theoretical starting point for a UWB antenna design. Finally, there is a variety of potential solutions to the problem of coupling unbalanced lines to balanced antennas (or vice versa). The best solution, however, is to design in compatibility from the beginning.

A well-matched antenna is an important part of a successful UWB system. Equally important, however, is an antenna that radiates and receives signals with a desired pattern and polarization. To evaluate these issues requires looking at an antenna from an alternate point of view, delving into currents and fields, and treating an antenna as a radiating and receiving structure. This will be the topic of Chapter 4.

Endnotes

[1]   Schelkunoff, Sergei A., Bell System Technical Journal, Vol. 17, 1938, p. 17.

[2]   Munk, Ben A., private communication with the author, 1999.

[3]   Munk, Ben A., *Finite Antenna Arrays and FSS*, New York: John Wiley and Sons, 2003.

[4]   Pozar, David M., *Microwave Engineering*, 2nd ed., New York: John Wiley and Sons, 1998.

[5] Schantz, Hans, "Planar elliptical element ultra-wideband dipole antennas," 2002 IEEE Antennas and Propagation Society International Symposium, San Antonio, Texas, June 16-21, 2002, Vol. 3, pp. 44–47.

[6] Agrawall, N. P., et al., "Wide-band monopole antennas," IEEE Transactions on Antennas and Propagation, Vol. 46, No. 2, February 1998, pp. 406–407.

[7] Bryant, J. H, "Coaxial transmission lines, related two-conductor transmission lines, connectors, and components: a U.S. historical perspective," IEEE Transactions on Microwave Theory and Techniques, Vol. 32, September 1984, pp. 970–983.

[8] Gupta, K. C., et al., Microstrip Lines and Slot Lines, Norwood, MA: Artech House, 1979. This is the ultimate reference on PCB transmission lines.

[9] Hilberg, Wolfgang, Electrical Characteristics of Transmission Lines, Norwood, MA: Artech House, 1979. Hilberg's work isn't as practical and useful as the text of Gupta et al., but it does have an extensive table of transmission line geometries with analytical expressions for impedance.

[10] Excellent free transmission line software is available from a variety of sources, including
Agilent AppCAD: www.hp.woodshot.com
UltraCAD: www.ultracad.com
Applied Wave Research: www.appwave.com/products/txline.html

[11] Carter, Phillip S., "Wide band, short wave antenna and transmission line system," U.S. Patent 2,181,870, December 5, 1939.

[12] Brillouin, Leon N., "Broad band antenna," U.S. Patent 2,454,766, November 30, 1948.

[13] Nester, William, "Microstrip notch antenna," U.S. Patent 4,500,887, February 19, 1985.

[14] Schantz, Hans, "Apparatus for establishing signal coupling between a signal line and an antenna structure," U.S. Patent 6,512,488, January 28, 2003.

[15] Smolders, A.B. and M.J. Arts, "Wide-band antenna element with integrated balun," Proceedings of the IEEE Antennas and Propagation Society International Symposium, 1998. Vol. 3, June 21–26, 1998, pp. 1394–1397.

[16] Jokanovic, Branka, and Velimir Trifunovic, "Double-Y Baluns for MMICs and Wireless Applications," Microwave Journal, Vol. 41, No. 1, January 1998, pp. 70–92.

[17] Trifunovic, Velimir, and Branka Jokanovic, "Four decade bandwidth uniplanar balun," Electronics Letters, Vol. 28, No. 6, March, 12 1992, pp. 534–535.

[18] Gu, Huifang and Wu Ke, "Broadband design considerations of uniplanar double-Y baluns for hybrid and monolithic integrated circuits," 1999 IEEE MTT-S Digest, 1999, pp. 863–866.

[19] Besser Associates have a fascinating online interactive Java applet that shows reflection and mismatch due to an impedance discontinuity. At the time of this writing, the applet was available at www.bessernet.com/Ereflecto/appletframe.htm.

[20] Collin, Robert E., Foundations for Microwave Engineering, New York: McGraw-Hill, 1966.

[21] Klopfenstein, R. W., "A Transmission Line Taper of Improved Design," Proceedings of the IRE, Vol. 44, January 1956, pp. 31–35. See also Ref. 4, pp. 291–293.

[22] Capcon® Suppressant Tubing; See www.capconemi.com. Capcon® is a registered trademark of Capcon International, Inc.

[23] Guanella, G., "High frequency balancing units," U.S. Patent 3,025,480, March 13, 1962.

[24] Sevick, Jerry, *Building and Using Baluns and Ununs: Practical Designs for the Experimenter*, Hicksville, NY: CQ Communications, Inc., 1994.

[25] Duncan, J. W. and V. P. Minerva, "100:1 Bandwidth Balun Transformer," Proceedings of the IRE, Vol. 48, 1960, pp. 156–164.

[26] Climer, B., "Analysis of suspended microstrip taper balun," IEE Proceedings, 135, Pt. H, No.2 April 1988, pp. 65–69.

[27] McCorkle, John W., "Microstrip DC-to-GHz field stacking balun," U.S. Patent 5,523,728, June 4, 1996.

[28] Knott, P., et al., "Coaxially-fed tapered slot antenna," Electronics Letters, Vol. 37, August 30, 2001, No. 18, pp. 1103–1104.

[29] Roberts, Willmar K., "A New Wide-Band Balun," Proceedings of the IRE, Vol. 45, December 1957, pp. 1628–1631.

[30] Siwiak, Kazimierez, "Microstrip balun-antenna apparatus," U.S. Patent 4,737,797, April 12, 1988.

Chapter 4

Antennas as Radiators

Electric and magnetic force. May they live forever, and never be forgot, if only to remind us that the science of electromagnetics, in spite of the abstract nature of the theory, involving quantities whose nature is entirely unknown at present, is really and truly founded upon the observation of real Newtonian forces, electric and magnetic respectively.

Oliver Heaviside, 1900

This chapter looks at antennas from a more traditional viewpoint: as radiators and receivers of electromagnetic fields. Then as now, electromagnetics is indeed an abstract subject, as Heaviside observed [1]. Its application to antennas provides a welcomed opportunity for electromagnetic science to demonstrate its practical value. First, this chapter compares and contrasts time domain and frequency domain analysis of signals. Then, it presents time domain electromagnetics for linear antennas. Next, it introduces a variety of approximations that make a direct (if qualitative) analysis of time domain antenna problems easier and it explains the fields around an ideal dipole source. Finally, this chapter summarizes and explains the advantages and limitations of thinking of antennas as radiators and receivers of electromagnetic fields.

4.1 TIME DOMAIN AND FREQUENCY DOMAIN

The time and frequency domains present two complementary and equivalent ways of looking at physical signals. Traditional electromagnetic practice has leaned heavily in the direction of frequency domain or "harmonic" signals. To be sure, the harmonic functions like sin ωt and cos ωt (or more generally, $e^{j\omega t}$) have many attractions. They offer an excellent approximation to the steady-state behavior of periodic systems or signals. Harmonic functions have their place in every engineer's toolbox. Some, however, have been so beguiled by the beauty of harmonic functions that they regard all physical signals as sine waves or as superpositions of sine waves. This is most emphatically not the case. In many situations, harmonic functions offer a potentially misleading description.

Unlike harmonic functions, any physically real signal has a definite beginning and a definite end. As one begins to consider nonsteady-state, or transient, systems, the shortcomings of harmonic analysis become increasingly obvious. For instance, any attempt to model a step function using superposition of harmonic functions yields overshoot and ringing, a phenomenon so well known that it has a special name ("Gibbs' Phenomenon"). As Henning Harmuth aptly quipped, "Hence, we must conclude that the Fourier series converges everywhere, except where it is needed" [2]. Harmuth has dubbed this overreliance on harmonic functions the "Dogma of the Circle" in a devastating critique [3].

In short, harmonic signals are only one member of a vastly larger family of functions. These functions become accessible through time domain analysis. Short duration, transient, or UWB processes are often better modeled using functions other than harmonic ones. Thus, the UWB engineer must be bilingual, conversant in both the time domain as well as in the frequency domain.

4.1.1 Impulses and Sine Waves

Begin by considering a sine wave. Representing a time domain function using a capital T and a frequency domain function using a capital F, a sine wave may be written

$$T(t) = \sin \omega t \qquad (4.1)$$

and

$$F(f) = \delta(f - f_C) \qquad (4.2)$$

in the time and frequency domains, respectively. Assume the sine wave has a center frequency $f_C = \omega/(2\pi)$ and a period $\tau = 1/f_C$. In the time domain, a sine wave is infinite in extent. In the frequency domain, it is a spike or Dirac delta function at its particular frequency. Figure 4.1(a, b) illustrates this behavior.

Now consider a time domain impulse occurring at time $t = \tau$. Such a function is a spike in time and of infinite span in frequency. This function may be written

$$T(t) = \delta(t - \tau) \qquad (4.3)$$

and

$$F(f) = 1 \qquad (4.4)$$

in the time and frequency domains, respectively. Figure 4.2(a, b) shows the time domain and frequency domain representations of a time domain impulse.

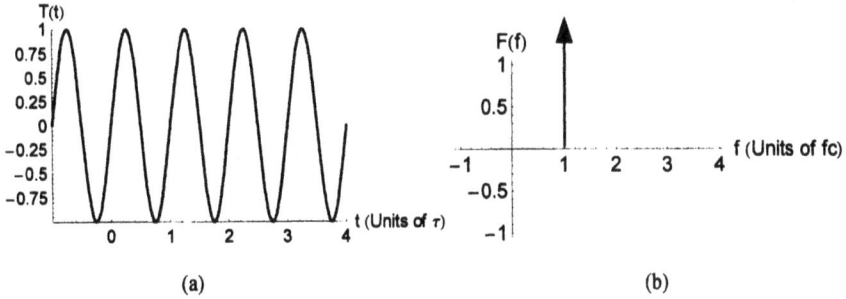

(a)   (b)

Figure 4.1   (a) A sine wave with period 1 and frequency 1 in the time domain, and (b) the same sine wave as a spike, or Dirac delta, in the frequency domain.

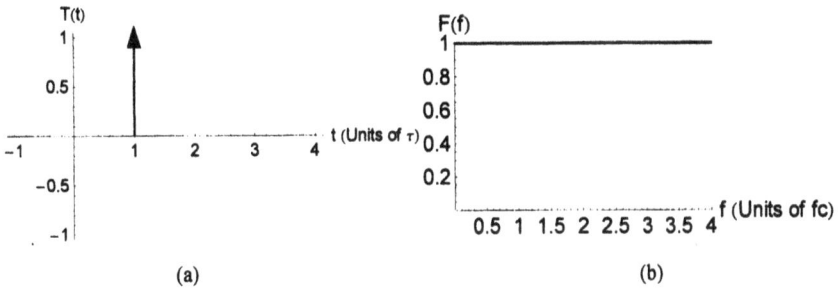

(a)   (b)

Figure 4.2   (a) A time domain impulse and, (b) the same impulse in the frequency domain.

4.1.2 Basic Principles of the Frequency and Time Domains

Analytic functions may be transformed between the time and frequency domains using integral relations. The Fourier Integral Transformation

$$F(\omega) = \int_{-\infty}^{+\infty} T(t) e^{-j\omega t} dt$$

$$(4.5)$$

yields the frequency domain spectral density, $F(\omega)$, of a time domain signal $T(t)$. The Inverse Fourier Integral Transformation

$$f(t) = \frac{1}{2\pi} \int_{-\infty}^{+\infty} F(\omega) e^{+j\omega t} d\omega$$

$$(4.6)$$

yields the time domain signal corresponding to a frequency domain spectral density. From a practical point of view, an engineer is usually more interested in the power spectral density (PSD), $u(\omega) = |F(\omega)|^2$, than in the signal spectral density, $F(\omega)$.

A key feature of time and frequency domain signals is the scaling property. A signal that is narrow in time is broad in frequency. Similarly, a signal that is broad in time is narrow in frequency. Expressed mathematically, if $T(t) \leftrightarrow F(\omega)$, then $f(at) \leftrightarrow \frac{1}{a} F(\omega/a)$. This property has many significant implications. Any attempt to create a "brick wall" filter, one with an abrupt and severe roll off in the frequency domain, will lead to a long duration ringing in the time domain. Similarly, a pulse of extremely short duration in time or an abrupt change in a time domain signal will have broad spectral content.

The shifting properties have similar importance. A shift in time (or, in other words, a time delay) yields a linear phase shift. Expressed mathematically, if $T(t) \leftrightarrow F(\omega)$, then $T(t - t_0) \leftrightarrow F(\omega) e^{-j\omega t}$. A frequency shift, multiplies a time domain signal by $e^{-j\omega_0 t}$. In other words, if $T(t) \leftrightarrow F(\omega)$, then $T(t) e^{-j\omega_0 t} \leftrightarrow F(\omega - \omega_0)$.

Finally, many time domain radiation processes involve differentiation, so it is important to understand the impact of time domain differentiation and integration on frequency domain spectra. Time differentiation multiplies frequency domain spectra by ω, while time integration divides frequency domain spectra by ω. Thus,

if $T(t) \leftrightarrow F(\omega)$, then $\dfrac{df(t)}{dt} \leftrightarrow j\omega F(\omega)$ and $\displaystyle\int_{-\infty}^{t} f(t)dt \leftrightarrow \dfrac{F(\omega)}{j\omega}$.

More properties of frequency domain signals may be found in any decent text on linear systems [4].

4.1.3 Time Domain Signals

Many families of time domain signals are useful in UWB engineering. "Gaussian" signals offer a good mix between time domain compactness and frequency domain compactness. "Edges" support modeling of transients. "Sincs," are abruptly bounded in one domain and broad in the other domain. Finally, truncated sine waves are of great significance because of the ease with which they may be generated.

4.1.3.1 Gaussian Signals

The family of Gaussian signals is defined by

$$T_n(t) = \frac{\tau^n \left(\frac{n}{2}\right)!}{n!} \frac{d^n}{dt^n} e^{-\frac{t^2}{\tau^2}}$$

(4.7)

Gaussian Signals

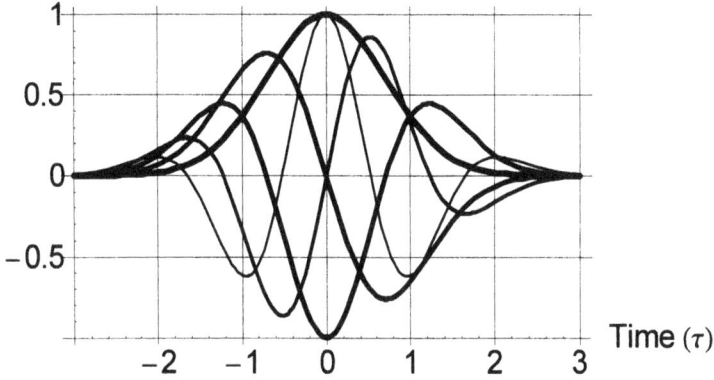

Figure 4.3 The first five members ($n = 0, 1,..., 4$) of the Gaussian family where $\tau = 1$.

where τ is a time constant, and index n is the integer number of derivatives. Note that the normalization of this function ensures that for all even n, the Gaussian function will have unit maximum amplitude. Figure 4.3 shows the family of Gaussian signals for $n = 0, 1,..., 4$. The first derivative ($n = 1$) is sometimes called a "Gaussian doublet." The second derivative ($n = 2$) is sometimes called a "Gaussian W." The Gaussian signal will be almost entirely contained within $\pm 3\tau$ of $t = 0$, even for arbitrarily large n.

In the frequency domain,

$$F_n(\omega) = \frac{\tau^n \left(\frac{n}{2}\right)!}{n!} (j\omega)^n \sqrt{\pi\tau^2} e^{-\frac{\tau^2\omega^2}{4}}$$

(4.8)

The center frequency follows from solving for the critical point

$$\frac{d}{d\omega} F_n(\omega) = 0 = n j^n \omega^{n-1} e^{-\frac{\tau^2\omega^2}{4}} - \frac{\tau^2\omega}{2} j^n \omega^n e^{-\frac{\tau^2\omega^2}{4}}$$

(4.9)

to yield

$$\omega_c = \frac{\sqrt{2n}}{\tau}$$

(4.10)

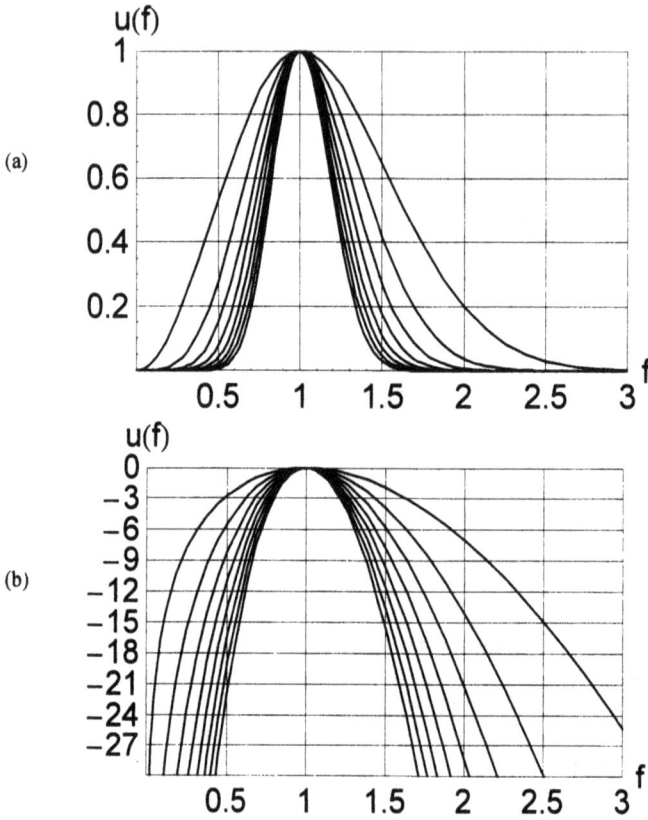

Figure 4.4 (a) Linear PSD for $n = 1, 2, \ldots, 8$, and (b) log PSD for $n = 1, 2, \ldots, 8$.

or

$$f_c = \frac{\omega_c}{2\pi} = \frac{\sqrt{2n}}{2\pi\tau}$$

(4.11)

Alternatively, one may solve for the time constant in terms of the desired center frequency

$$\tau = \frac{\sqrt{2n}}{2\pi f_c}$$

(4.12)

Normalizing to unit peak, the PSD is

Q (Inverse Bandwidth) of Gaussian Signals

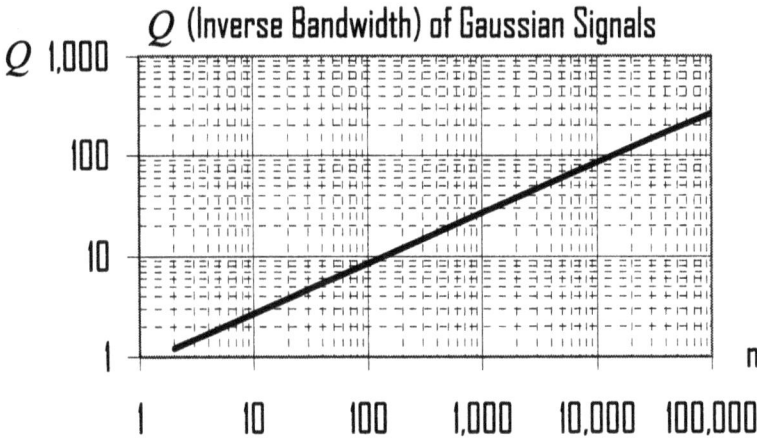

Figure 4.5   Inverse bandwidth of Gaussian signals as a function of n.

$$u(\omega) = \begin{cases} e^{-\frac{\tau^2 \omega^2}{2}} & n = 0 \\ \left(\dfrac{e}{4\pi^2 f_c^2}\right)^n \omega^{2n} e^{-\frac{n\omega^2}{4\pi^2 f_c^2}} & n > 0 \end{cases} \tag{4.13}$$

Figure 4.4(a, b) portrays this PSD in both linear and logarithmic form. Figure 4.4(a, b) makes obvious the slow narrowing of the bandwidth with increasing n. In fact, this slow convergence is one of the major shortcomings of Gaussian signals. Many signals of practical use require a significant number of derivatives. This results in unwieldy analytic expressions. For instance, a 0.2 fractional bandwidth UWB signal ($Q = 5$) requires $n \sim 35$. Figure 4.5 shows the result of a numeric calculation of the inverse fractional bandwidth (or Q) of Gaussian signals as a function of n. These bandwidth results assume −3dB bandwidth.

4.1.3.2 Edge Signals

The family of step functions, or "edge" signals, was first introduced by Heaviside. An excellent parameterization for a rising or falling edge with a characteristic time τ (or, in other words, a Heaviside step function) is

$$T_\pm^0(t) = \tfrac{1}{2}\left(1 \pm \tanh \tfrac{t}{\tau}\right) \tag{4.14}$$

where a rising edge is associated with the "+" sign, and a falling edge is associated with the "–" sign [5]. The derivatives of this function form an orthogonal family of signals that can be useful in analyzing transient, or time domain, signals. The first few derivatives are

$$T_\pm^1(t) = \pm \frac{\operatorname{sech}^2 \frac{t}{\tau}}{2\tau} \tag{4.15a}$$

$$T_\pm^2(t) = \mp \frac{1}{\tau^2} \operatorname{sech}^2 \frac{t}{\tau} \tanh \frac{t}{\tau} \tag{4.15b}$$

$$T_\pm^3(t) = \frac{1}{\tau^3} \left(\mp \operatorname{sech}^4 \frac{t}{\tau} \pm 2 \operatorname{sech}^2 \frac{t}{\tau} \tanh^2 \frac{t}{\tau} \right) \tag{4.15c}$$

$$T_\pm^4(t) = \frac{1}{\tau^4} \left(\pm 8 \operatorname{sech}^4 \frac{t}{\tau} \tanh \frac{t}{\tau} \mp 4 \operatorname{sech}^2 \frac{t}{\tau} \tanh^3 \frac{t}{\tau} \right) \tag{4.15d}$$

and

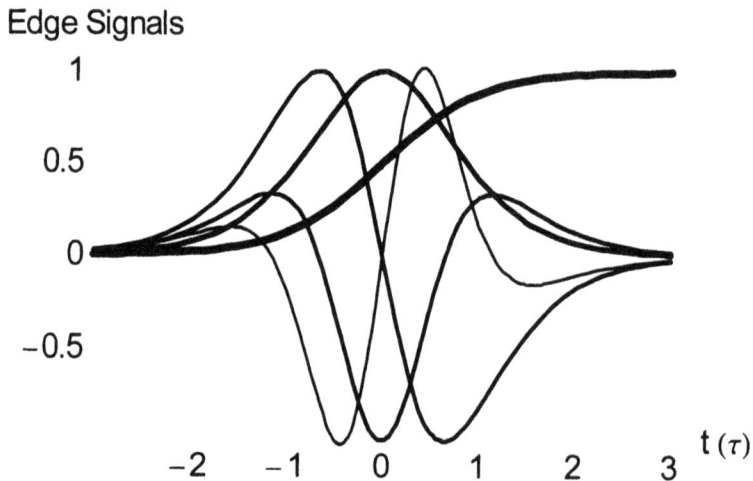

Figure 4.6   Family of signals related to a rising edge and first four derivatives.

$$T_\pm^5(t) = \frac{1}{\tau^5}\left(\pm 8\,\text{sech}^6\tfrac{t}{\tau} \mp 44\,\text{sech}^4\tfrac{t}{\tau}\tanh^2\tfrac{t}{\tau} \pm 8\,\text{sech}^2\tfrac{t}{\tau}\tanh^4\tfrac{t}{\tau}\right)$$

(4.15e)

Figure 4.6 shows this family of signals. Increasing derivatives yield a higher-frequency-content, narrower-band signal.

This behavior becomes more obvious in considering the frequency response. The frequency response follows from the Fourier Integral Transform (4.5). The challenge in applying the Fourier Integral Transform to a family of time domain signals is in selecting a particular member of that family whose integral is easy to evaluate. In this case, (4.15a) provides a good starting point

$$\int_{-\infty}^{+\infty} \frac{\text{sech}^2\tfrac{t}{\tau}}{2\tau}\exp(-j\omega t)dt = \omega\,\text{csch}\frac{\pi\tau\omega}{4}\,\text{sech}\frac{\pi\tau\omega}{4}$$

(4.16)

Remaining members of the frequency domain family are readily found by using the appropriate power of ω

$$F_0(\omega) = \text{csch}\tfrac{\pi\tau\omega}{4}\,\text{sech}\tfrac{\pi\tau\omega}{4}$$

(4.17a)

$$F_1(\omega) = \omega\,\text{csch}\tfrac{\pi\tau\omega}{4}\,\text{sech}\tfrac{\pi\tau\omega}{4}$$

(4.17b)

$$F_2(\omega) = \omega^2\,\text{csch}\tfrac{\pi\tau\omega}{4}\,\text{sech}\tfrac{\pi\tau\omega}{4}$$

(4.17c)

$$F_3(\omega) = \omega^3\,\text{csch}\tfrac{\pi\tau\omega}{4}\,\text{sech}\tfrac{\pi\tau\omega}{4}$$

(4.17d)

and

$$F_4(\omega) = \omega^4\,\text{csch}\tfrac{\pi\tau\omega}{4}\,\text{sech}\tfrac{\pi\tau\omega}{4}$$

(4.17e)

Figure 4.7(a, b) shows the linear and logarithmic power spectra of these signals.

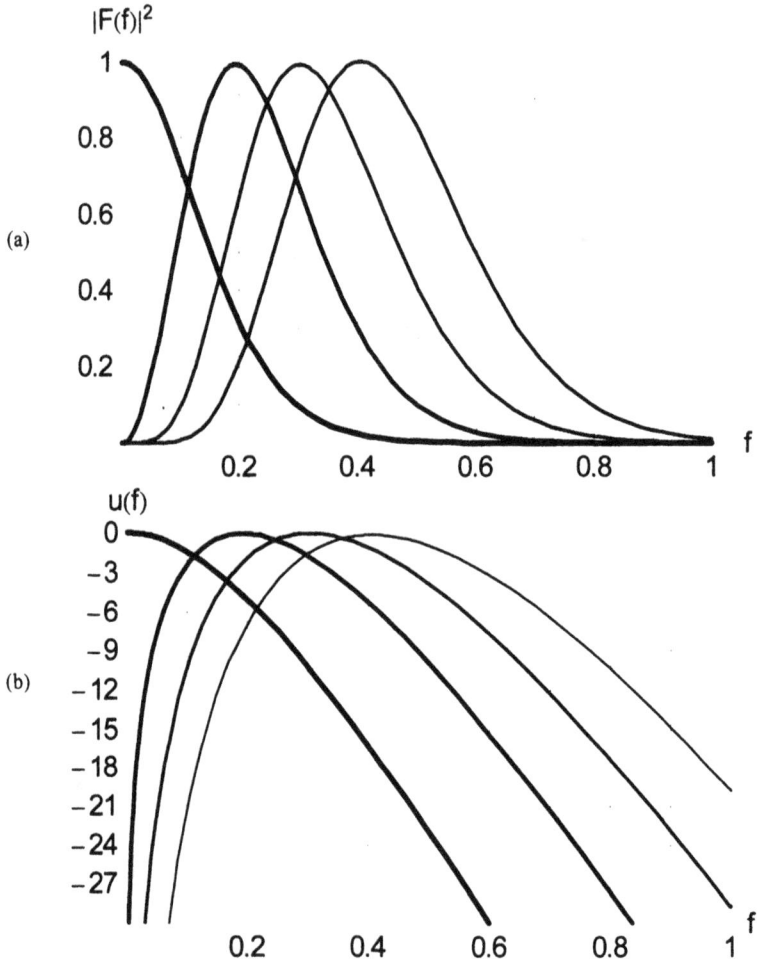

Figure 4.7   (a) Linear PSD of edge signals, and (b) logarithmic PSD of edge signals.

4.1.3.3 Sinc Signals

Another useful time domain function is a "sinc" signal

$$T(t) = \text{sinc}\frac{\pi t}{\tau} = \frac{\sin\dfrac{\pi t}{\tau}}{\dfrac{\pi t}{\tau}}$$

(4.18)

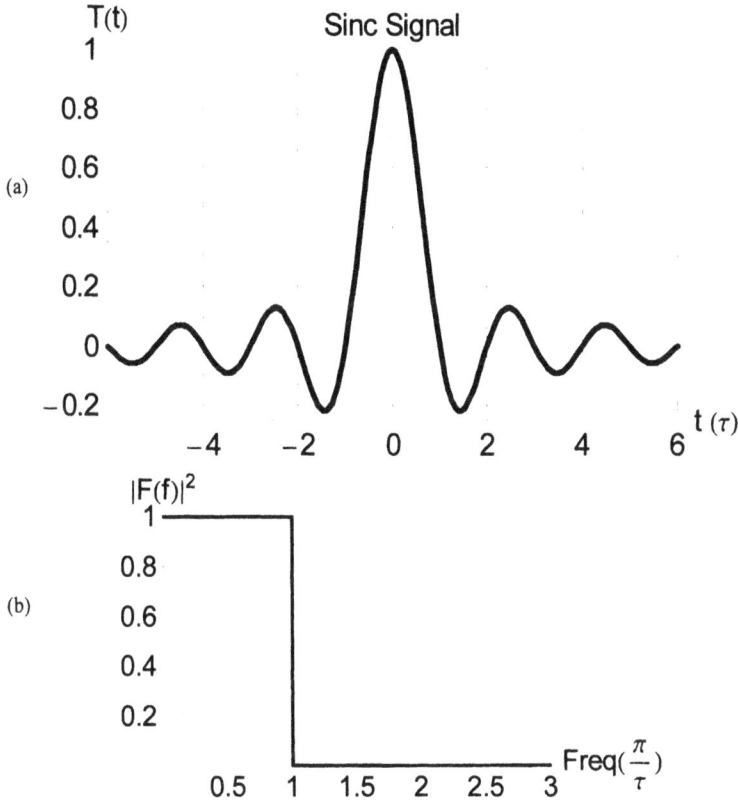

Figure 4.8 (a) Sinc signal, and (b) frequency distribution of a sinc signal.

A sinc function is effectively a Dirac delta smeared out over a characteristic time $\pm\tau$. A sinc function is an excellent illustration of the scaling property of Fourier transforms. A sinc function has a brick-wall edge in the frequency domain, making it relatively short in frequency span. A sinc function has a correspondingly very long extent ringing in the time domain. Figure 4.8(a) presents a sinc function plotted with respect to time in units of τ, and Figure 4.8(b) shows the frequency domain representation of this signal.

The advantage of sinc functions is that they are narrowly bounded in the frequency domain. The disadvantage of these signals is their dc content. Radiated signals cannot have dc content for reasons explained later in this chapter. Thus, a sinc signal is not a very useful representation of time domain fields, unless one were to consider the family of derivatives of a sinc function. An alternate and easier way to consider families of signals that are well defined in the frequency domain follows from applying the shifting theorem in the time domain.

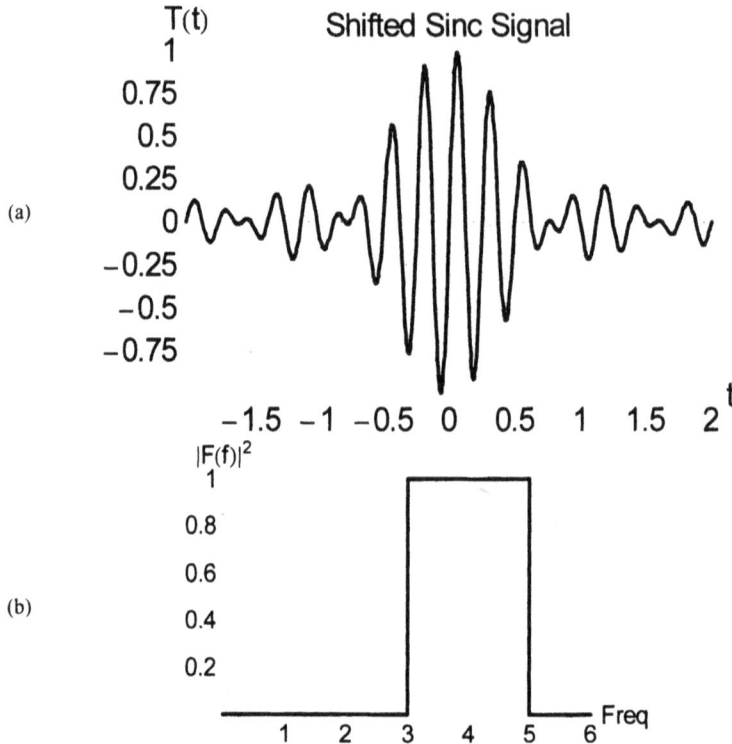

Figure 4.9   (a) Shifted sinc signal (in phase and quadrature) with bandwidth $BW = 2$ and center frequency $f_C = 4$, and (b) PSD of a shifted sinc signal.

A "shifted sinc" function has a defined bandwidth BW about a center frequency f_C

$$T(t) = \frac{\sin(2BW\ t)}{2BW\ t}\left(\begin{array}{c}\sin 2\pi f_C t \\ \cos 2\pi f_C t\end{array}\right)$$

(4.19)

Note that the sine and cosine define an in-phase signal (I) and a corresponding quadrature signal (Q). For instance, consider a quadrature pair of signals with bandwidth $BW = 2$ and center frequency $f_C = 4$. Figure 4.9(a, b) shows these signals and their power spectrum.

Shifted sinc signals are a good example of signals tightly bounded in the frequency domain. By the scaling theorem however, these signals are characterized by the long time domain ringing evident in Figure 4.9(a). Filtering a broadband impulse signal through a brick-wall filter yields a shifted sinc signal. Also, applying an Inverse Fourier Transform to a swept frequency system (like a network analyzer) yields a shifted sinc signal. This process (with a more gentle windowing function) is illustrated in Figure 2.5. Nevertheless, applications are few where one can tolerate the time-dispersed ringing inherent to shifted sinc functions.

4.1.3.4 Truncated Sine Signals

"Truncated sine" signals are a very useful family of signals that combine a short time domain duration and a reasonably compact representation in the frequency domain [6]. They have the additional advantage of being easily generated from relatively straightforward hardware [7, 8, 9]. A truncated sine signal is a truncated sine wave of n cycles such as may be generated by rapidly switching (or flipping) a carrier at time intervals that are integer multiples of the carrier period (T_0). One definition of a truncated sine signal is

$$T(t) = \begin{cases} \sin 2\pi f_C t & 0 < t < nT_0 \\ 0 & else \end{cases}$$

(4.20)

The spectra of truncated sine signals is

$$F(f) = \frac{2f_c}{n} \frac{f_c\left(1 - \exp\left[-\dfrac{jn2\pi f}{f_C}\right]\right)}{2\pi(f - f_C)(f + f_C)}$$

(4.21)

Figure 4.10(a) shows a time domain truncated sine signal with center frequency $f_C = 4$ and $n = 2$. Figure 4.10(b) shows the PSD of the family of truncated sine signals centered at $f_C = 4$.

Truncated sine signals exhibit sinclike spectral behavior with spectral sidelobes on the order of 10–15dB below the main lobe. These sidelobes may be readily filtered out in practical implementations.

4.1.3.5 Summary of Signals

Gaussian signals and edge signals offer a balance between time domain and frequency domain compactness. Shifted sinc signals are preferred when frequency domain compactness is critical, as in meeting a sharp spectral mask. The worst of the time domain ringing associated with shifted sinc signals may be removed by applying modest frequency domain windowing. Truncated sine signals are preferred where time domain compactness is critical, as in a high-data-rate UWB system. The frequency domain sidelobes of truncated sine signals may be improved by filtering.

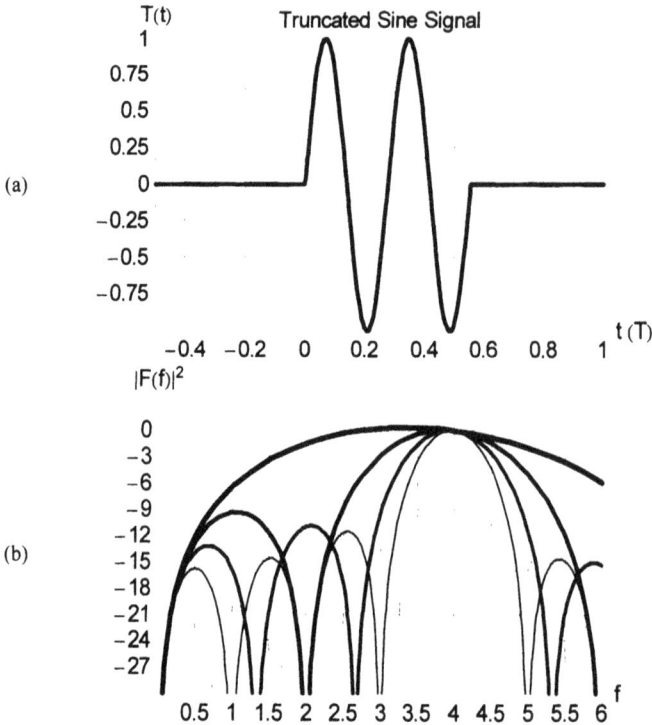

Figure 4.10 (a) Truncated sine signal with $f_C = 4$ and $n = 2$, and (b) Frequency spectra of truncated sine signals with center frequency $f_C = 4$ and $n = 1, 2,...,6$.

4.1.4 Time Domain versus Frequency Domain

The time domain and the frequency domain offer two complementary and equivalent ways of looking at physical signals. A proficient UWB antenna designer needs to be "bilingual," able to translate between, or operate within, either of these domains as the context demands.

The frequency domain is suited for studying narrow bandwidth, or steady state systems. These are the usual systems considered in traditional RF practice, hence the overwhelming predominance of frequency domain thinking in RF theory. The time domain is preferred for studying wide bandwidth, or transient, systems. Thus, the time domain is of particular interest to those who work with UWB systems.

The time domain has significant advantages over the frequency domain when considering wide bandwidth, or transient, systems. The frequency domain obscures time-dependent processes behind the opaque cloak of Fourier transforms.

Conversely, in the time domain, one may directly examine the natural evolution of time-dependent processes from start to finish. The frequency domain assumes a steady-state process in which each cycle is followed by an identical cycle ad infinitum. The time domain assumes a transient process with a definite beginning and a definite end. Finally, and most importantly, operating in the time domain allows one to track fields and energy in time, following their course throughout a physical process. This ability yields valuable insight into the functioning of physical systems.

4.2 MAXWELL'S EQUATIONS

In 1864, Maxwell presented his *A Dynamical Theory of the Electromagnetic Field*, summarizing the behavior of electric and magnetic fields in about 20 equations [10, 11]. Two decades later, Heaviside derived a more compact form of Maxwell's equations, using his novel vector notation and introducing vector differential operators [the gradient (∇), the curl ($\nabla\times$), and the divergence ($\nabla\cdot$)] [12]. The differential form of Maxwell's equations has the advantage of being localized so it pertains to the fields at a particular point in space:

Gauss's Law		Gauss's Law for Magnetism	
$\nabla \cdot \mathbf{E} = \dfrac{\rho}{\varepsilon_0}$	(4.22a)	$\nabla \cdot \mathbf{B} = 0$	(4.22b)
Faraday's Law		**Ampere's Circuit Law**	
$\nabla \times \mathbf{E} = -\dfrac{\partial \mathbf{B}}{\partial t}$	(4.22c)	$\nabla \times \mathbf{B} = \varepsilon_0 \mu_0 \dfrac{\partial \mathbf{E}}{\partial t} + \mu_0 \mathbf{J}$	(4.22d)

Although this form of Maxwell's equations pertains to the fields at a particular point in space, these fields also depend upon the behavior of sources distributed throughout space. Application of Maxwell's equations to time domain problems involves some concepts familiar to traditional frequency domain calculations, like "generalized coordinates." There are also wrinkles unique to the time domain, like the concept of "retardation."

The traditional form of Maxwell's equations suffers from a particularly serious difficulty: this form expresses electric fields in terms of magnetic fields and vice versa, making a simultaneous solution difficult. Fortunately there is an alternate formation of Maxwell's equations in which the fields are explicit functions of the currents and charges. This alternate form lends itself to a right-hand rule that makes understanding radiation fields relatively straightforward.

Maxwell's equations allow a calculation of the radiated fields from a known charge or current distribution: a description of the behavior of the fields around a known current distribution or a known geometry. This is simultaneously their strength and their weakness. Although Maxwell's equations enable an understanding of a known antenna geometry, they do not directly allow us to derive a novel geometry with useful properties.

This section will review generalized coordinates and retardation. Then, this section will present generalizations of the Coulomb and Biot-Savart laws which allow a calculation of the electric and magnetic fields as a function of the charge and current densities. Finally, this section presents a simplified right-hand rule that makes understanding electromagnetic radiation much easier.

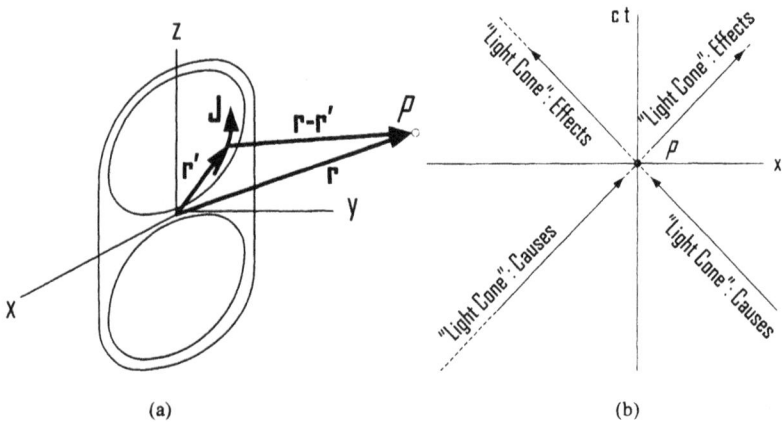

(a)   (b)

Figure 4.11   (a) A current filament is a source on an antenna. The fields at a point P are a function of the vector difference between the source point located at $\mathbf{r'}$ and the field point located at $\mathbf{r}$. (b) A space-time diagram shows the location in space and time of causes and effects.

4.2.1 Generalized Coordinates and Retardation

The fields at a field point P depend on the distance between the field point and the source point. If the source point is located at vector position $\mathbf{r'}$ and the field point is located at vector position $\mathbf{r}$, then the fields depend upon the magnitude $x = |\mathbf{r} - \mathbf{r'}|$. Figure 4.11(a) illustrates this principle.

Electromagnetic influences propagate at the speed of light, so only those events that happened at time t in the past at a distance $x = ct$ away can exert a causal influence at a field point P at time $t = 0$. Physicists depict this behavior using a "space-time diagram" like the one shown in Figure 4.11(b). The vertical axis of a space-time diagram denotes time, and the horizontal axis denotes distance. Typically, the axes are scaled so that influences propagating at the speed of light trace paths at a 45° angle. These paths converging on point P form a "causal light cone." This light cone consists of the locations of all the potential causes in space-time that happened at time t in the past a distance $x = ct$ away. Similarly, what happens at a field point P at time $t = 0$ can exert an effect only at locations lying a distance $x = ct$ away at a time t in the future.

This accounting for cause-and-effect relationships requires use of "retarded" time

$$t_{ret} = t - {}^x/_c \tag{4.23}$$

A time domain function $T(t)$ evaluated at retarded time may be denoted using square brackets

$$[T(t)] = T(t - {}^x/_c) \tag{4.24}$$

Tracking the causal chain of events in an electromagnetic system is among the most challenging aspects of time domain electromagnetics.

4.2.2 Jefimenko Form of the Biot-Savart and Coulomb Laws

The classical Coulomb's Law relates an electrostatic field to a static charge distribution. Similarly, the classical Biot-Savart Law relates a magnetostatic field to a static current distribution. Both these laws may be generalized to fully time varying charge and current distributions [13]. These generalizations were popularized by Jefimenko [14] but have been derived elsewhere in earlier literature [15]. The Jefimenko form of Coulomb's Law is

$$\mathbf{E} = \frac{1}{4\pi\varepsilon_0} \int \left(\frac{[\rho]}{x^2}\hat{\mathbf{x}} + \frac{1}{xc}\left[\frac{\partial\rho}{\partial t}\right]\hat{\mathbf{x}} - \frac{1}{xc^2}\left[\frac{\partial\mathbf{J}}{\partial t}\right] \right) dV \tag{4.25}$$

where ρ is the time-dependent charge distribution, $\partial\rho/\partial t$ is the time derivative of the time-dependent charge distribution, and $\partial\mathbf{J}/\partial t$ is the time derivative of the time-dependent current density. All three of these quantities are enclosed in the square brackets that denote evaluation at retarded time. This expression for the time-dependent electric field makes clear that the electric field comprises three terms: an electrostatic term proportional to ρ, an electroinduction term proportional to $\partial\rho/\partial t$, and a radiation term proportional to $\partial\mathbf{J}/\partial t$.

The Jefimenko form of the Biot-Savart Law is

$$\mathbf{H} = \frac{1}{4\pi} \int \left(\frac{[\mathbf{J}]\times\hat{\mathbf{x}}}{x^2} + \frac{1}{xc}\left[\frac{\partial\mathbf{J}}{\partial t}\right]\times\hat{\mathbf{x}} \right) dV \tag{4.26}$$

where **J** is the time-dependent current distribution, and $\partial J/\partial t$ is the time derivative of the time-dependent current density. This expression for the time-dependent magnetic field shows that the magnetic field comprises two terms: a magnetic-induction term proportional to **J**, and a radiation term proportional to $\partial J/\partial t$. Interestingly, the geometry by which the current distribution gives rise to the magnetic-induction field is identical to the manner in which the rate of change of the current distribution gives rise to the radiation term. The next section will exploit this interesting fact to derive a right-hand rule for radiation.

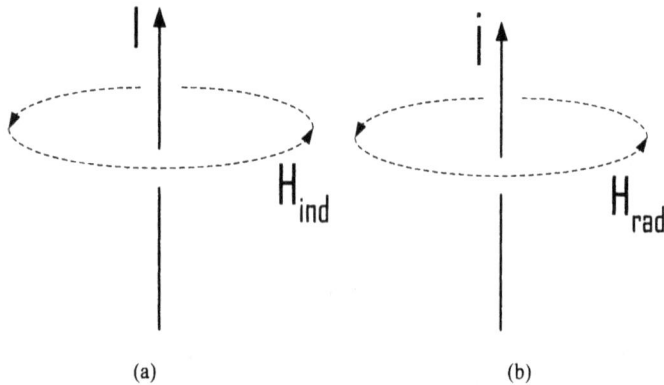

Figure 4.12 (a) The right-hand rule for magnetic induction fields about currents. (b) The right-hand rule for magnetic radiation fields about changing currents (After [16]).

4.2.3 Right-Hand Rule for Radiation

Place the thumb of your right hand in the direction of an electric current and the fingers of your right hand will curl in the direction of the induction magnetic field. Generations of electromagnetics students have learned this simple rule illustrated in Figure 4.12(a). There is an analogous rule for the radiation magnetic field, little appreciated until recently [16]. Place the thumb of your right hand in the direction of the rate of change of an electric current, and the fingers of your right hand will curl in the direction of the radiation magnetic field. Figure 4.12(b) depicts this rule.

Near a source current (typically within about $\lambda/2\pi$), static and induction components of the electric and magnetic fields dominate. These near fields are coupled to the source and have components radial or longitudinal to the source current. By the time the fields have decoupled to form radiation, they are almost entirely transverse. The electric-radiation field vector **E** and the magnetic-radiation field vector **H** lie in the transverse plane. The transverse plane is orthogonal to the direction $\hat{k}$ in which the wave propagates. Figure 4.13 shows this geometry.

In free space, the ratio between the electric-field intensity and the magnetic-field intensity is given by the free space impedance

$$Z_S = \frac{|\mathbf{E}|}{|\mathbf{H}|} = \sqrt{\frac{\mu_0}{\varepsilon_0}} = 376.6\ \Omega \approx 120\pi\ \Omega$$

(4.27)

where ε_0 = 8.854pF/m is the free space permitivity, and μ_0 = 1.257μH/m ($4\pi \times 10^{-7}$H/m) is the free space permeability. By numerical coincidence, the value for the free space impedance is approximately equal to 120π.

Combining these three principles—the right-hand rule for radiation, the transverse nature of electromagnetic waves, and the definition of field impedance—yields a relation for the electric-radiation field in terms of the magnetic-radiation field,

$$\mathbf{E} = -Z_S \hat{k} \times \mathbf{H}$$

(4.28)

and a relation for the magnetic-radiation field in terms of the electric-radiation field,

$$\mathbf{H} = \frac{1}{Z_S} \hat{k} \times \mathbf{E}$$

(4.29)

These principles are very useful for understanding radiation from simple linear antennas.

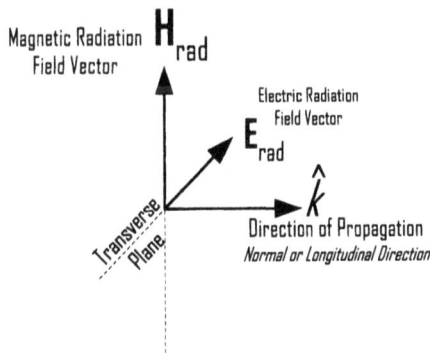

Figure 4.13 Orientation of fields and direction of propagation for an electromagnetic wave.

4.3 LINEAR ANTENNAS

The time-dependent Biot-Savart and Coulomb Laws allow a detailed calculation of the fields around a linear antenna given an assumed source current distribution. In many cases, however, a few simplifying rules and assumptions can yield significant insight into antenna behavior. This section will present some fundamental generalizations that, once understood, make antenna behavior much easier to grasp. Then this section will apply these generalizations to a few simple antenna examples.

4.3.1 Linear Antenna Behavior

The complexity of electromagnetic calculations can be both intimidating and overwhelming at times. An antenna designer may be tempted to resort to numerical modeling as a means of acquiring answers without the necessity of developing an understanding of reality. Ultimately though, without an intuitive understanding of a physical system, numerical analysis alone is unlikely to lead a designer to optimal answers. A few simple qualitative rules, once understood, pay great dividends in mastering an antenna problem.

First, one needs some idea of the bandwidth and frequency response of the antenna under analysis. This allows one to select a tentative time domain response that should be characteristic of what the design can generate. Second is the observation that RF currents tend to concentrate on the edges of planar structures. Third is understanding that radiation occurs roughly orthogonally to the direction in which the current is changing. Fourth is the idea that the current distribution on an antenna is roughly sinusoidal with a periodicity dependent on the center frequency. The final topic is tracking the differences in path length from sources distributed on an extended object. These concepts and the limits of their application are addressed in turn.

4.3.1.1 Time Domain Waveforms Follow From Frequency Response

The first step is to assume a tentative time domain waveform. If analyzing an existing antenna, a time domain waveform may be selected from empirical measurements. Alternatively, one may use an assumed waveform with whatever frequency response one desires for a particular application. The time domain signals of Section 4.3.2 offer a variety of possibilities.

4.3.1.2 Currents Concentrate on Edges (Hertz's Principle)

RF currents tend to concentrate on the sharp edges of a planar antenna element. In fact, Hertz recognized this phenomenon, saying "Just as electricity when distributed by electrostatic induction would tend to accumulate on the sharp edge of the strip, so here the current seems by preference to move along the edge" [17].

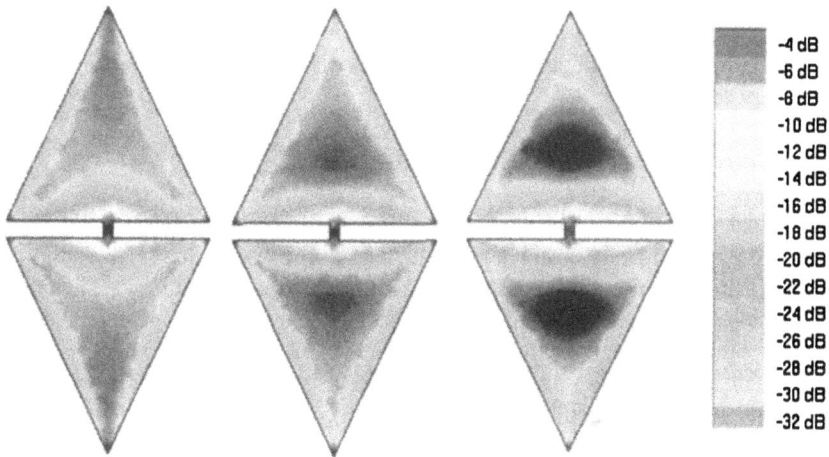

Figure 4.14 The current distribution on a diamond dipole at (a) 3 GHz, (b) 6 GHz, and (c) 8 GHz ([18]; © 2004 IEEE, courtesy of Guofeng Lu).

This tendency of currents to concentrate on edges becomes more pronounced with increasing frequency. Figure 4.14 shows the current concentration on a diamond dipole at a variety of frequencies. Edge shape is the critical parameter in planar antenna design, far more so than the area or interior of the element.

There are two important corollaries to current concentration on edges. First, to a good approximation, one may model a small planar antenna element by replacing the outline of the antenna with linear sources [18, 19]. Second, a planar antenna design replaced with an outline will offer virtually identical performance to that of the original antenna. Thus, the center or interior of an antenna element is an ideal place for support circuitry, mechanical attachments, or transmission-line feeds.

4.3.1.3 Radiation Is Approximately Orthogonal to Change of Current

The radiation power pattern of an infinitesimal current source is $\sin^2 \theta$, where θ is measured from the axis of the current change vector. The radiation power pattern is at a maximum when θ is 90°, in the plane transverse to current change. Thus, to a reasonable approximation, one may assume that the radiation is emitted approximately orthogonally to the current change vector. This approximation works best for curved edges. If an antenna has long linear segments, the full $\sin \theta$ treatment may be necessary.

As will be seen in Section 4.4.3.1, for a small electric antenna the radiated signal time dependence is the second time derivative of an applied signal. This effect can be of significance for systems with a decade of bandwidth or more. For UWB systems with a couple of octaves bandwidth or less, a reasonable approximation is to assume that the second time derivative is just an inverted copy of the applied signal.

4.3.1.4 Current Densities Follow Cosine Tapers

H. C. Pocklington first observed that the current distribution on a thin-wire antenna is approximately cosinusoidal with a maximum at the feed point, a null at the end of the wire, and a wavelength approximately equal to the wavelength of the excitation [20]. Antenna engineers have long exploited this phenomenon to simplify calculation and analysis. The cosinusoidal taper approximation often does an adequate job for pattern prediction. However, it can lead to woefully inaccurate results if applied to a calculation of antenna impedance. As Schelkunoff and Friis observed, "Some radio engineers call it 'a practical engineering approximation,' and some theoreticians once called it 'a colossal fraud'" [21].

4.3.1.5 Relative Delays and Phase Offsets Depend Upon Path Length

The final topic is determining the relative path length from an antenna feed to a source point, and then to a radiated signal. Begin by determining the source points responsible for radiation in a particular direction. These source points will be the locations on the edges where the edge runs perpendicular to the radiation direction, described by a radiation direction angle (for instance, θ). Then, determine the path length from the feed point along the edge to the source point as a function of the radiation direction angle. Finally, determine the distance from the source point to a common reference plane. Usually, it is convenient to select a plane running through the feed point of the antenna (or other point of symmetry) and normal to the radiation direction. This technique is analogous to ray tracing used in the context of geometric optics.

Although the techniques of this section are approximations and not an exact method, these approximations usually allow one to evaluate antenna dispersion and other uniquely time domain behaviors that arise from the distributed sources of a typical antenna. The following section presents a couple of illustrative examples.

4.3.2 Examples

This section offers a pair of simple examples to illustrate the procedure for analyzing linear and planar antenna elements. The first and simplest example is a thin-wire dipole. The second and more elaborate example is a circular element planar dipole antenna.

4.3.2.1 Thin-Wire Dipole

The transient response of thin-wire antennas has long been a topic of considerable interest [22]. This subject has also been well examined in more recent treatments [23]. As a first example, this section presents a simplified model that captures the essential physical characteristics of the radiation process.

An impulsively excited thin-wire dipole has three source points: the feed and the two tips. The signal from the source point is upright, while the signals from the tips, involving reflections, are inverted. Furthermore, the signals from the tips are delayed by the time required for signals to propagate along the length L of the elements. The net effect is that the signals radiated from a thin-wire dipole consist of these three components, with delays depending upon the radiation angle. Figure 4.15 shows the behavior of these three overlapping signal components. This being a linear element antenna, a sin θ dependence on waveform amplitude must be included for each of the three signals.

By way of example, consider the Gaussian W waveform of Figure 4.16(a). Suppose this waveform is the feedpoint signal, and the thin-wire dipole element length L is scaled to superimpose the reflected tip signals on main lobe of the feed signal. Then, the resulting waveform in the azimuthal plane ($\theta = 90°$) is given by Figure 4.16(b). This azimuthal waveform appears at first glance to be the derivative of the original Gaussian waveform but is actually a composite of the original waveform and two delayed inverted copies.

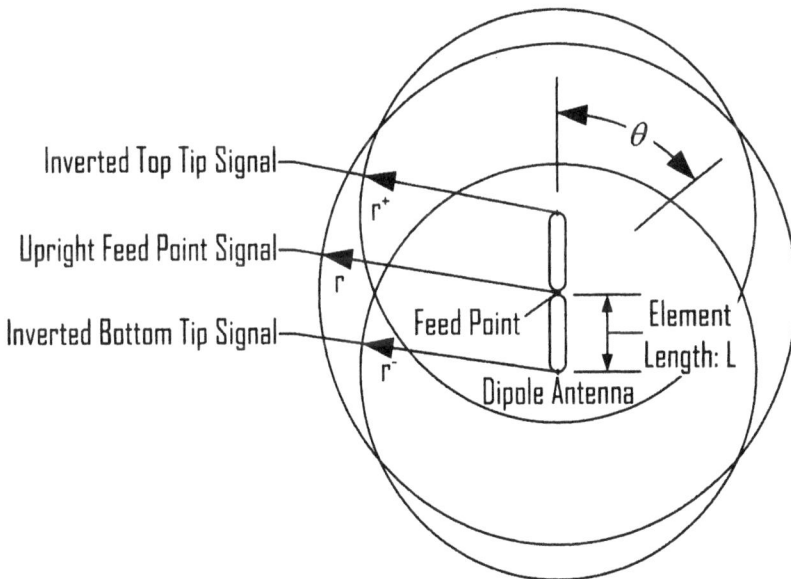

Figure 4.15 Radiation signals from thin-wire dipoles make up an upright feed signal and two inverted tip signals. Relative signal delays depend upon radiation angle.

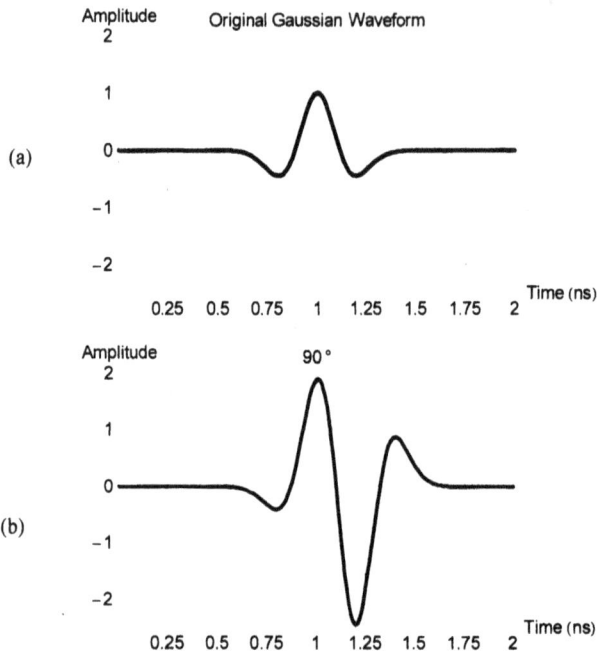

Figure 4.16 (a) A Gaussian *W* waveform contributed by the feed region, and (b) the composite waveform (upright feed and two delayed inverted tip signals) radiated in the azimuthal plane.

Although the composite waveform seen in the azimuthal plane ($\theta = 90°$) roughly corresponds to a time derivative of the original Gaussian *W* feed point signal, the relative delays between the signals vary significantly with radiation angle. As $\theta \rightarrow 0°$, the "top-tip" signal becomes relatively advanced and the "bottom-tip" signal becomes relatively retarded. Conversely as $\theta \rightarrow 180°$, the bottom-tip signal becomes relatively advanced, and the top-tip signal becomes relatively retarded.

Because the tip signals are identical, symmetry is preserved about the azimuthal plane. Nevertheless, the original azimuthal waveform quickly becomes distorted. At $\theta = 75°$, 15° from the azimuthal plane, the waveform is largely unchanged. At $\theta = 60°$, just 30° from the azimuthal plane, the waveform is significantly distorted. At $\theta = 45°$, the waveform is completely mangled, and it becomes difficult to see any remnant of the original waveform structure. Figure 4.17 shows normalized radiated waveforms as a function of angle, and Figure 4.18 shows a planar cross section plot of the waveform dispersion around the impulsively excited dipole. Results closely match those obtained using a more elaborate finite different time domain (FDTD) analysis [24].

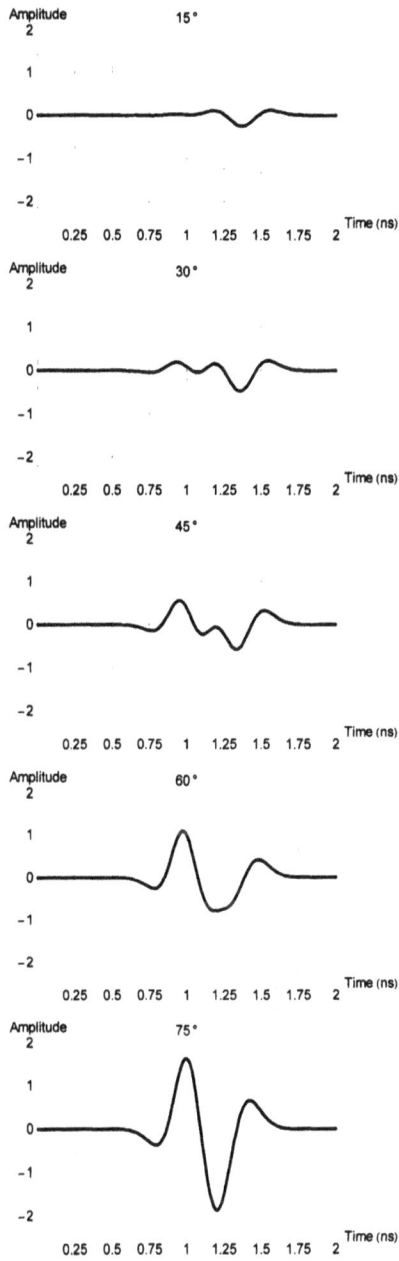

Figure 4.17 Waveforms as a function of polar angle about an impulsively excited thin-wire dipole antenna.

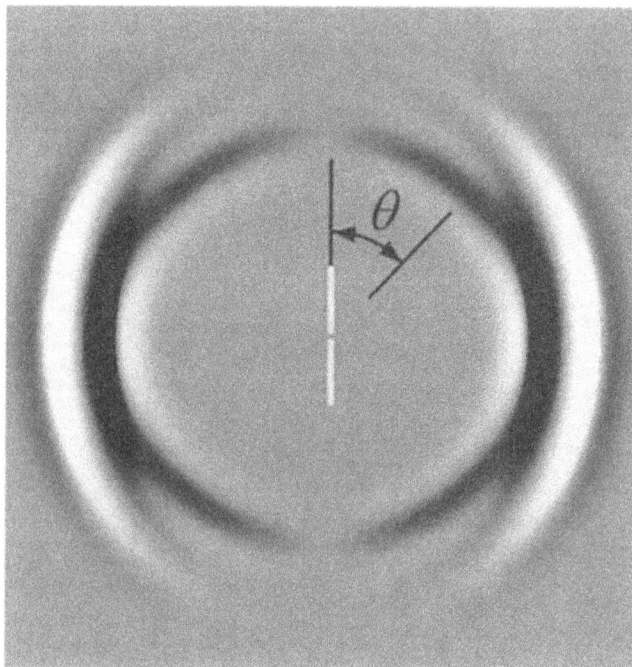

Figure 4.18 Radiation from a thin wire linear dipole at any radiation angle θ is due to the contribution of the upright feed point signal and delayed, inverted tip signals.

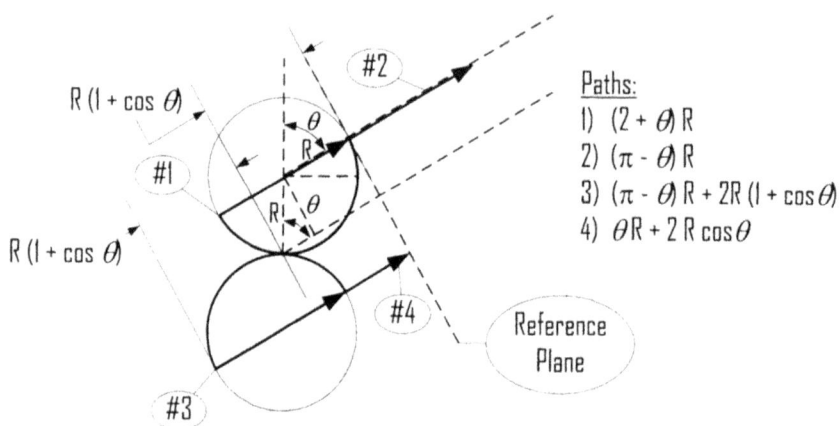

Figure 4.19 Radiation from planar circle dipoles at any radiation angle θ is due to the contribution of four source points with path lengths that depend upon the radiation angle.

4.3.2.2 Circle Dipole

Circular planar element antennas are of significant interest in UWB antenna practice. First described by Mike Thomas and Ronald Wolfson [25], their excellent performance and compact size have drawn much attention. Section 6.2.4.2 provides a further description of circle dipole antennas.

A circle dipole provides a more elaborate and realistic example of a planar small-element antenna from actual UWB antenna practice. Unlike a thin-wire antenna—characterized by only three radiation sources—a circle dipole is an example of a distributed source antenna. Currents distributed along the periphery of this antenna are the source of radiated signals. For any given radiation direction, there are two points on each circle where the edge runs perpendicular to the radiation direction. There is a "near" point and a corresponding "far" point on the other side of the antenna element. A pair of these source points for each of the two elements ("top" and "bottom") yields a total of four sources for signals radiated in each direction. Figure 4.19 designates these four source points as, first, a top-element far source; second, a top element near source; third, a bottom element far source; and fourth, a bottom-element near source. Additionally, Figure 4.18 lists the relative path length of each source from the feed point to the reference plane.

Furthermore, this model includes a cosine tapering of the source strength with a maximum at the feed between the circular elements and a minima at the end points or tips of the circular elements. Unlike the thin-wire dipole, the four signals from the circle dipole add up largely coherently in all directions. Starting with the original assumed Gaussian W waveform of Figure 4.16(a), the superposition of signals from the four sources yields the Gaussian W azimuthal ($\theta = 90°$) signal of Figure 4.20.

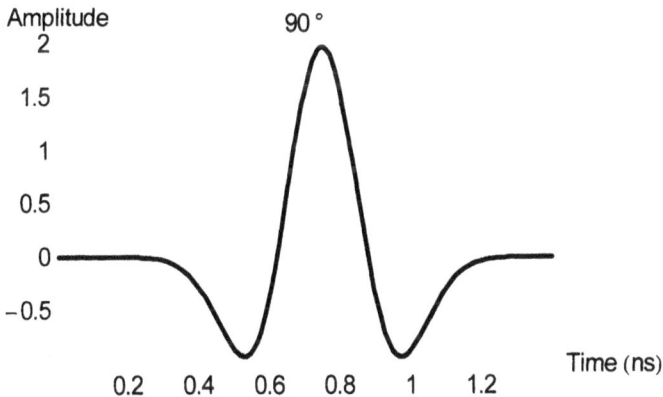

Figure 4.20 The composite waveform for a circle dipole (from the four sources) radiated in the azimuthal plane.

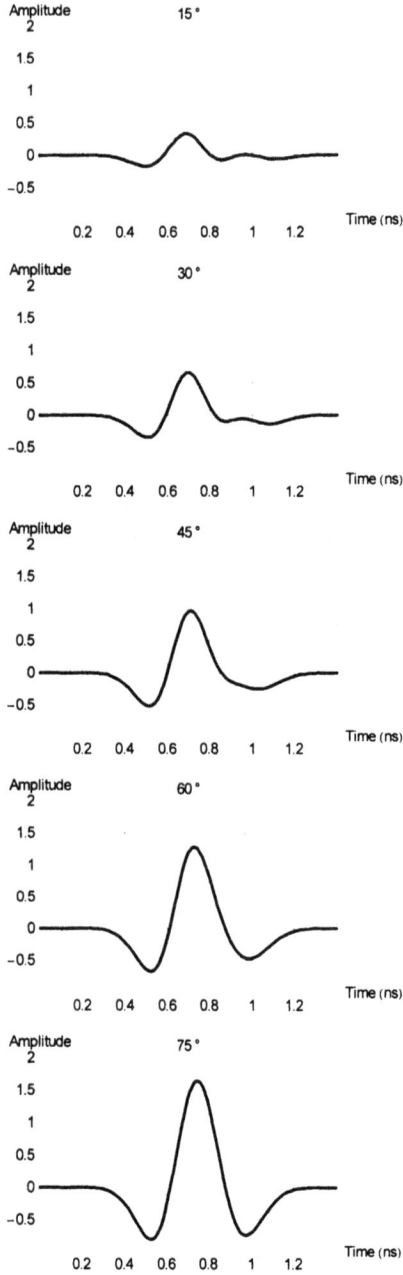

Figure 4.21   Waveforms as a function of polar angle about an impulsively excited circle dipole antenna.

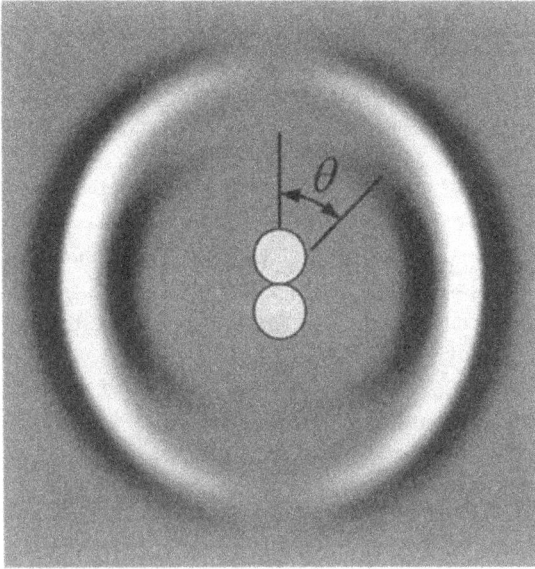

Figure 4.22 Radiation from planar circle dipoles is much less dispersive than that from thin linear elements.

Also, unlike the thin-wire dipole, the azimuthal waveform remains relatively undistorted as a function of radiation angle. Figure 4.21 shows normalized radiated waveforms as a function of angle, and Figure 4.22 shows a planar cross-section plot of the waveform dispersion around the impulsively excited circle element dipole

4.3.3 Summary

A few basic assumptions allow construction of quick and simple analytical models of small-element antennas. First, assume an appropriate time domain response. Second, assume currents are concentrated on the edges of planar elements. Third, assume radiation is orthogonal to curved edges and varies as $\sin\theta$ for linear segments. Fourth, assume cosine tapered currents. Fifth, find the appropriate relative path lengths for radiation in each direction of interest.

This simple approach, properly applied can yield qualitatively correct results. These results provide a designer with the appropriate guidance to make informed design decisions and help illuminate potentially troubling aspects of antenna behavior. Although a formal computational electromagnetic calculation may yield more accurate results for a specific geometry of interest, the simple analytical model discussed in this section allows a designer to focus much more quickly on candidate designs worth the investment of a computational effort.

They [the dipole equations] are the most important equations in the theory and practice of antennas.

Sergei Schelkunoff and Harald Friis, 1952

4.4 DIPOLE FIELDS

Hertz first derived the equations for the dipole fields in the 1880s [26], and they remain at the foundation of antenna practice [27]. The classical dipole is more properly called an "electric dipole," or, alternatively, a "Hertzian dipole."

An electric dipole may be thought of as equal and opposite charges Q separated by a distance d, as shown in Figure 4.23(a). Figure 4.23(b) shows an equivalent definition of an electric dipole as an infinitesimal current element I of length ℓ. The time-dependent electric-dipole moment is defined as

$$\mathbf{p} = Qd\,\hat{\mathbf{z}} = p_0 T(t)\,\hat{\mathbf{z}}$$

(4.30)

where p_0 is the magnitude of the electric-dipole moment, and $T(t)$ is the time dependence. The direction of the electric-dipole moment lies along the axis of the dipole, pointing in the direction from the negative to positive charge or, alternatively, in the direction of current flow. The definition of an electric-dipole moment as static offset charges connects to the "current-segment" definition of an electric-dipole moment via a time derivative

$$I\ell = \dot{p} = \frac{d}{dt}(Qd)$$

(4.31)

An electric dipole provides an excellent analytic model of a small electric antenna.

Finally, there is a magnetic analog to the electric dipole. A "magnetic dipole" may be thought of as an infinitesimal current loop with area A and current I. The time-dependent magnetic-dipole moment is defined as

$$\mathbf{m} = IA\,\hat{\mathbf{z}} = m_0 T(t)\,\hat{\mathbf{z}}$$

(4.32)

where m_0 is the magnitude of the magnetic-dipole moment. The direction of the magnetic-dipole moment lies along the normal to the plane containing the current loop. Figure 4.23(c) shows a magnetic dipole. A magnetic dipole provides an excellent analytic model of a small magnetic antenna. This section will present the electric- and magnetic-dipole fields and discuss some basic antenna lessons dipoles can teach.

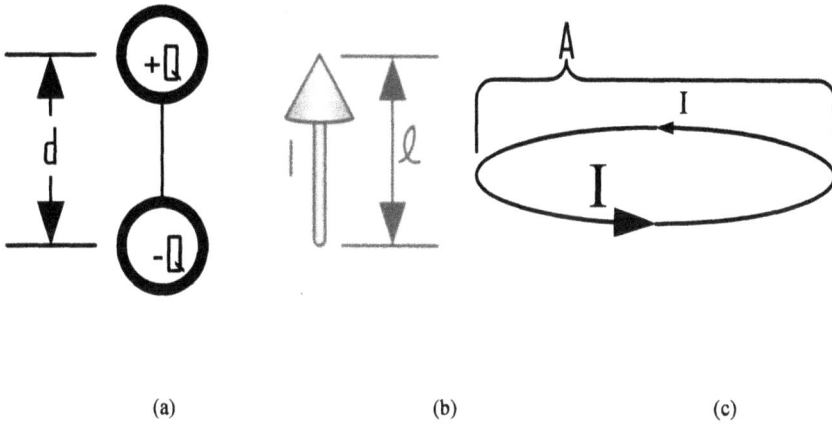

(a) (b) (c)

Figure 4.23 (a) An electric dipole may be thought of as equal and opposite charges Q separated by a distance d. (b) Alternatively an electric dipole may be thought of as an infinitesimal current I of length ℓ. (c) A magnetic dipole may be thought of as an infinitesimal loop with current I and area A.

4.4.1 Electric Dipole Fields

The magnetic field of an electric dipole follows from the time-dependent Biot-Savart Law (4.26)

$$\mathbf{H} = \frac{p_o}{4\pi r}\left(\frac{[\dot{T}]}{r} + \frac{[\ddot{T}]}{c}\right)\sin\theta\,\hat{\phi}$$

$$(4.33)$$

Here, the first term is the magnetic induction field, and the second term is the magnetic radiation field. The electric field may be obtained by a time integration using Ampere's Circuit Law (4.22d)

$$\mathbf{E} = \frac{1}{4\pi\varepsilon_0}\frac{p_o}{r^2}\left(\frac{[T]}{r} + \frac{[\dot{T}]}{c}\right)\left(2\cos\theta\,\hat{\mathbf{r}} + \sin\theta\,\hat{\theta}\right) + \frac{1}{4\pi\varepsilon_0}\frac{p_o[\ddot{T}]\sin\theta}{c^2 r}\hat{\theta}$$

$$(4.34)$$

Here, the first two terms are the electrostatic and the electric-induction fields respectively. The final term is the electric-radiation field. Figure 4.24 shows the coordinates around an electric dipole oriented along the positive z-axis.

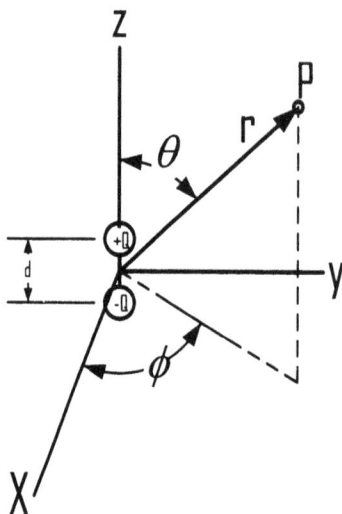

Figure 4.24 Coordinates around an electric dipole oriented along the positive z-axis.

4.4.2 Magnetic Dipole Fields

The magnetic dipole fields follow from a principle called the "duality" of the electric and magnetic fields. Making the substitution: $\mathbf{E}' = -\frac{1}{\sqrt{\varepsilon_0 \mu_0}}\mathbf{B}$ and $\mathbf{B}' = +\sqrt{\varepsilon_0 \mu_0}\,\mathbf{E}$, to Maxwell's equations (4.22a–d) in a source-free ($\rho = 0$, $\mathbf{J} = 0$) region yields:

$\nabla \cdot \mathbf{B}' = 0$	(4.35a)	$\nabla \cdot \mathbf{E}' = 0$	(4.35b)
$\nabla \times \mathbf{B}' = \varepsilon_0 \mu_0 \dfrac{\partial \mathbf{E}'}{\partial t}$	(4.35c)	$\nabla \times \mathbf{E}' = -\dfrac{\partial \mathbf{B}'}{\partial t}$	(4.35d)

Thus, Maxwell's equations are invariant under the duality transformation.

Applying the duality transformation and taking $p_0 \to \sqrt{\varepsilon_0 \mu_0}\, m_0$ yields the magnetic-dipole fields. The magnetic field of a magnetic dipole is

$$\mathbf{H} = \frac{m_0}{4\pi r^2}\left(\frac{[T]}{r} + \frac{[\dot{T}]}{c}\right)\left(2\cos\theta\,\hat{\mathbf{r}} + \sin\theta\,\hat{\theta}\right) + \frac{m_0[\ddot{T}]\sin\theta}{4\pi c^2 r}\,\hat{\theta}$$

$$(4.36)$$

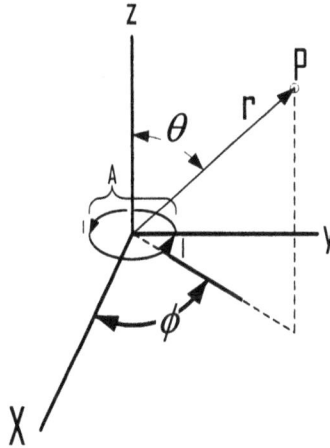

Figure 4.25 Coordinates around a magnetic dipole oriented along the positive *z*-axis.

The first two terms are the magnetostatic and magnetic-induction fields, respectively. The final term is the magnetic-radiation field. The electric field of a magnetic dipole is

$$\mathbf{E} = \frac{m_0}{4\pi\varepsilon_0 c^2 r} \left(\frac{[\dot{r}]}{r} + \frac{[\ddot{r}]}{c} \right) \sin\theta \; \hat{\phi}$$

(4.37)

Figure 4.25 shows the coordinates around a magnetic dipole oriented along the positive *z*-axis.

4.4.3 Dipole Behavior

Hertzian dipoles, of both the electric and the magnetic varieties, provide a good first-order model of small-element antennas. Understanding the behavior of dipoles allows us to better understand the behavior of small-element antennas. This section will discuss some of the lessons dipoles can teach us about fundamental antenna physics.

4.4.3.1 Antenna Differentiation

If a voltage signal is applied to a small electric-dipole antenna, it induces a dipole moment proportional to the applied voltage. The radiated fields have a time dependence proportional to the second time derivative of the voltage signal. This general principle is almost as well known as it is widely abused.

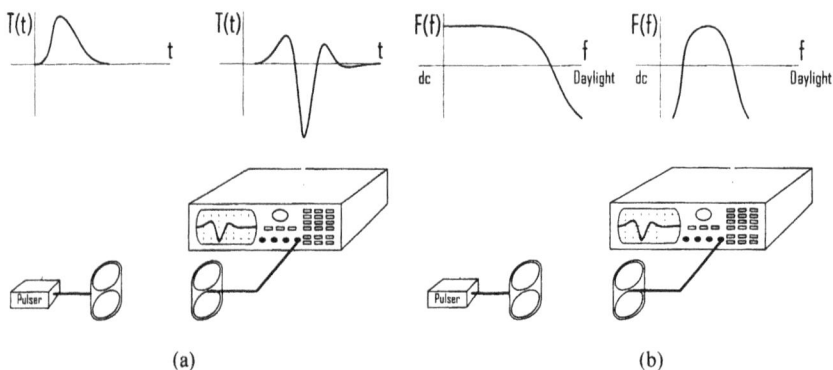

Figure 4.26 (a) Time domain response of a UWB antenna excited by an even broader band impulse, and (b) corresponding frequency domain response.

The first common abuse is to confuse differentiation with band-pass filtering. If one applies a broadband impulse to a less-broadband antenna, the effect of the antenna filtering will be to add zero-crossings, or lobes, to the waveform. By coincidence, this may yield a waveform qualitatively similar to a differentiated copy of the applied impulse. Unless the frequency content of the antenna excitation fits comfortably within the antenna bandwidth, one will not observe a true differentiation. Figure 4.26(a, b) illustrates what might happen when a broadband antenna is excited by an even broader-band impulse.

A second common abuse is to look for signs of differentiation in waveforms with bandwidths too narrow to support any obvious features. To see a differentiation clearly requires a bandwidth in excess of a couple of octaves, preferably a decade or more. For instance, see the example of Figure 2.5 in which a two octave (1.5–6.0GHz) excitation is applied to a UWB dipole antenna. The received waveform at the other end of the link is virtually an identical copy of the transmitted waveform. The "double differentiation" predicted by the dipole model manifests itself as a 180° phase shift in the transmitted signal, just as in the narrowband or frequency domain context.

For typical UWB practice in the 3.1–10.6GHz band authorized by the FCC, differentiation of time domain waveforms is well hidden and barely perceptible if it manifests itself at all. An important corollary follows from this observation. Elaborate link models and transfer functions involving time derivatives of transmitted waveforms may be of academic or theoretical interest, but their practical utility is lacking. The subtleties captured by such models are unlikely to survive contact with the high-multipath propagation environments of real-world applications. In short, antenna differentiation is one of the best known, yet least needed, principles of UWB antenna engineering.

4.4.3.2 Field Impedance

Conventional practice in electromagnetics is to deal in the frequency domain or time-harmonic fields. The general time-dependent dipole fields transform to the frequency domain fields by taking

$$T \to 1, \tag{4.38a}$$

$$\dot{T} \to j\omega \tag{4.38b}$$

and

$$\ddot{T} \to -\omega^2 \tag{4.38c}$$

This yields time-harmonic expressions for the dipole fields

$$\mathbf{H} = \frac{p_0}{4\pi r}\left(\frac{j\omega}{r} - \frac{\omega^2}{c}\right)\sin\theta\,\hat{\phi} = \frac{p_0\omega k^2}{4\pi}\left(\frac{j}{k^2 r^2} - \frac{1}{kr}\right)\sin\theta\,\hat{\phi} \tag{4.39a}$$

and

$$\mathbf{E} = \frac{1}{4\pi\varepsilon_0}\frac{p_0}{r^2}\left(\frac{1}{r} + \frac{j\omega}{c}\right)\left(2\cos\theta\,\hat{\mathbf{r}} + \sin\theta\,\hat{\theta}\right) - \frac{1}{4\pi\varepsilon_0}\frac{p_0\omega^2\sin\theta}{c^2 r}\hat{\theta}$$

$$= \frac{p_0 k^3}{4\pi\varepsilon_0}\left(\frac{1}{k^3 r^3} + \frac{j}{k^2 r^2}\right)\left(2\cos\theta\,\hat{\mathbf{r}} + \sin\theta\,\hat{\theta}\right) - \frac{p_0 k^3}{4\pi\varepsilon_0}\frac{\sin\theta}{kr}\hat{\theta} \tag{4.39b}$$

where $\omega p_0 = I\ell$ and $k = 2\pi/\lambda$ is the wave number.

The field impedance is the ratio of the electric to the magnetic field amplitude. In the far field, the field impedance is defined in terms of the radiation fields

$$Z = E_\theta/H_\phi = E_\phi/H_\theta = 1/(c\varepsilon_0) = 376.7\,\Omega \tag{4.40}$$

Extrapolating this formula into the near field and using the time-harmonic expressions for the dipole fields yield

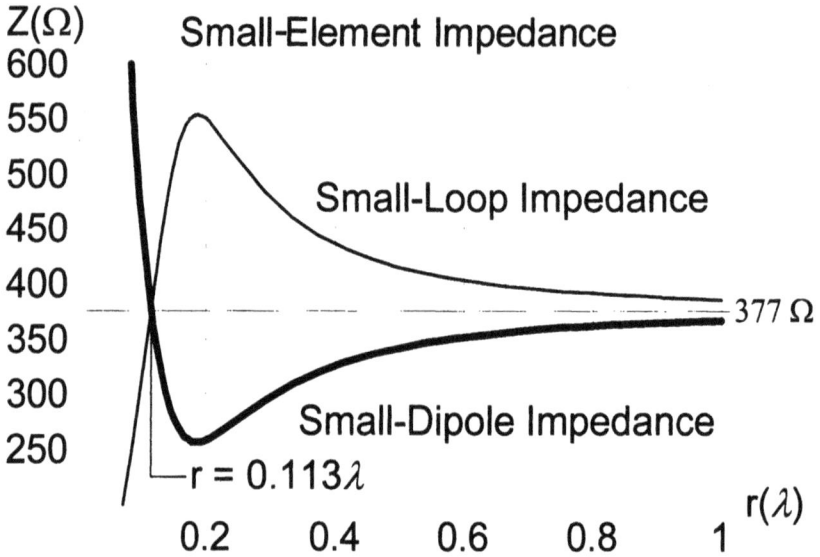

Figure 4.27 Field impedance of the time-harmonic dipole fields as a function of range in units of wavelength (After [28]).

$$Z_E = \frac{E_\theta}{H_\phi} = Z_S \frac{\dfrac{1}{k^3 r^3} + \dfrac{j}{k^2 r^2} - \dfrac{1}{kr}}{\dfrac{j}{k^2 r^2} - \dfrac{1}{kr}} \tag{4.41a}$$

for the electric-dipole impedance and

$$Z_H = \frac{E_\theta}{H_\phi} = Z_S \frac{\dfrac{j}{k^2 r^2} - \dfrac{1}{kr}}{\dfrac{1}{k^3 r^3} + \dfrac{j}{k^2 r^2} - \dfrac{1}{kr}} \tag{4.41b}$$

for the magnetic-dipole impedance. Figure 4.27 shows the field impedance of the time-harmonic dipole fields. Not surprisingly, at very close range, a small dipole looks like a high-impedance capacitive open and a small magnetic dipole looks like a low-impedance inductive short. An excellent analysis is available elsewhere on the definition of near field versus far field from the time-harmonic point of view [28].

4.4.3.3 Basic Antenna Physics

This section will use the dipole field relations to establish some basic principles of antenna physics. First, this section will take a close look at the shortcomings of the usual radiation-field approximation.

The Radiation Approximation

Conventional practice in electromagnetics identifies field terms with an inverse r ($1/r$) dependence as the "radiation fields." For instance, in the case of an electric dipole, the electric-radiation field is

$$\mathbf{E}_{rad} = \frac{1}{4\pi\varepsilon_0} \frac{p_0 \left[\ddot{T}\right] \sin\theta}{c^2 r} \hat{\theta}$$

(4.42)

Far away from a source, the traditional concept of a radiation field is an excellent approximation, but it is neither a complete nor a valid solution to Maxwell's equations.

By way of example, assume a harmonic dipole with

$$\left[\ddot{T}\right] = -\omega^2 \sin\omega[t] = -\omega^2 \sin\omega\left(t - \frac{r}{c}\right) = -\omega^2 \sin(\omega t - kr)$$

(4.43)

Consider the divergence of the electric-radiation field term for a harmonic electric dipole

$$\nabla \cdot \mathbf{E}_{rad} = -\nabla \cdot \frac{1}{4\pi\varepsilon_0} \frac{p_0\omega^2 \sin(\omega t - kr)\sin\theta}{c^2 r} \hat{\theta}$$

$$= \frac{2p_0\omega^2}{4\pi\varepsilon_0} \frac{\sin(\omega t - kr)\cos\theta}{c^2 r^2} \neq 0$$

(4.44)

The divergence of the radiation field term is not zero as would be expected from Gauss's Law (4.22a) for a source-free region. Either the electromagnetic radiation described above carries with it some sort of charge distribution as it propagates through free space, or, at the very least, it is not a complete and valid solution to Maxwell's equations.

One might argue (with considerable justice) that since this nonzero divergence goes as the inverse square of distance (as $1/r^2$), the error becomes increasingly negligible at large distances from the source. But before dismissing this minor anomaly out of hand as an inconsequential error in the radiation approximation, consider what would be required to remedy the discrepancy.

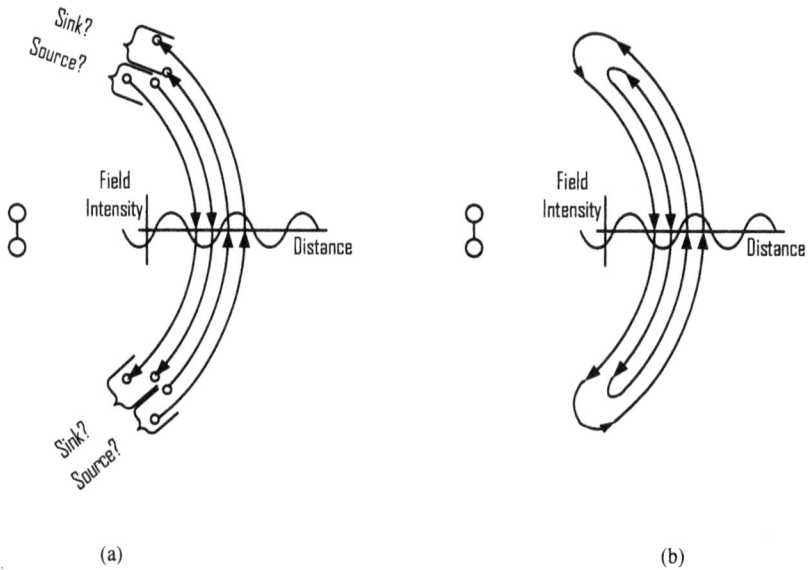

Figure 4.28   (a) The usual approximation for radiation leaves field lines hanging from implied sources and sinks. (b) Radiation fields actually form closed loops with neither sinks nor sources.

Suppose we were to add an additional term to the radiation field to create an alternate expression for the radiation

$$\mathbf{E}^*_{rad} = \frac{1}{4\pi\varepsilon_0}\frac{p_0}{r^2}\frac{[\dot{T}]}{c}2\cos\theta\,\hat{\mathbf{r}} + \frac{1}{4\pi\varepsilon_0}\frac{p_0[\ddot{T}]\sin\theta}{c^2 r}\,\hat{\theta}$$

$$(4.45)$$

Then, evaluating for the harmonic time dependence yields

$$\nabla\cdot\mathbf{E}^*_{rad} = \nabla\cdot\frac{p_0}{4\pi\varepsilon_0}\left[\frac{2\omega\cos(\omega t - kr)\cos\theta}{cr^2}\,\hat{r} - \frac{\omega^2\sin(\omega t - kr)\sin\theta}{c^2 r}\,\hat{\theta}\right]$$

$$= \frac{p_0}{4\pi\varepsilon_0}\left[-\frac{2k\omega\sin(\omega t - kr)\cos\theta}{cr^2} + \frac{2\omega^2\sin(\omega t - kr)\cos\theta}{c^2 r^2}\right] = 0 \quad (4.46)$$

where the wave number is $k = \omega/c$. This alternate expression for the radiation now satisfies Gauss's law.

Radiation as a Closed Field Line

The previous mathematical example probes the fundamental difference between bound, or reactive, fields and unbound, or radiation, fields. An electric field is bound when it begins and ends on an electric charge. The positive charge where a field line begins is a "source," and the negative charge where a field line ends is a "sink." Unbound or radiation fields, on the other hand, form a closed loop with neither source nor sink. The extra term in the alternate expression for the radiation field above is needed to close the loop. Figure 4.29(a, b) illustrates this point for a particular cycle of a harmonic electric dipole.

Radiation of a DC Signal?

Understanding that radiation fields have neither source nor sink leads relatively painlessly to another basic principle of UWB systems. The complete time integral of a radiated waveform must be equal to zero [29]. Expressed mathematically,

$$\int_{-\infty}^{+\infty} \ddot{T}(t)\,dt = 0$$

$$(4.47)$$

There must be as much positive excursion of the radiated waveform as there is negative excursion. For every radiation field line pointing in one direction, there must be an equal and opposite radiation field line pointing in the other direction. A corollary to this rule might be stated, "You can't radiate dc."

Suppose one attempted to excite an antenna so as to create a radiated waveform whose field intensity had a net positive excursion. Such an excitation would require an antenna current $[\,I \sim \dot{T}(t)\,]$ that did not return to zero at the end of the radiation $[\,\lim_{t\to\infty} \dot{T}(t) > 0\,]$. Thus, the antenna voltage $[\,V \sim T(t)]$ would have to continue to grow, a process that could not continue forever. Figure 4.29(a) illustrates what radiating a "dc impulse" implies for antenna current and voltage.

Of course, this "no dc" rule still allows considerable flexibility for designing a UWB system with a radiated impulse that is almost, but not quite, dc. For instance, one might quickly increase antenna current, then allow it to decay gradually. The net positive antenna current flow leaves a net positive antenna voltage in its wake. Figure 4.29(b) shows how a quasi-dc impulse may be created with a physically realizable antenna current and voltage. Note how the total positive field intensity excursion in the quasi-dc impulse is balanced by a long, gradual, negative field intensity precursor. Figure 4.29(c) provides a qualitative diagram of the radiation field lines that might be associated with a quasi-dc impulse.

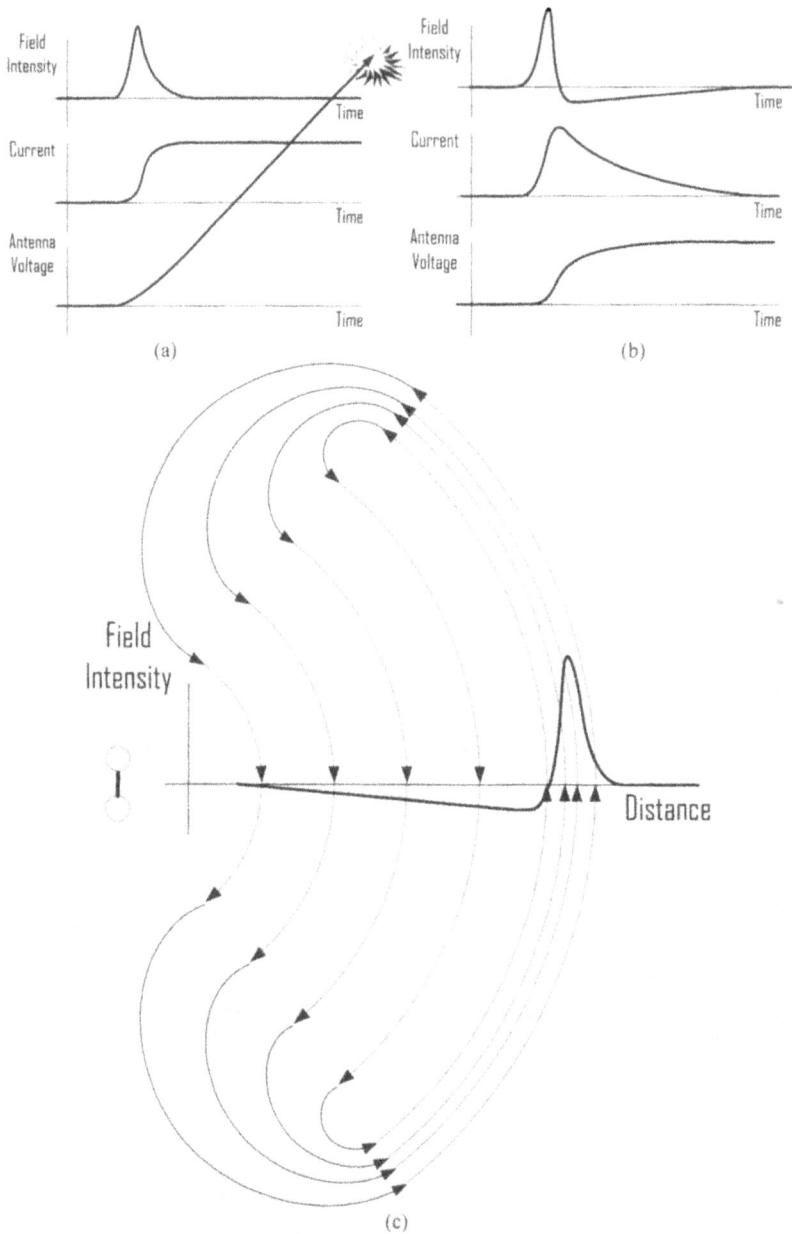

Figure 4.29 (a) Radiating a true dc impulse (top) requires a step current (middle) that implies an unrealizable ever-increasing antenna voltage (bottom). (b) Radiating a quasi-dc impulse with zero complete time integral (top) requires an asymmetric current impulse (middle) that implies an antenna voltage charging to a final non-zero value (bottom). (c) A quasi-dc impulse with closed field lines is physically possible.

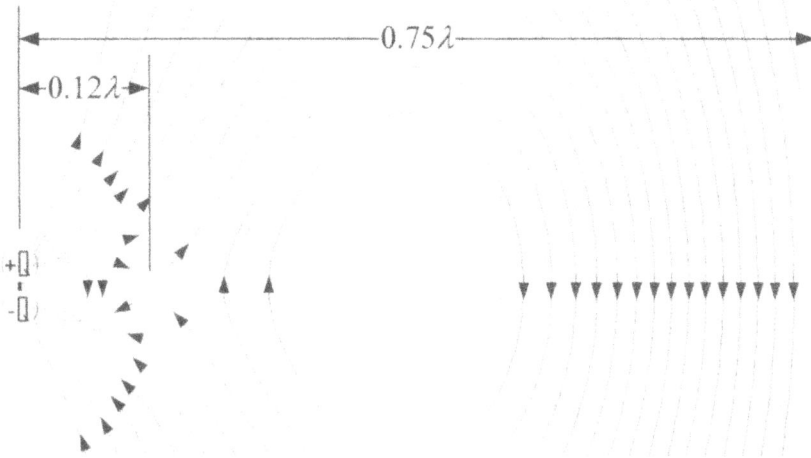

Figure 4.30 Detail of electric field lines in the vicinity of an ideal point electric dipole at $t = 0.1975T$ after dipole voltage maximum.

Radiation as a Decoupling of Field Lines from a Source

The "magic" of radiation lies in the transformation of bound, or reactive, electromagnetic fields to unbound, or radiation, fields. This transformation is particularly clear and striking in the electric field of an ideal point electric dipole. Assume a harmonic dipole with maximum voltage at time zero (i.e., $[T(t)] = \cos \omega(t - r/c)$). As the dipole voltage decreases, the electric fields collapse. These field lines pinch off at about $\lambda/8$ to $\lambda/6$ from the dipole source. Field lines inside this pinch point collapse back into the dipole source. Field lines outside this pinch point form the closed field line loop geometry characteristic of radiation fields. Deprived of their umbilical chord back to the dipole charges, they decouple and radiate away. Figure 4.30 shows a detailed view of the field lines as a radiated impulse is being born.

4.5 ANTENNAS AS RADIATORS

Frequency domain analysis provides an excellent tool kit for understanding and dealing with systems that involve steady-state, harmonic, or narrowband signals. This chapter has introduced a few of the time domain analytic tools useful in understanding transient, or broadband, antennas. These tools include functions well suited for analyzing time domain signals, like Gaussian, edge, sinc, and truncated sine signals.

Time domain electromagnetics enables prediction of radiated fields around an antenna of known geometry. Maxwell's equations, as written, can be difficult to use in analysis because electric fields are a function of magnetic fields and vice versa. This chapter presented an alternate formation of Maxwell's equations, the time-dependent Coulomb and Biot-Savart laws in which the fields are a function of the charge and current distribution. The concept of generalized coordinates is common to the usual frequency domain approach to electromagnetics, but the time domain approach has its own peculiar wrinkles, like causality and retardation.

Some simple rules of thumb, assumptions, and approximations can make understanding a time domain system much easier. These include:

- Right-hand rule for radiation;

- Geometry and ratio (or impedance) of fields in an electromagnetic wave;

- Concentration of currents on the edges of a planar antenna;

- Cosinusoidal distribution of currents around an antenna;

- Accounting for path-length differences from sources distributed around an antenna.

These principles allow understanding and analysis of linear antennas as well as complicated planar structure antennas.

Finally, the electric- and magnetic-dipole fields provide an analytically simple, yet physically profound, model of small-antenna behavior. The dipole fields can be used to understand such concepts as the impedance of small antennas, the field-line model of radiation where radiation fields are treated as closed loops, and the requirement that the total time integral of a radiated signal be zero.

This chapter has applied time domain analysis and electromagnetics, as well as a variety of engineering approximations useful to understanding the behavior of UWB antennas as radiators of electromagnetic waves. Given a known structure, these techniques allow prediction of the fields radiated by an antenna and such useful properties as pattern, polarization, and gain.

The perspective of an antenna as a radiator is descriptive, not normative: this point of view is useful for understanding existing antennas, but it does not describe how to design a new antenna to meet a particular specification. Antenna design requires further insight—insight that can be gained by an understanding of electromagnetic energy flow. This topic is the subject of the next chapter.

Endnotes

[1] Heaviside, Oliver, *Electromagnetic Theory*, Vol. 3, London: "The Electrician" Printing and Publishing Company, Limited, 1912, p. 1. Heaviside's remarkable three volume work on electromagnetic theory was recently reprinted in facsimile edition (2003) and is now available from Elibron Classics at www.elibron.com.

[2] Harmuth, Henning, *Sequence Theory*, New York: Academic Press, 1977, p. 6.

[3] Harmuth, Henning, *Op. Cit.*, pp. 1-17.

[4] McGillem, Clare D., *Continuous and Discrete Signal and System Analysis*, New York: Holt Rinehart, and Winston, Inc., 1984.

[5] Arfken, George, *Mathematical Methods for Physicists*, 2nd ed., New York: Academic Press, 1970, p. 419.

[6] Harmuth, Henning, *Antennas and Waveguides for Nonsinusoidal Waves*, New York: Academic Press, Inc., 1984, pp. 17-22.

[7] Siwiak, Kazimierz, "Ultra-wideband high data-rate communication apparatus and associated methods," Pub. No. US 2004/0017841, January 29, 2004.

[8] Siwiak, Kazimierz, "High data-rate communication apparatus and associated methods," Pub. No. US 2004/0017840, January 29, 2004.

[9] Siwiak, Kazimierz, James Richards, and Hans Schantz, "Transmitting and receiving spread spectrum signals using continuous waveforms," Pub. No. US 2004/0174928, September 9, 2004.

[10] Maxwell, James Clerk, "A Dynamical Theory of the Electromagnetic Field," Royal Society Transactions Vol. CLV (1864), as reprinted in W. D. Niven, ed., *The Scientific Papers of James Clerk Maxwell*, New York: Dover, 1952, pp. 526-597.

[11] Bork, Alfred M., "Maxwell and the Electromagnetic Wave Equation," American Journal of Physics Vol. 35 (1967), pp. 844-849. Bork provides an excellent analysis linking Maxwell's original Cartesian formulation to the currently accepted vector form.

[12] Nahin, Paul J., *Oliver Heaviside: Sage in Solitude*, New York: IEEE Press, 1988. See in particular Chapter 7 of this delightful book for an explanation of Heaviside's contributions to electrodynamics.

[13] Griffiths, David J. and Mark A. Heald, "Time-dependent generalizations of the Biot-Savart and Coulomb laws," American Journal of Physics Vol. 59, 1991, pp. 111-117.

[14] Jefimenko, Oleg D., *Electricity and Magnetism*, New York: Appleton-Century-Crofts, 1966, §15-7, pp. 515-518.

[15] Panofsky, Wolfgang K.H., and Melba Phillips, *Classical Electricity and Magnetism*, 2nd ed., New York: Dover, 2005.

[16] Schantz, Hans Gregory, "Electromagnetic Radiation Made Simple," APS/AAPT Joint Meeting, Washington, D. C., April 18-21, 1997.

[17] Hertz, Heinrich, *Electric Waves*, London: Macmillan and Co., 1893, p. 164.

[18] Lu, Guofeng, I. Korisch, L. Greenstein, and P. Spasojevic, "Antenna modelling using linear elements, with applications to UWB," 2004 IEEE Antennas and Propagation Society

International Symposium, Monterey, California. June 20-25, 2004, Vol. 3, pp. 2544-2547. I am indebted to Guofeng Lu for permission to reproduce Figure 4.14.

[19] Pele, I., Y. Mahe, A. Chousseaud, S. Toutain, and P.Y. Garel, "Antenna design with control of radiation pattern and frequency bandwidth," 2004 IEEE Antennas and Propagation Society International Symposium, Monterey, California. June 20-25, 2004, Vol. 1, pp. 783-786.

[20] Pocklington, H. C., "Electrical Oscillations in Wires," Cambridge Philosophical Society Proceedings, Vol. 9, 1897, pp. 324-332.

[21] Schelkunoff, Sergei, and Harald Friis, *Antennas: Theory and Practice*, New York: John Wiley and Sons, 1952, p. ix.

[22] Wu, T. T., "Transient Response of a Dipole Antenna," Journal of Mathematical Physics, Vol. 2, 1961, p. 892.

[23] Smith, Glenn S., *An introduction to classical electromagnetic radiation*, Cambridge: Cambridge University Press, 1997. See in particular Chapter 8.

[24] Maloney, J. G., G. S. Smith, and W. R. Scott, Jr., "Accurate computation of radiation from simple antennas using the finite-difference time-domain method," IEEE Transactions on Antennas and Propagation, Vol. 38, July 1990, pp. 1059-1068.

[25] Thomas, Mike and Wolfson, Ronald, "Wideband arrayable planar reflector," U.S. Patent 5,319,377, June 7, 1994.

[26] Hertz, Heinrich, *Op. Cit.*, pp. 141-150.

[27] Schelkunoff, S. and H. Friis, *Op. Cit.*, p. 120.

[28] Capps, Charles, "Near Field or Far Field?," EDN, August 16, 2001, pp. 95-102. At present, this article is available at www.edn.com/contents/images/150828.pdf. Originally "Electrical Design News," the title of this publication is now officially the three letter acronym, "EDN."

[29] Baum, Carl E., "Some Limiting Low-Frequency Characteristics of a Pulse-Radiating Antenna," Sensor and Simulation Note #65, Air Force Research Lab Directed Energy Directorate, October 28, 1968.

Chapter 5

Antennas as Energy Converters

> *The energy in electro-magnetic phenomena is mechanical energy. The only question is, Where does it reside? On the old theories it resides in electrified bodies, conducting circuits, and magnets, in the form of an unknown quantity called potential energy, or the power of producing certain effects at a distance. On our theory it resides in the electro-magnetic field, in the space surrounding the electrified and magnetic bodies, as well as in those bodies themselves....*
>
> James Clerk Maxwell, 1865

This chapter treats antennas as converters of electromagnetic energy. From this perspective, an antenna is a device that converts guided, bound, or reactive electromagnetic energy into unbound, or radiation, electromagnetic energy. Electromagnetic energy analysis is not commonly applied in antenna theory. Thus, this chapter begins by motivating the subject, discussing the value of examining fundamental antenna physics. Then, this chapter considers localization of electromagnetic energy and develops an understanding of how bound electromagnetic energy uncouples and radiates away. Unlike the more traditional analytic techniques presented in the last chapter, analysis of electromagnetic energy flow leads to novel antenna designs that conform to these flows. Finally this chapter uses energy-flow techniques to derive fundamental limits to antenna size and performance.

5.1 MOTIVATION

Traditional thinking about antennas treats them as radiators and receivers of electromagnetic fields as described by Maxwell's equations. This approach does a remarkable job of analyzing real-world antenna problems and providing correct answers. Why then should one study a seemingly esoteric subject like electromagnetic energy flow?

This question may be answered on a couple of levels. First, models that more closely correspond to the underlying physical reality yield not only correct answers, but also profound physical insights that open the door to a deeper and more valuable understanding. Second, traditional "field-based" thinking about electromagnetics provides an incomplete and often confusing model of physical behavior. This section will consider these two topics in turn.

5.1.1 Models and Reality

The Ptolemaic, or geocentric, model of the solar system yielded a remarkably accurate mathematical model of the relative motion of the planets with respect to the celestial firmament. In the Ptolemaic model, a planet follows a circular path, or "epicycle," about a center of rotation that itself follows a circular path, or "deferent," about the Earth. The Ptolemaic superposition of circular motions models the periodic retrograde motion of planets against the fixed stars to a respectable degree of accuracy. Of course, an arbitrarily close match to observations may be obtained by superimposing additional epicycles. Figure 5.1 presents a diagram of planetary motion in the Ptolemaic model.

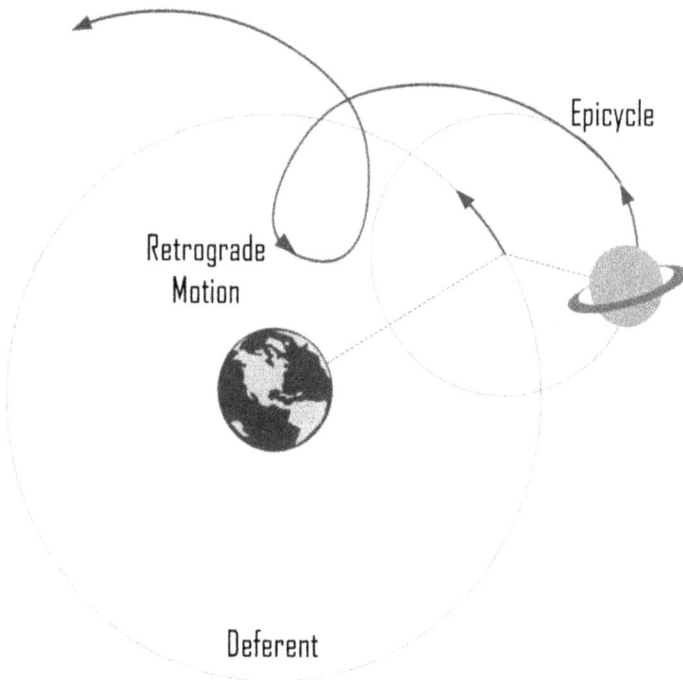

Figure 5.1 Ptolemaic model of planetary motions.

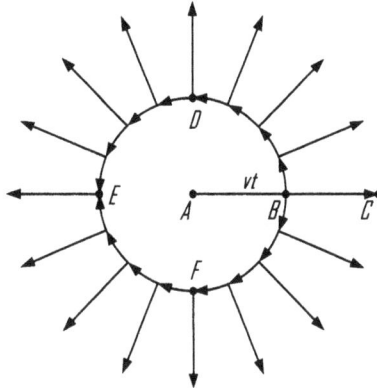

Figure 5.2 Field lines around a suddenly accelerated charge (After [1]).

The Ptolemaic model of planetary motions is adequate if mere calculation of planetary positions against the field of stars is the only goal. The competing Copernican (or heliocentric) model ultimately yields more accurate results with less complexity. The value of the Copernican model lies not merely in its simplification of astronomical calculation. The principal benefit of this model is that by being a closer, better description of the underlying physical reality, it opened the door to Kepler's discovery of elliptical orbits and his laws of planetary motion. Kepler's discoveries in turn led Newton to devise his Universal Law of Gravitation. All these discoveries arose from adopting a model that more closely reflected the underlying reality.

The story of how Copernican astronomy eclipsed Ptolemaic astronomy has an important moral: correct models yield profound insights. The science behind UWB antenna design is the science of how electromagnetic energy radiates from antennas. To understand how to design a good antenna, one must understand how this process works. In this spirit, the following section considers the difficulties with the traditional view of electromagnetic radiation as a kink, or bend, in a field line.

5.1.2 The Fallacy of the Kinked Field Line

The usual explanation of how radiation works begins with treating radiation as a transverse kink, or bend, in a field line. In 1893, Heaviside pioneered this view by imagining a charge originally at a point *A* suddenly accelerated to the speed of light in a direction *AC* [1]. Heaviside argued that the field lines would be analogous to lines of longitude on the sphere described by *B*, *D*, *E*, and *F*, centered at *A* and with an axis traversing *B* and *E*. Figure 5.2 is based upon Heaviside's 1893 depiction of radiation as a transverse kink in a field line.

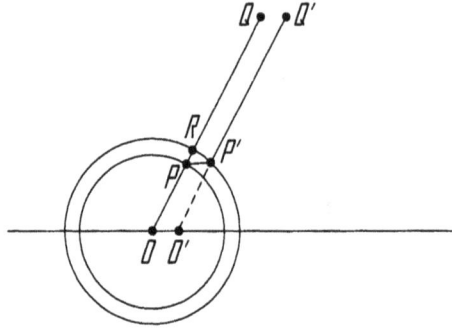

Figure 5.3 Field lines around a suddenly decelerated charge (After [2]).

Not long thereafter, J. J. Thompson presented a similar depiction of electromagnetic radiation as a kinked field line. Imagine a charge originally at a point O' with a field line $O'\,P'\,Q'$ moving with a constant velocity. Suddenly the charge is jerked to a stop at point O. What would have been the $O'\,P'\,Q'$ field line adapts to the change, becoming field line $O\,P\,Q$. The transition is not instantaneous, however, because electromagnetic fields propagate at the speed of light. Thus, if we examine the field line not long after the charge halts, we will observe a transverse component $P\,P'$ propagating out. This transverse field line $P\,P'$ is the locus where the $O'\,P'\,Q'$ field line becomes the $O\,P\,Q$ field line. Figure 5.3 is based on Thompson's 1904 illustration of this phenomenon [2].

This perspective of radiation as a kinked field line is deeply inscribed in the textbooks and training of physicists and engineers alike. Working from this point of view, it is easy to envision accelerating charges as the source of radiation energy. The idea that charges are the source of energy leads to two significant problems, however.

First, there is no limit to the amount of energy a charge can radiate. A charge accelerated around and around in a cyclotron will radiate an arbitrarily large amount of energy from a seemingly bottomless reservoir. If one assumes that an accelerating charge is the source of radiation energy, an accelerating charge becomes a moving violation of the Law of Conservation of Energy.

Second, the kinked field line model assumes static, radial field lines propagating out from a point charge to infinity. In reality, static field lines originate at positive charges and terminate at negative charges. Furthermore (as noted in Section 4.4.3.3), radiation field lines form closed loops. Thus, the kinked field line model of radiation is at least incomplete if not physically misleading.

Given the importance of a correct physical model, the difficulties with the usual kinked-field-line model of radiation mean that a deeper examination of electromagnetic physics is called for. The following section begins an examination of where electromagnetic energy resides and how it works.

5.2 LOCALIZATION OF ELECTROMAGNETIC ENERGY

This examination of electromagnetic energy localization begins with the historical development of John Henry Poynting's and Heaviside's theory of electromagnetic energy transfer [3]. Then, the problems and alleged paradoxes of this theory are examined. Finally, some simple examples illustrate methods of tracking electromagnetic energy.

5.2.1 A Brief History of Electromagnetic Energy

Hermann von Helmholtz discussed electromagnetic energy in his 1847 treatise on energy conservation. Helmholtz introduced the concept that electric energy density (u_E) depends upon charge density (ρ_e) and the electromagnetic potential (Φ) [4]

$$u_{E(static)} = \frac{1}{2}\rho_e\Phi \tag{5.1}$$

Helmholtz's charge-based energy density provides a mathematical foundation for the idea that electromagnetic energy is localized with charges. Helmholtz's theory runs into difficulty when considering transfers of electromagnetic energy. If energy is localized on charges, then when one charge gains energy at the expense of another, the energy must transfer from one charge to the other without crossing the intervening space. This "action-at-a-distance" approach to physics was generally accepted in the early nineteenth century.

The primacy of action-at-a-distance in physical thought began to crumble at the hands of Faraday. Faraday introduced the concept of a field. The implication of Faraday's field approach is that electric and magnetic processes are distributed throughout space, not localized with sources. In 1853, William Thompson (later Lord Kelvin) introduced the idea that energy itself might be localized with the fields [5]. In the context of the electric field, Thompson's idea meant that the field carries an energy

$$u_{E(static)} = \frac{1}{2}|\mathbf{D} \cdot \mathbf{E}| = \frac{1}{2}\varepsilon_0|\mathbf{E}|^2 \tag{5.2}$$

Maxwell extended this field-based approach to energy density. In fact, localization of energy in electric and magnetic fields was a cornerstone of Maxwell's thinking. His thoughts on the subject grace this chapter's epigraph [6, 7].

The equivalence of the Helmholtz charge-based energy density to the Kelvin-Maxwell field-based energy density follows from Green's First Identity

$$\iiint_V \left[\Phi \nabla^2 \Psi + (\nabla \Phi) \cdot (\nabla \Psi) \right] dV = \oiint_S (\Phi \nabla \Psi) \cdot d\mathbf{S} \qquad (5.3)$$

under the assumption of electrostatic fields. In this context, the electric field is the gradient of the electric potential (Φ)

$$\mathbf{E} = -\nabla \Phi \qquad (5.4)$$

By Gauss's Law [(4.22a)],

$$\nabla \cdot \mathbf{E} = -\nabla \cdot \nabla \Phi = -\nabla^2 \Phi = \frac{\rho_e}{\varepsilon_0} \qquad (5.5)$$

Substituting into Green's First Identity (5.3) and evaluating over a region so large that in the limit the surface integral goes to zero yields

$$\iiint_V \left[\Phi \nabla^2 \Phi + (\nabla \Phi) \cdot (\nabla \Phi) \right] dV = \overset{\to 0}{\oiint_S (\Phi \nabla \Phi) \cdot d\mathbf{S}}$$
$$\iiint_V \Phi \nabla^2 \Phi \, dV = -\iiint_V (\nabla \Phi) \cdot (\nabla \Phi) dV \qquad (5.6)$$
$$\frac{1}{2} \iiint_V \rho_e \Phi \, dV = \frac{1}{2} \varepsilon_0 \iiint_V |\mathbf{E}|^2 \, dV.$$

Identifying the integrands with each other establishes the mathematical equivalence of the Helmholtz charge-based energy density to the Kelvin-Maxwell field-based energy density.

Not long thereafter, Poynting [8] and Heaviside [9] independently arrived at what came to be known as the "Poynting vector" to describe the flow of electromagnetic energy

$$\mathbf{S} = \mathbf{E} \times \mathbf{H} \qquad (5.7)$$

The Poynting vector and its relation to electromagnetic energy follow directly from Maxwell's equations. Taking the scalar product of the electric field with Ampere's Circuit Law [(4.22d)] yields

$$\mathbf{E} \cdot \nabla \times \mathbf{H} = \varepsilon_0 \mathbf{E} \cdot \frac{\partial \mathbf{E}}{\partial t} + \mathbf{E} \cdot \mathbf{J} \qquad (5.8)$$

Similarly, taking the scalar product of the electric field with Faraday's Law [(4.22c)] yields

$$\mathbf{H} \cdot \nabla \times \mathbf{E} = -\mu_0 \mathbf{H} \cdot \frac{\partial \mathbf{H}}{\partial t} \qquad (5.9)$$

Subtracting these two relations one from the other yields the following result

$$\mathbf{H} \cdot \nabla \times \mathbf{E} - \mathbf{E} \cdot \nabla \times \mathbf{H} = -\left(\varepsilon_0 \mathbf{E} \cdot \frac{\partial \mathbf{E}}{\partial t} + \mu_0 \mathbf{H} \cdot \frac{\partial \mathbf{H}}{\partial t} + \mathbf{E} \cdot \mathbf{J} \right) \qquad (5.10)$$

Noting that

$$\nabla \cdot (\mathbf{E} \times \mathbf{H}) = (\mathbf{H} \cdot \nabla \times \mathbf{E} - \mathbf{E} \cdot \nabla \times \mathbf{H}) \qquad (5.11)$$

and

$$\frac{\partial}{\partial t} \left(\tfrac{1}{2} \varepsilon_0 |\mathbf{E}|^2 + \tfrac{1}{2} \mu_0 |\mathbf{H}|^2 \right) = \varepsilon_0 \mathbf{E} \cdot \frac{\partial \mathbf{E}}{\partial t} + \mu_0 \mathbf{H} \cdot \frac{\partial \mathbf{H}}{\partial t} \qquad (5.12)$$

the mathematical relation known as Poynting's Theorem follows

$$\nabla \cdot (\mathbf{E} \times \mathbf{H}) + \frac{\partial}{\partial t} \left(\tfrac{1}{2} \varepsilon_0 |\mathbf{E}|^2 + \tfrac{1}{2} \mu_0 |\mathbf{H}|^2 \right) + \mathbf{E} \cdot \mathbf{J} = 0 \qquad (5.13)$$

As a mathematical relation, Poynting's Theorem follows unequivocally from Maxwell's equations. Interpreted as the fundamental law of electromagnetic energy transfer and conservation, Poynting's Theorem,

$$\nabla \cdot \mathbf{S} + \frac{\partial}{\partial t} (u_E + u_H) + \mathbf{E} \cdot \mathbf{J} = 0 \qquad (5.14)$$

establishes that the divergence, or outward flux, of electromagnetic energy from a particular point $[\nabla \cdot \mathbf{S} = \nabla \cdot (\mathbf{E} \times \mathbf{H})]$, the time rate of change of the local electromagnetic energy density $(\partial u/\partial t = \partial/\partial t(u_E + u_H))$, and the ohmic losses $(\mathbf{E} \cdot \mathbf{J})$ must balance out. This interpretation has not always been without controversy.

With Hertz's discovery of radio waves (1888-1889) however, the triumph of Faraday's field point of view was generally considered complete [10]. Having demonstrated that radio waves move at the speed of light and have optical properties, Hertz proved that they are decoupled from their source: "In the sense of our theory we more correctly represent the phenomena by saying that fundamentally the waves which are being developed do not owe their formation solely to processes at the origin, but arise out of the conditions of the whole surrounding space, which latter, according to our theory, is the true seat of the energy [11]."

5.2.2 Puzzles and Paradoxes of Electromagnetic Energy

Later investigators noticed difficulties with the energy-flow interpretation advocated by such pioneers as Poynting, Heaviside, and Hertz. Under certain circumstances, the Poynting-Heaviside theory yields seemingly nonsensical results, such as closed loops of energy in otherwise static systems. For instance, consider a static point charge q with field

$$\mathbf{E} = \frac{1}{4\pi\varepsilon_0}\frac{q}{r^2}\hat{\mathbf{r}}$$ (5.15)

superimposed with a static magnetic dipole with field

$$\mathbf{H} = \frac{m_0}{4\pi r^3}\left(2\cos\theta\,\hat{\mathbf{r}} + \sin\theta\,\hat{\theta}\right)$$ (5.16)

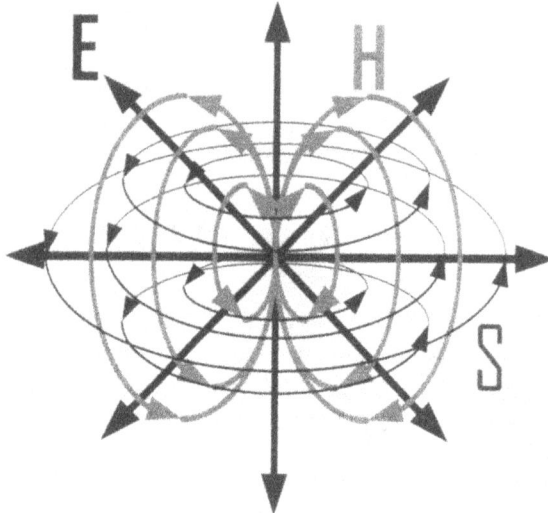

Figure 5.4   Circular Poynting flux around a superposition of a static charge and static magnetic dipole.

This "static" system has a nonzero Poynting vector

$$\mathbf{S} = \mathbf{E} \times \mathbf{H} = \frac{m_0 q}{16\pi^2 \varepsilon_0 r^5} \sin\theta \, \hat{\phi} \qquad (5.17)$$

corresponding to azimuthal loops of flux. Figure 5.4 shows the pattern of radial electric field lines, dipolelike magnetic field lines, and circular flux of electromagnetic energy.

Further, the Poynting vector is ambiguous to a solenoidal (i.e., divergenceless) term. In other words, if $\mathbf{S}' = \mathbf{S} + \mathbf{G}$, where $\nabla \cdot \mathbf{G} = 0$, the new resulting Poynting vector, $\mathbf{S}'$, will still satisfy Poynting's Theorem. Thus, it is often argued that the Poynting vector has no physical significance unless integrated over a completely closed surface [12].

Some observers reject the idea that the Poynting vector represents a localized flow of energy while at least accepting that the integral of the Poynting vector over a closed surface has a physical significance—the rate of change of energy in the bounded volume [13]. Others, like R. W. P. King, reject the idea that energy has any physical significance whatsoever, beyond being a mathematical quantity that may be useful in calculations [14].

More recent investigation has tended to uphold the original vision of Poynting and Heaviside. The seemingly implausible "loops" of Poynting flux are now generally recognized as required by the demands of angular momentum conservation. A simple example (adapted from one of Richard Feynman's [15]) serves to illustrate the point.

Consider a charged dielectric hoop at rest with a bar magnet along its axis, a physical system whose fields will be similar to the point charge, point magnetic dipole described above. By Lenz's Law, when the magnet is removed, an electromotive force will be induced in the hoop, setting it spinning. Since the bar magnet is removed along the axis of the hoop, no angular momentum will be imparted to the system, and, yet, it spins. Figure 5.5 shows this configuration. A detailed analysis of an analogous system shows that the electromagnetic angular mometum of the original "static" fields, as predicted by the closed loops of Poynting flux, is the same as the imparted mechanical angular momentum [16, 17]. In fact, the torque associated with circularly polarized light has been experimentally measured. The results are also in accord with the Poynting-Heaviside interpretation [18].

The seeming ambiguity of the Poynting vector is thus subject to the constraint that any modification must not only satisfy conservation of angular momentum, but also yield the same correct value for the angular momentum. Similarly, there is a linear momentum (source of the so-called light pressure) known to be correctly described by the Poynting vector [19]. Naturally, any modification must further satisfy conservation of energy. All these constraints severely limit alternatives to the Poynting vector for describing the local electromagnetic energy flow [20]. In fact, it has long been realized that any additional solenoidal term added to the Poynting vector cannot be a function of the electromagnetic fields (or their time derivatives) and still satisfy Poynting's Theorem [21]. Since it is reasonable to assume that any measure of local electromagnetic energy flow must be a function of the electromagnetic fields, it is very difficult to imagine how there could be a physically meaningful alternative to the Poynting vector.

A final consideration has largely been neglected in the debate on the physical interpretation of the Poynting vector. Every communication link ever designed relies on the Friis transmission formula to predict the received power. The Friis transmission formula in turn relies on our ability to predict the power flux using the Poynting vector, not integrated over a closed surface surrounding the transmitter but, rather, over the very tiny piece of that surface intercepted by a receive antenna. The undeniable success of the Friis formula suggests that the Poynting vector is the correct local measure of electromagnetic energy flow or at least places very stringent limitations on alternatives. Open questions may remain, but there is no good reason to reject unilaterally the insights available from localizing and tracking electromagnetic energy. The next section will apply some of these insights to reevaluate the canonical example of an accelerating charge.

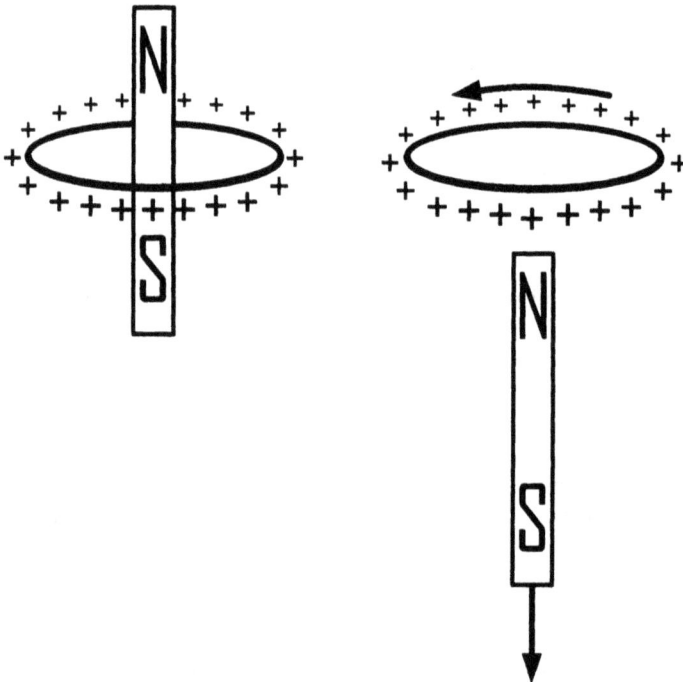

Figure 5.5   Conversion of electromagnetic angular momentum to mechanical angular momentum.

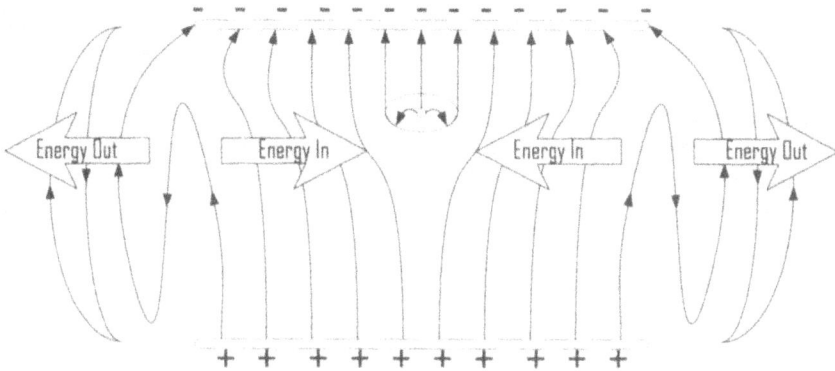

Figure 5.6 Behavior of fields and energy around an accelerating charge.

5.2.3 Example: Accelerating Charge

Consider the energy flow around an accelerating charge. An accelerating charge (q) with an increasing velocity (v) is like an increasing infinitesimal current ($I \sim qv$). The magnetic field energy associated with the charge is proportional to I^2, so an accelerating charge is actually absorbing energy to store in an ever-increasing magnetic field about the charge. From a mechanical point of view, as the velocity of the charge increases, its kinetic energy increases, so here again the charge must be absorbing not losing energy. However one looks at accelerating charges, they are absorbing energy from their immediate environment, not losing energy to radiation.

If an accelerating charge is not the source of radiation energy, what is the true source? The answer is that radiation energy comes from the applied field. A charge does not accelerate by itself—it accelerates in response to an applied field. Suppose a charge is placed between the plates of a capacitor and accelerates in response to the applied field as shown in Figure 5.6. Appealing to the right-hand rule for radiation (see Section 4.3.3), the orientation of the magnetic fields must be up out of the page to the left of the charge and down into the page to the right of the charge. In between the plates, the applied field dominates, and the net flow of energy is in. At the boundary of the plates, the applied field weakens, and finally, the charge's radiation field dominates the fringing fields of the capacitor. In this boundary region, the energy flow reverses direction finally to flow outward and radiate away. Therefore, the true source of the radiation energy is in the fringe region of the applied field.

The traditional view of radiation as a transverse kink in the fields of a point charge is highly misleading because it only considers the fields of the point charge and not the applied field responsible for the acceleration. When one considers the applied field, a more complete and informative picture of radiation emerges.

A quick calculation in a comoving reference frame helps establish boundaries for regions with different energy flows. On a sufficiently local scale, the external field must always be uniform and constant. The applied field cannot dominate arbitrarily close to the accelerating charge. In a frame comoving with respect to an accelerating charge, the electric field due to an accelerating charge q with acceleration $\dot{v}$ and with mass m is

$$\mathbf{E} = \frac{q}{4\pi\varepsilon_0}\left(\frac{\hat{\mathbf{r}}}{r^2} + \frac{\dot{v}\sin\theta}{c^2 r}\,\hat{\theta}\right) \tag{5.18}$$

The external field will look something like this

$$\mathbf{E}_{ext} = E_0\hat{\mathbf{z}} = E_0\left(\cos\theta\,\hat{\mathbf{r}} - \sin\theta\,\hat{\theta}\right) \tag{5.19}$$

Summing the tangential ($\hat{\theta}$) fields and noting that the magnitude of the force on the charge is

$$F = qE_0 = m\dot{v} \tag{5.20}$$

the tangential fields will go to zero on a spherical surface of radius

$$r_e = \frac{q^2}{4\pi\varepsilon_0 mc^2} \tag{5.21}$$

Since the E_θ field is zero, and since the radial component of the Poynting vector is

$$S_r = E_\theta H_\phi \tag{5.22}$$

no energy passes through this surface of radius r_e.

Interestingly, this result is entirely independent of the magnitude of the external field. In fact, the result of (5.21) is exactly the so-called classical electron radius. Since the net tangential electric field is zero, energy cannot be extracted from within this surface. The surface serves as an electromagnetic analog to the event horizon of a black hole. Whatever energy is absorbed by or radiated from an accelerating charge must reside outside the sphere defined by the classical electron radius. Of course, this is strictly a nonrelativistic result, but since the fields in an instantaneously comoving reference frame should reduce to the low velocity result here, a similar result should hold for the more general case. Further, it should be noted that this is strictly a classical calculation and neglects any quantum mechanical effects. Figure 5.7 shows the detailed behavior of the fields and energy flow in the immediate vicinity of the charge.

Distances on the scale of the classical electron radius are not of practical relevance to most antenna problems. Thus, for the purposes of understanding antenna physics, one is fully justified in assuming that accelerating charges absorb energy, and radiation energy comes from the fringing region of the external or applied field.

To summarize, electromagnetics is not exempt from the Law of Energy Conservation. Radiation energy does not come into being out of nothing, rather it is converted from other forms of electromagnetic energy. In the case of accelerating charges, the external field loses energy—it is the source of both the radiated energy as well as the magnetic field energy gained by the now more quickly moving charge. In the case of decelerating charges, the more slowly moving charge loses magnetic field energy—it is the source of both the radiated energy as well as the energy gained by the external field.

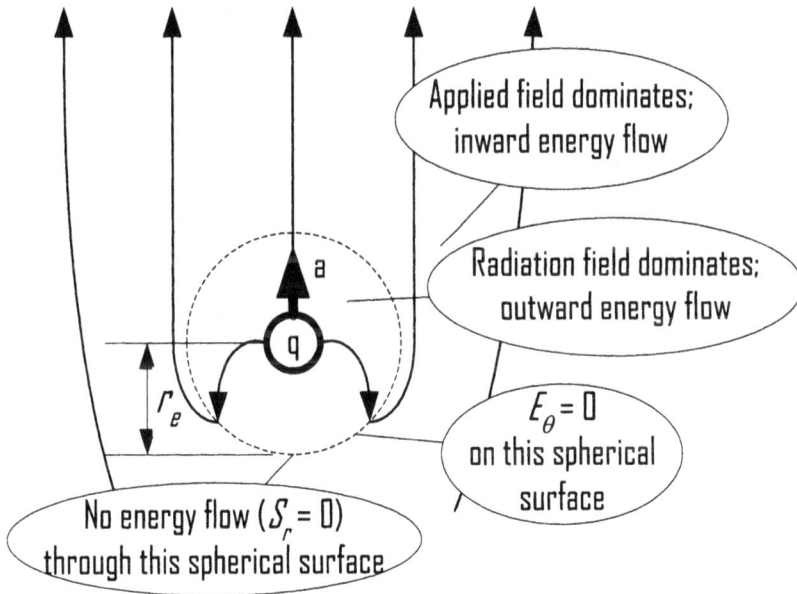

Figure 5.7 Detailed behavior of fields and energy in the immediate vicinity of an accelerating charge.

5.2.4 Causal Surfaces

A detailed examination and understanding of electromagnetic energy flow can involve some complicated analysis. This analysis can be simplified by looking for surfaces on which the normal component of the Poynting vector goes to zero. There is no net energy flow through such a surface. These surfaces bound distinct regions of energy flow and reveal the cause-and-effect relationships at the root of electromagnetic phenomena. Thus, these surfaces have been dubbed "causal surfaces." An analysis of causal surfaces yields great insights into the location and behavior of electromagnetic energy.

In the previous example of an accelerating charge, an identification of these causal surfaces allows one to determine the boundaries at which the original electrostatic energy of a capacitor cleaves to yield outgoing radiation energy and in-flowing kinetic, or magnetic, energy of the accelerating charge. Although the preceding treatment of an accelerating charge was general and qualitative, these same techniques can be applied with mathematical rigor and precision, as will be demonstrated in Section 5.3.

This section has discussed the localization of electromagnetic energy from a historical viewpoint and addressed some of the alleged puzzles and paradoxes of the Poynting-Heaviside theory of electromagnetic energy transfer. The example of an accelerating charge demonstrated that conventional thinking about radiation as a kinked field line is misleading and fails to capture the subtleties of electromagnetic energy flow at the root of the radiation process. In the following section, these ideas of energy flow and localization will be applied to the dipole fields to yield insights of value to a UWB antenna designer.

> *If we believe in the continuity of the motion of energy, that is, if we believe that when it disappears at one point and reappears at another it must have passed through the intervening space, we are forced to conclude that the surrounding medium contains at least a part of the energy, and that it is capable of transferring it from point to point.*

John Henry Poynting, 1885

5.3 DIPOLE FIELD ENERGY

This section will explain how to understand and analyze the time domain energy flow around a Hertzian dipole. First, this section will show how to localize energy and track the flow of energy in the dipole fields. Then, this section will apply these methods to a variety of examples including, exponential, damped harmonic, harmonic, and transient dipole excitations.

The dipole field equations, (4.31) and (4.32), may be combined with the Kelvin-Maxwell theory that the energy density is proportional to the square of the field intensity [(5.2)]. For a dipole then, the electric field energy density (assuming free space) is

$$
\begin{aligned}
u_E = \frac{\varepsilon_0}{2}|\mathbf{E}|^2 &= \frac{p_0^2}{32\pi^2\varepsilon_0 r^2}\left[\begin{array}{l} 4\cos^2\theta\left(\dfrac{T}{r^2}+\dfrac{\dot{T}}{cr}\right)^2 \\[3mm] +\sin^2\theta\left(\dfrac{T}{r^2}+\dfrac{\dot{T}}{cr}+\dfrac{\ddot{T}}{c^2}\right)^2 \end{array}\right] \\[4mm]
&= \frac{p_0^2}{32\pi^2\varepsilon_0 r^2}\left[\begin{array}{l} 4\cos^2\theta\left(\dfrac{T^2}{r^4}+\dfrac{2T\dot{T}}{cr^3}+\dfrac{\dot{T}^2}{c^2 r^2}\right) \\[3mm] +\sin^2\theta\left(\dfrac{T^2}{r^4}+\dfrac{2T\dot{T}}{cr^3}+\dfrac{\dot{T}^2+2T\ddot{T}}{c^2 r^2}+\dfrac{2\dot{T}\ddot{T}}{c^3 r}+\dfrac{\ddot{T}^2}{c^4}\right) \end{array}\right] \\[4mm]
&= \frac{p_0^2}{32\pi^2\varepsilon_0 r^2}\left[\begin{array}{l} \left(3\cos^2\theta+1\right)\left(\dfrac{T^2}{r^4}+\dfrac{2T\dot{T}}{cr^3}+\dfrac{\dot{T}^2}{c^2 r^2}\right) \\[3mm] +\sin^2\theta\left(\dfrac{2T\ddot{T}}{c^2 r^2}+\dfrac{2\dot{T}\ddot{T}}{c^3 r}+\dfrac{\ddot{T}^2}{c^4}\right) \end{array}\right]
\end{aligned}
\tag{5.23}
$$

The corresponding magnetic field energy density (again, assuming free space) is

$$
\begin{aligned}
u_H = \frac{\mu_0}{2}|\mathbf{H}|^2 &= \frac{p_0\mu_0}{32\pi^2 r^2}\left(\frac{\dot{T}}{r}+\frac{\ddot{T}}{c}\right)^2\sin^2\theta \\[3mm]
&= \frac{p_0\mu_0}{32\pi^2 r^2}\left(\frac{\dot{T}^2}{r^2}+\frac{2\dot{T}\ddot{T}}{cr}+\frac{\ddot{T}^2}{c^2}\right)\sin^2\theta
\end{aligned}
\tag{5.24}
$$

The energy flow is given by the Poynting vector

$$
\mathbf{S} = \mathbf{E}\times\mathbf{H} = S_r\hat{\mathbf{r}}+S_\theta\hat{\boldsymbol{\theta}}
\tag{5.25}
$$

where

$$
S_r = E_\theta H_\phi = \frac{p_0^2\sin^2\theta}{\varepsilon_0(4\pi)^2}\left(\frac{T\dot{T}}{r^5}+\frac{1}{r^4 c}\left(T\ddot{T}+\dot{T}^2\right)+2\frac{\dot{T}\ddot{T}}{r^3 c^2}+\frac{\ddot{T}^2}{c^3}\right)
\tag{5.26}
$$

and

$$S_\theta = -E_r H_\phi = -\frac{p_0^2 \sin 2\theta}{\varepsilon_0 (4\pi)^2} \left(\frac{T\dot{T}}{r^5} + \frac{1}{r^4 c}\left(T\ddot{T} + \dot{T}^2\right) + \frac{\ddot{T}\dot{T}}{r^3 c^2} \right) \qquad (5.27)$$

The following sections apply these relations in a variety of examples.

5.3.1 Exponentially Decaying Dipoles

An electric dipole is analogous to a small capacitor. If this dipole capacitor C is charged up, then discharged rapidly through a resistance R, the resulting discharge has an exponentially decaying time dependence

$$T(t) = \exp\left(-\frac{t}{\tau}\right) \qquad (5.28)$$

where the time constant of the decay is $\tau = RC$. Consider the magnetic field around an exponentially decaying dipole. The time derivatives are

$$\dot{T}(t) = -\frac{1}{\tau}\exp\left(-\frac{t}{\tau}\right) \qquad (5.29)$$

and

$$\ddot{T}(t) = \frac{1}{\tau^2}\exp\left(-\frac{t}{\tau}\right) \qquad (5.30)$$

The magnetic fields [given by (4.32)] around an exponentially decaying dipole go to zero for

$$\frac{[\dot{T}]}{r} + \frac{[\ddot{T}]}{c} = 0 \rightarrow -\frac{1}{r\tau} + \frac{1}{c\tau^2} = 0 \rightarrow r = c\tau \qquad (5.31)$$

Since the magnetic field is zero everywhere on a spherical surface of radius $R_S = c\tau$ throughout the exponential decay, the Poynting vector is also identically equal to zero. Thus, this spherical shell is a causal surface defining the boundary between energy hurtling inward to be dissipated as heat in the resistor, and uncoupled energy radiating away. Figure 5.8 depicts this flow of energy around an exponentially decaying dipole.

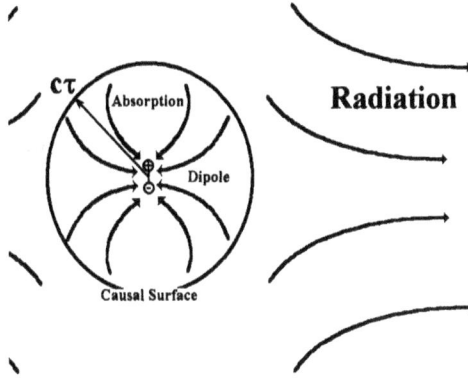

Figure 5.8   Energy flow around an exponentially decaying dipole ([22], © 1995 H. Schantz).

In fact, an exact calculation shows the equivalence of the radiated energy
(U_{rad}) with the loss in energy ($\Delta U_F = U_{F1} - U_{F0}$) from the region outside the causal
surface during a time period ($\Delta t = t_1 - 0 = t_1$). Consider an arbitrary spherical
surface of radius R outside the causal surface ($R > R_S$). Figure 5.9 shows the
space-time geometry of this configuration. The radiated energy through the sphere
of radius R is identically equal to the change in the stored energy outside the
causal surface during the period

$$U_{rad} = \Delta U_F = \frac{p_0^2(R - c\tau)(R^2 - Rc\tau + \tau^2 c^2)}{12\pi\varepsilon_0 R^3 \tau^3 c^3} \tag{5.32}$$

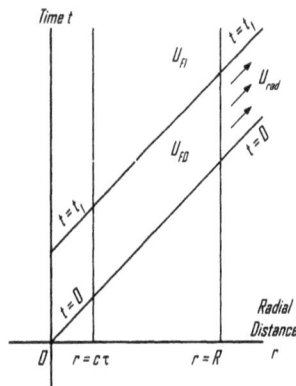

Figure 5.9   Space-time geometry of energy storage around an exponentially decaying dipole (After
[23]).

5.3.2 Damped Harmonic Dipoles

The exponentially decaying dipole described in the preceding section is enclosed by a spherical surface on which the magnetic field is equal to zero—a causal surface that partitions the energy flow. This causal surface denotes the boundary between the in-falling reactive energy and the outgoing radiation energy. A similar surface will exist where the tangential (or θ) component of the electric field goes to zero. Solving for $E_\theta \to 0$ yields the differential equation

$$\ddot{T} + \frac{c}{r}\dot{T} + \frac{c^2}{r^2}T = 0 \tag{5.33}$$

Setting the radial distance to a constant ($r = R_S$), solutions to this equation have the form

$$T(t) = \begin{cases} 1 & t \le 0 \\ e^{-\frac{ct}{2R_S}}\left(C_1 \sin\frac{\sqrt{3}c}{2R_S}t + C_2 \cos\frac{\sqrt{3}c}{2R_S}t\right) & t \ge 0 \end{cases} \tag{5.34}$$

under the assumption that the radiation process begins at time $t = 0$. These are also the solutions for the characteristic modes of a spherical conducting oscillator, a problem studied as early as 1884 by J. J. Thompson [24, 25]. In any event, the solution of (5.34) must be continuous in time at $t = 0$, so $C_2 = 0$. The other constant may be found by applying a conservation-of-energy boundary condition—equating the radiated energy with the original electrostatic energy stored outside the sphere of radius R_S.

The total radiated energy follows from integrating the far-field Poynting flux over a distant sphere for all time. This yields Larmor's dipole radiation energy formula [26]

$$U_{rad} = \frac{p_0^2}{6\pi\varepsilon_0 c^3}\int_0^\infty \ddot{T}^2(t)\,dt$$

$$= \frac{p_0^2}{6\pi\varepsilon_0 c^3}\int_0^\infty \frac{c^4}{4R_S^4} e^{-\frac{ct}{2R_S}}\left[\begin{matrix}\left(C_1 - \sqrt{3}C_2\right)\sin\frac{\sqrt{3}c}{2R_S}t \\ +\left(\sqrt{3}C_1 + C_2\right)\cos\frac{\sqrt{3}c}{2R_S}t\end{matrix}\right]^2 dt \tag{5.35}$$

$$= \frac{p_0^2}{24\pi\varepsilon_0 R^3}\left(3C_1^2 + C_2^2\right)$$

The energy density of the original electrostatic field is

$$u_E = \frac{\varepsilon_0}{2}\left|\mathbf{E}_{static}\right|^2 = \frac{p_0^2}{32\pi^2\varepsilon_0 r^6}\left(3\cos^2\theta+1\right) \tag{5.36}$$

Therefore the energy stored in the original electrostatic field outside the causal surface is

$$
\begin{aligned}
U_{static} = \int_v u_E\, dV &= \frac{p_0^2}{16\pi\varepsilon_0}\int_{R_S}^{\infty}\int_0^{\pi}\frac{3\cos^2\theta+1}{r^6}\sin\theta\, r^2\, d\theta\, dr \\[2mm]
&= -\frac{p_0^2}{16\pi\varepsilon_0}\int_{R_S}^{\infty}\frac{dr}{r^4}\left[\cos\theta+\cos^3\theta\right]_0^{\pi} = \frac{p_0^2}{4\pi\varepsilon_0}\int_{R_S}^{\infty}\frac{dr}{r^4} \\[2mm]
&= -\frac{p_0^2}{4\pi\varepsilon_0}\left[\frac{1}{3r^3}\right]_{R_S}^{\infty} = \frac{p_0^2}{12\pi\varepsilon_0 R_S^3}
\end{aligned}
\tag{5.37}
$$

The conservation-of-energy boundary condition requires that $U_{static} = U_{rad}$. Since $C_2 = 0$, this means that $C_1 = {}^1/\sqrt{3} \cong 0.577$. Figure 5.10 shows the resulting damped harmonic time dependence and its first two derivatives.

There is a variety of ways to depict and follow the energy flow for this damped harmonic decay process. Figures 5.11 through 5.14 provide snapshots at 1ns intervals of the energy density and energy flow around a damped harmonic decay process with $R_S = 30$cm. A point dipole is located at $(y, z) = (0, 0)$ aligned along the z-axis. The background shading shows the local energy density, and the arrows denote the direction of the energy flow as given by the Poynting vector direction. The snapshots, at 1ns intervals, show how the leading edge of the radiated signal gathers energy from the original electrostatic energy distribution. The energy inside the $R_S = 30$cm causal surface sphere collapses into the dipole while the energy outside the causal surface decouples and radiates away.

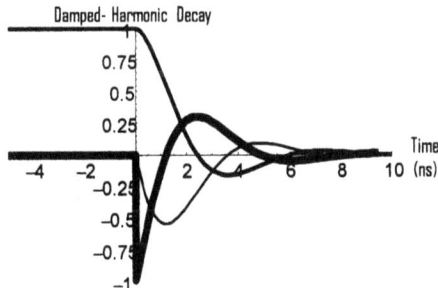

Figure 5.10 Normalized damped harmonic time dependence (thick line), first derivative (medium line), and second derivative (thin line) assuming $R_S = 30$cm (After [23]).

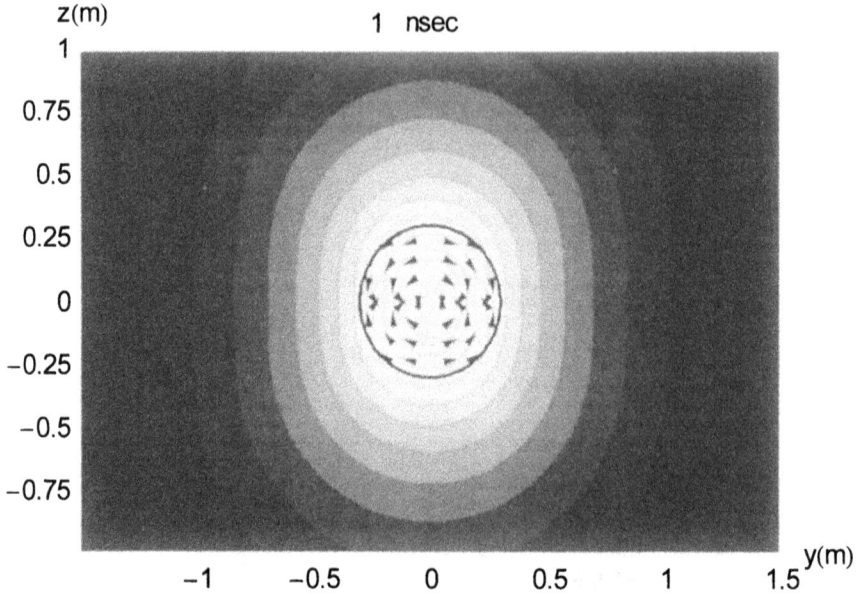

Figure 5.11 At 1ns, the leading edge has moved 30cm. Inside the ($R_S = 30$)–cm sphere, the energy flows into the dipole (After [23]).

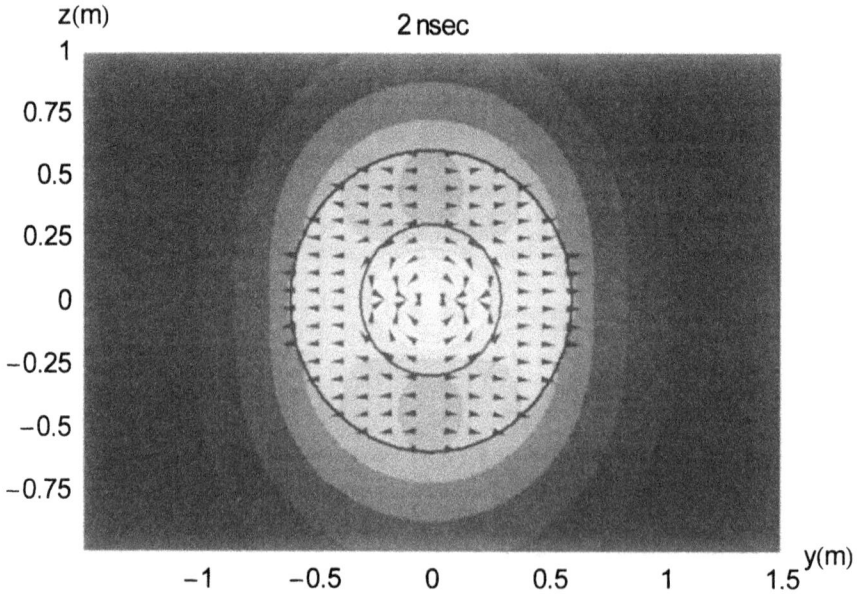

Figure 5.12 At 2ns, the leading edge has moved 60cm away from the dipole. As the leading edge sweeps past the ($R_S = 30$)–cm causal surface, it gathers up the original electrostatic field energy to form the radiated signal (After [23]).

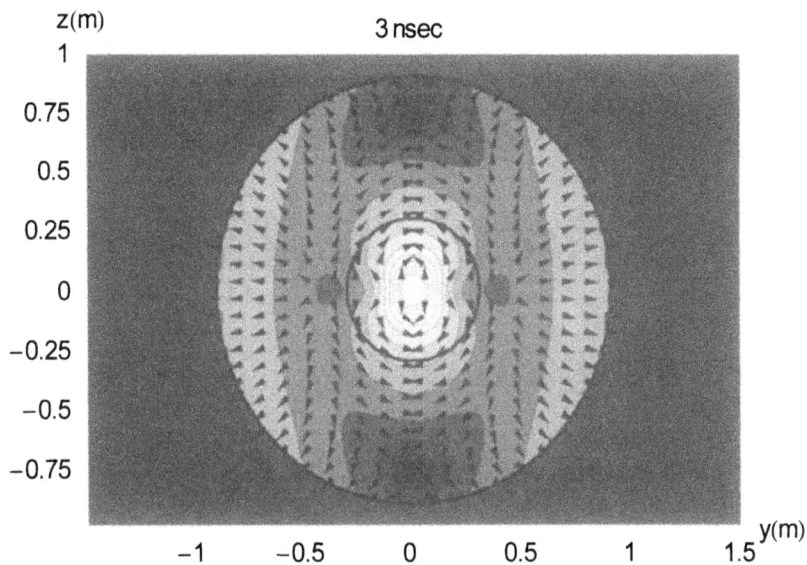

Figure 5.13 At 3ns, the leading edge has moved 90cm. The damped harmonic oscillation gives rise to periodic reversals in energy flow, but the causal surface still partitions the energy flow (After [23]).

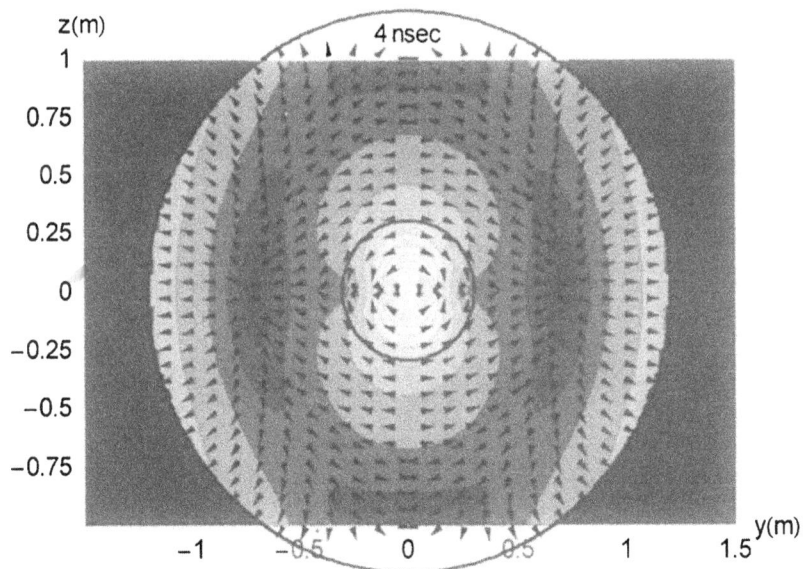

Figure 5.14 At 4ns, the leading edge has moved 120cm. Energy outside the causal surface that was not included in the leading edge collects along the z-axis and flows into the second radiated impulse (After [23]).

An alternate way to depict the flow of energy is to use a space-time energy-flow diagram. The horizontal axis shows radial distance (in meters), and the vertical axis depicts time (in nanoseconds). A radial space-time energy velocity follows from averaging the radial local energy velocity over the spherical shell at the radial distance in question

$$v(r,t) = \frac{\langle S_r(r,t) \rangle}{c \langle u_r(r,t) \rangle} = \frac{1}{c} \frac{\displaystyle\oiint_{S@r} \mathbf{S} \cdot \hat{\mathbf{r}} \, d\Omega}{\displaystyle\oiint_{S@r} u \, d\Omega} \tag{5.38}$$

The plot is scaled so that radiation energy traveling at the speed of light follows a $45°$ trajectory. Interestingly, the signal traverses the causal surface without any corresponding energy flow through the surface. Figure 5.15 presents a space-time energy-flow diagram of the damped harmonic decay process depicted in Figures 5.11 to 5.14.

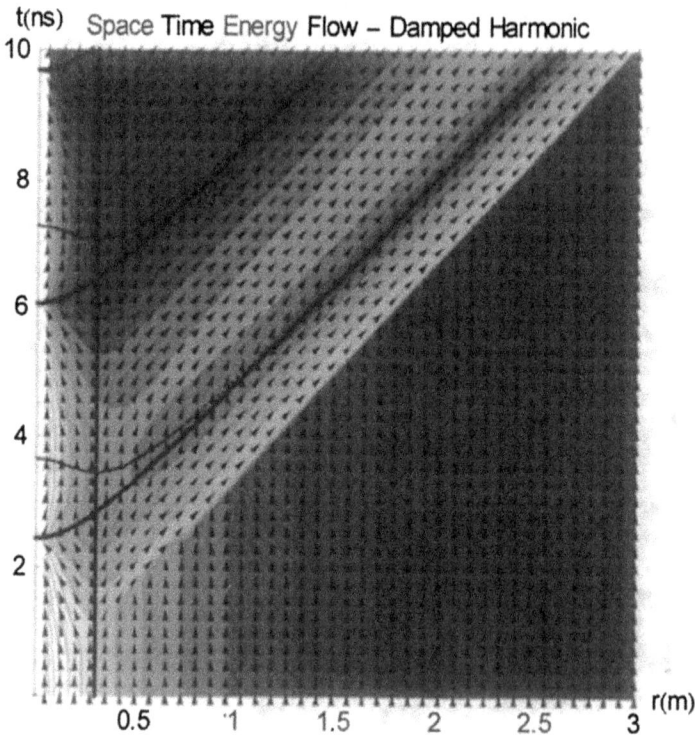

Figure 5.15 A space-time energy-flow diagram for a damped harmonic dipole with $R_S = 30$cm. The dark lines show the surfaces on which there is no net radial energy flow (After [23]).

5.3.3 Harmonic Dipoles

The two preceding examples have analyzed transient processes. Traditional electromagnetics uses frequency domain analysis. Frequency domain analysis assumes steady-state, or harmonic, behavior. This section will connect the frequency domain and the time domain by presenting a time domain analysis of a harmonic excitation. Consider a time dependence

$$T(t) = \sin \omega t \qquad (5.39)$$

A first derivative

$$\dot{T}(t) = \omega \cos \omega t \qquad (5.40)$$

A second derivative

$$\ddot{T}(t) = -\omega^2 \sin \omega t \qquad (5.41)$$

Since traditional practice looks at the energy of a harmonic dipole from a time-average basis, bridging the gap from time domain analysis to time-average analysis is a good place to start.

A time domain analysis of a harmonic dipole reveals subtle and important details of radiation physics. Consider a point dipole with time dependence $T(t) = \sin \omega t$. A new pulse of energy decouples and radiates away every half-period. Figures 5.16 through 5.19 illustrate this process. The gray scale denotes energy density with regions of highest energy density being lightest and areas of lowest energy density being darkest. The arrows denote the local direction of the energy flow as determined by the direction of the Poynting vector. The thick black contours show the locations of causal surfaces. All these figures are the result of numerical analysis of theoretical formulas, and thus results are subject to some minor inaccuracies due to numerical precision and the density of calculation points.

The energy flow around a harmonic dipole may be partitioned by considering two sets of causal surfaces: a first magnetic set where $H_\phi \rightarrow 0$ and a second electric set where $E_\theta \rightarrow 0$. The first magnetic set comprises solutions of

$$\frac{[\dot{T}]}{r} + \frac{[\ddot{T}]}{c} = 0 \rightarrow r = -c\frac{[\dot{T}]}{[\ddot{T}]} = \frac{c}{\omega}\cot \omega t \qquad (5.42)$$

and the second electric set comprises solutions of

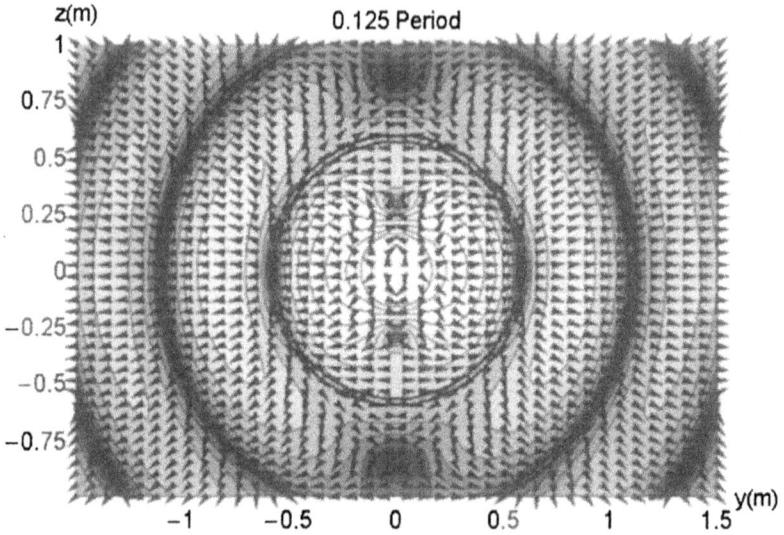

Figure 5.16 Energy density and energy flow around a point dipole with wavelength $\lambda = 1$m and time dependence $T(t) = \sin \omega t$ at 0.125 period. Magnetic field energy transforms to electric field energy. Some is stored near the dipole "poles;" some begins decoupling along the dipole "equator" (After [23]).

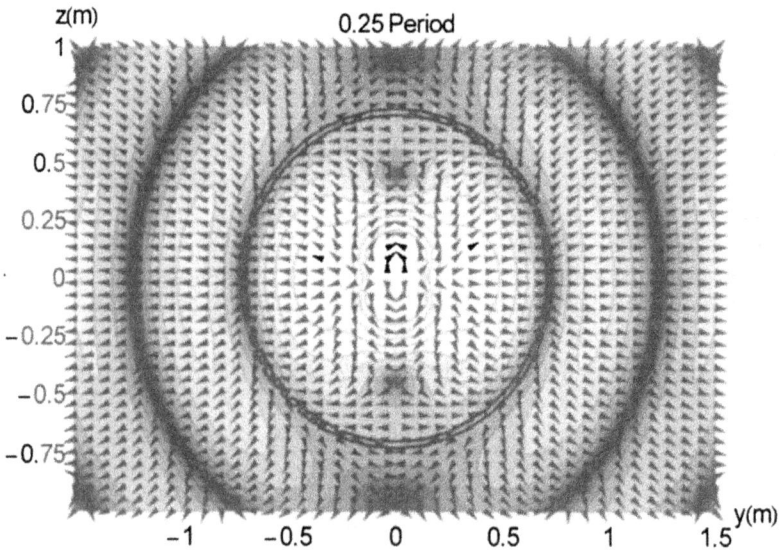

Figure 5.17 Energy density and energy flow around a point dipole with wavelength $\lambda = 1$m and time dependence $T(t) = \sin \omega t$ at 0.250 period. The dipole has reached maximum charge separation, and stored electric field energy begins converting to magnetic field energy. The equatorial energy is about to decouple and radiate away (After [23]).

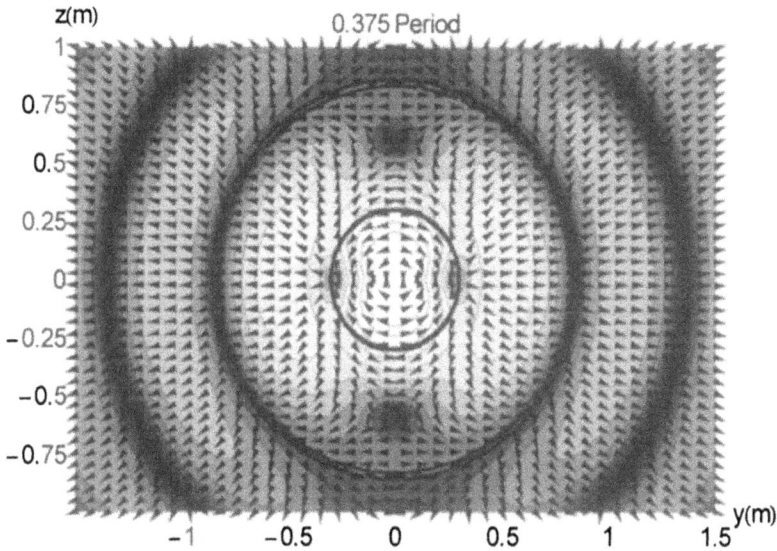

Figure 5.18 Energy density and energy flow around a point dipole with wavelength $\lambda = 1$m and time dependence $T(t) = \sin \omega t$ at 0.375 period. The dipole is discharging, and electric field energy transforms to magnetic field energy. Equatorial energy has decoupled and is radiating away (After [23]).

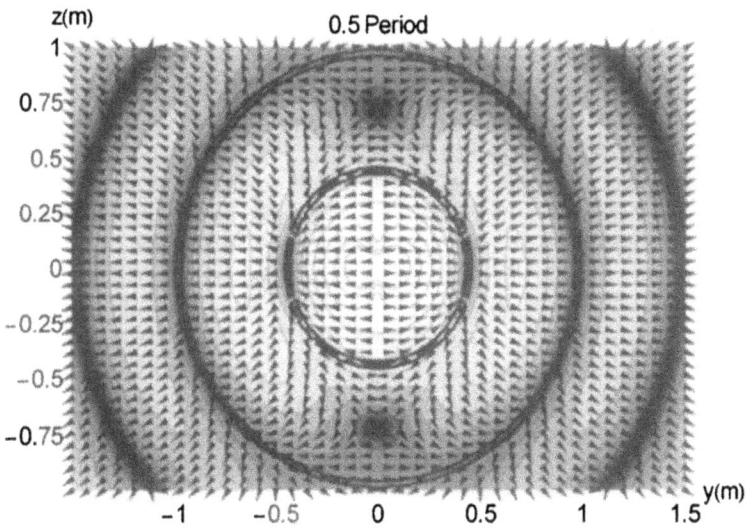

Figure 5.19 Energy density and energy flow around a point dipole with wavelength $\lambda = 1$m and time dependence $T(t) = \sin \omega t$ at 0.500 period. The dipole discharge current reaches a maximum, charge separation is a minimum, and stored magnetic field energy begins converting to electric field energy (After [23]).

$$\left[\ddot{T}\right]+\frac{c}{r}\left[\dot{T}\right]+\frac{c^2}{r^2}\left[T\right]=0 \rightarrow r^2-\frac{c}{\omega}r\cot\omega t-\frac{c^2}{\omega^2}=0 \qquad (5.43)$$

Taking into account retardation ($t \rightarrow t - \frac{r}{c}$) and solving for the location of the causal surfaces in time as a function of distance yield for the magnetic case

$$t_m = \frac{r}{c}+\frac{1}{\omega}\left(\cot^{-1} kr + n\pi\right) \qquad (5.44)$$

and

$$t_e = \frac{r}{c}+\frac{1}{\omega}\left(\cot^{-1}\left(kr-\frac{1}{kr}\right)+n\pi\right) \qquad (5.45)$$

for the electric causal surfaces. Note that n = 0, ±1, ±2, ±3... and that there is a branch cut in (5.45). A space-time energy-flow diagram can show these two relations. Because the surfaces are not stationary with respect to time, they no longer bound distinct regions with fixed quantities of energy. Nevertheless, they still denote surfaces on which the radial Poynting flux goes to zero. Thus, on one side of the surface, the flux of energy is radially in; on the other side, the flux is radially out. In the far field, both E_θ and H_ϕ go to zero every half-period, and both sets of causal surfaces converge asymptotically. Figure 5.20 shows the space-time energy flow around a harmonic dipole.

Each half-period, a pulse of energy is radiated. The dipole itself emits energy for the first quarter of a period, then spends the next quarter of a period absorbing some, but not all, of the energy it emitted earlier. The radiation energy is the extra energy that is not reabsorbed. Although an observer in the far field perceives radiation energy coming from the dipole for the entire half-period, that energy was emitted by the dipole *only* during the first quarter of the period. The seeming origin of radiation energy from the dipole during the second quarter of the period when the dipole is actually absorbing is an effect strikingly similar to the concept of a virtual image in optics. From the far field, it looks as though the radiation energy may be traced back to the dipole, but the source of the radiation energy is actually the stored energy in the reactive fields around the dipole.

If one defines the source of the radiation as that point where static, or reactive, energy is finally converted to radiant energy (in other words, where the radiation begins an uninterrupted trip to the far field), then the source of the radiation is the causal surface on which $E_\theta \rightarrow 0$ (the upper of the two causal surfaces that merge together in the far field). Interestingly *none* of the energy "radiated by the dipole" comes directly from the dipole itself. At one point or another, it must pass through the $H_\phi \rightarrow 0$ surface, begin to be reabsorbed (if only briefly), then pass through the $E_\theta \rightarrow 0$ surface and be reradiated.

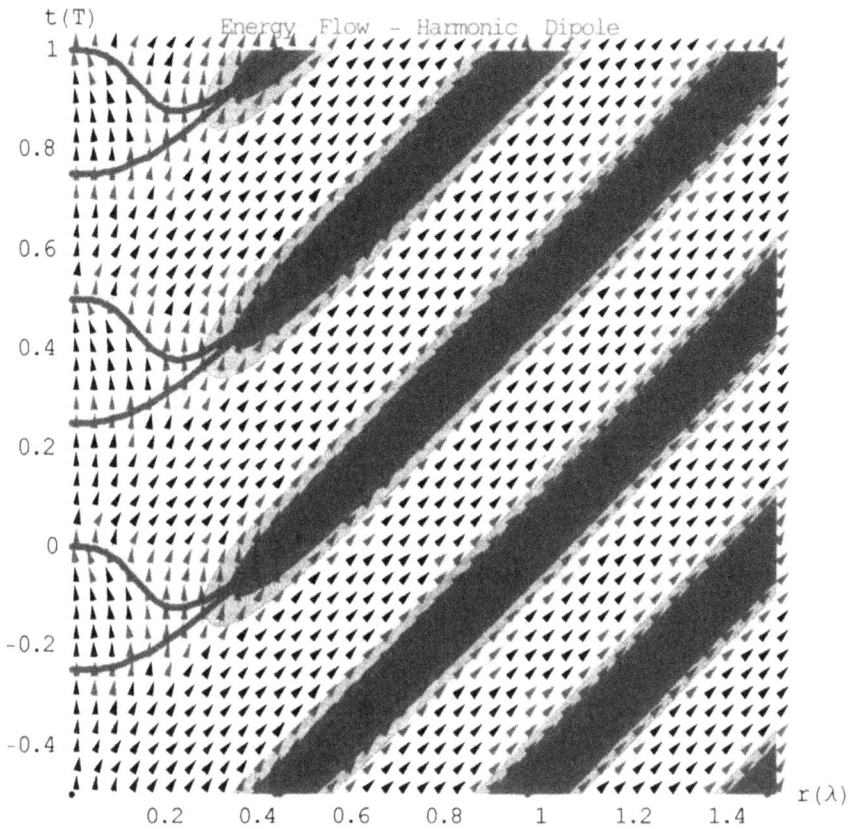

Figure 5.20 Space-time diagram of the energy flow around a steady-state Hertzian dipole with harmonic time dependence. Time is in units of period; radial distance is in units of wavelength. Radiation travels with a slope of one (i.e., one wavelength per period). Light shading indicates a higher energy density; dark shading indicates a lower energy density (After [23]).

Additional support for this interpretation comes from the field line perspective. The surface where $E_\theta \to 0$ is the surface on which the electric field lines reverse themselves to form the closed field loops that characterize radiated fields. The causal surface thus shows us exactly where and when the electric field becomes uncoupled from the reactive fields around the dipole. This process was described in detail by Hertz [27]. The $H_\phi \to 0$ and $E_\theta \to 0$ surfaces are also surfaces of constant phase. In fact, Hertz's own diagrams of dipole phase relationships are strikingly similar to the results of Figure 5.20 [28].

To determine the average origin of the radiation energy, we can consider the average location of surface on which $E_\theta \to 0$. Solving for this radial distance as a function of time,

$$r = \frac{c}{2\omega}\left(\cot \omega t + \sqrt{4 + \cot^2 \omega t} \right) \qquad (5.46)$$

where the other root is unphysical. Normalizing according to the $\sin^2 \omega t$ time dependence of the dipole radiation energy, and solving the integral for a half-period pulse numerically (using *Mathematica*), the average distance from the dipole at which the radiation energy originates is

$$\langle r \rangle = \frac{\frac{\omega}{\pi} \int_0^{\frac{\pi}{\omega}} \frac{c}{2\omega}\left(\cos \omega t + \sqrt{3\sin^2 \omega t + 1} \right) \sin \omega t \, dt}{\frac{\omega}{\pi} \int_0^{\frac{\pi}{\omega}} \sin^2 \omega t \, dt} = 0.173\lambda \qquad (5.47)$$

This result is roughly consistent with Harold Wheeler's concept of a "radian-sphere" [29, 30].   The radiansphere was Wheeler's identification that in a frequency domain analysis of a dipole, the reactive fields are dominant for $r < \lambda/(2\pi) = 0.159\lambda$, and outside this distance, the radiation fields are dominant. The method of causal surfaces allows Wheeler's time-harmonic radiansphere concept to be extended to dipoles with an arbitrary time dependence. The analysis of fundamental antenna limits in Section 5.4.1 further discusses the radiansphere concept.

Since the present discussion pertains to an infinitesimal "point" dipole source whose aperture (d) is zero, the usual $2d^2/\lambda$ phase front uniformity criterion to determine the far-field limit does not apply. The far-field limit of a point source is determined by finding that distance where the radiation fields are dominant. Wheeler's radiansphere marks the transition distance from the reactive to radiative region, but to be in the far-field limit, one must be even further away from the source. At $r = \lambda/2$, the peak reactive field terms are about 10 dB below the radiated, so $r = \lambda/2$ seems a good criterion for the "far-field limit" of a harmonic Hertzian dipole source. It should be noted, however, that since the radiated field of a time-harmonic dipole goes to zero every half-period, at that instant, the reactive field will be the dominant term, even when arbitrarily far away. Of course, there are a great many ways to define the distinction between near field and far field [31].

The steady-state behavior of a harmonic dipole makes understanding the conversion of energy into radiation more difficult. The next section examines transient excitations with a definite beginning and end.

5.3.4 Time Domain Excitations

This section presents some examples of the energy flow around Hertzian dipoles excited by a variety of time domain excitations. An external source or sink of energy must be assumed to be connected to the dipole. This section looks at three different types of excitations.

- ✲ **Transient**: A Gaussian impulse is an example of a transient time domain excitation. A transient time domain excitation begins and ends with no stored energy in the field but nevertheless radiates energy.

- ✲ **Charging**: A rising Heaviside step (or rising edge) is an example of a charging time domain excitation. A charging excitation not only yields radiation energy but also leaves in its wake the energy of a static electric field.

- ✲ **Discharging**: A falling Heaviside step (or falling edge) is an example of a discharging time domain excitation. A discharging excitation starts with the stored energy of a static electric field. Some of this stored energy decouples and radiates away; some is absorbed back into the dipole source.

In each case, the time parameter is taken to be $\tau = 1\text{m}/c = 3.24\text{ns}$. The causal surfaces are shown by the solid black lines.

5.3.4.1 Transient Time Domain Excitation

The time dependence of a Gaussian impulse (see Section 4.1.3.1) is

$$T(t) = \exp\left[-\frac{t^2}{\tau^2}\right] \tag{5.48}$$

Figure 5.21 shows a Gaussian impulse and its time derivatives.

With a Gaussian impulse, there is no reactive field energy before ($t \to -\infty$) or after ($t \to +\infty$) the pulse. The dipole moment starts and ends at zero. An external source of energy drives the dipole, providing the energy to create the dipole moment and then reabsorbing some of the energy as the dipole collapses. There are three distinct pulses of radiation. The first one is emitted directly from the local vicinity of the dipole and leaves in its wake a residue of reactive field energy that becomes the source of the radiant energy in the next two pulses. Note that the dipole (rather, the external source attached to it) is emitting energy for the first half of the process (i.e., for $t < 0$) and absorbing energy for the entire second half of the process (i.e., for $t > 0$). Figure 5.22 presents a space-time energy-flow diagram for a Gaussian impulse.

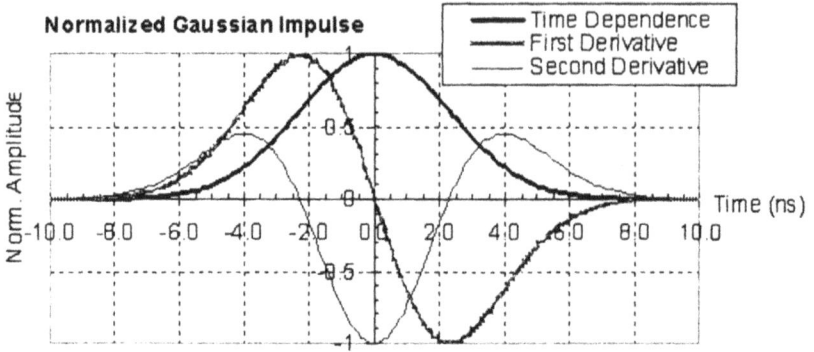

Figure 5.21 Gaussian excitation and time derivatives (After [23]).

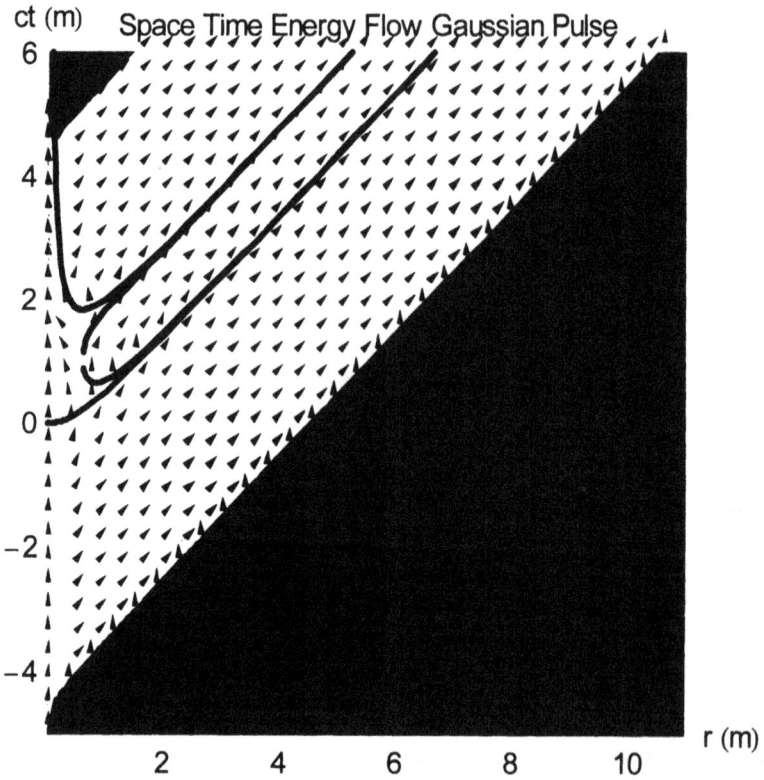

Figure 5.22 A space-time energy-flow diagram for a dipole excited by a Gaussian transient impulse (After [23]).

5.3.4.2 Charging Time Domain Excitation

The time dependence of a rising edge, or a charging (Heaviside), dipole is

$$T(t) = \tfrac{1}{2}\left(1 + \tanh \tfrac{t}{\tau}\right) \tag{5.49}$$

Figure 5.23 shows a rising edge and its time derivatives. This time domain response is further described in Section 4.1.3.2.

A dipole with the charging Heaviside step pulse has no near-field energy before ($t \to -\infty$) the pulse begins, but afterwards ($t \to +\infty$), the dipole is surrounded by a static field. The dipole starts at zero and ends up with a net moment, again, courtesy of the external source of energy. There are two distinct pulses of radiation. The first one is emitted directly from the local vicinity of the dipole and leaves in its wake a residue of near-field energy that becomes the source of both the energy radiated by the second pulse as well as the energy left behind in the static field of the charged dipole. Here again, the second pulse is slowly gaining energy from the first pulse, and in the far-field limit, will have an identical energy profile. Note that the dipole does not reabsorb any energy; it emits energy for the entire process. Figure 5.24 presents a space-time energy-flow diagram for a rising edge impulse.

5.3.4.3 Discharging Time Domain Excitation

The time dependence of a falling edge, or a discharging (Heaviside), dipole is

$$T(t) = \tfrac{1}{2}\left(1 - \tanh \tfrac{t}{\tau}\right) \tag{5.50}$$

Figure 5.25 shows a falling edge and its time derivatives. This time domain response is also further described in Section 4.1.3.2.

In the case of the Heaviside decay pulse, the dipole has been charged by some external source in the distant past ($t \to -\infty$) and begins already surrounded by the energy in its static electric field. Afterwards ($t \to +\infty$), the dipole moment is gone. The dipole begins with a net moment and ends with none. As the dipole decays, there are two distinct pulses of radiation, which are fed by the energy preexisting around the dipole. Note also the asymmetry in the energy of the two pulses shown. The second pulse is slowly gaining energy from the first pulse and becomes identical in the far-field limit. Further note that the dipole itself (rather, the external source attached to it) is absorbing energy throughout the process. Figure 5.26 presents a space-time energy-flow diagram for a falling edge impulse.

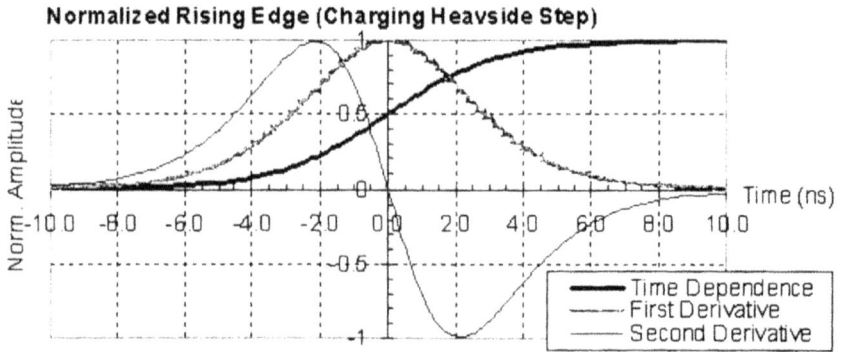

Figure 5.23 Charging Heaviside step excitation and time derivatives (After [23]).

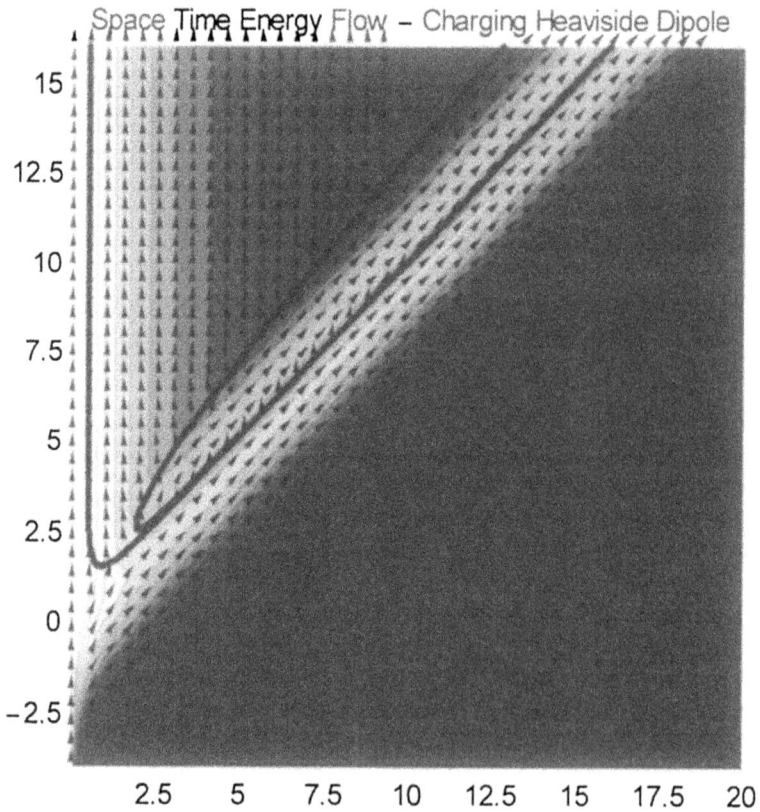

Figure 5.24 A space-time energy-flow diagram for a charging dipole (After [23]).

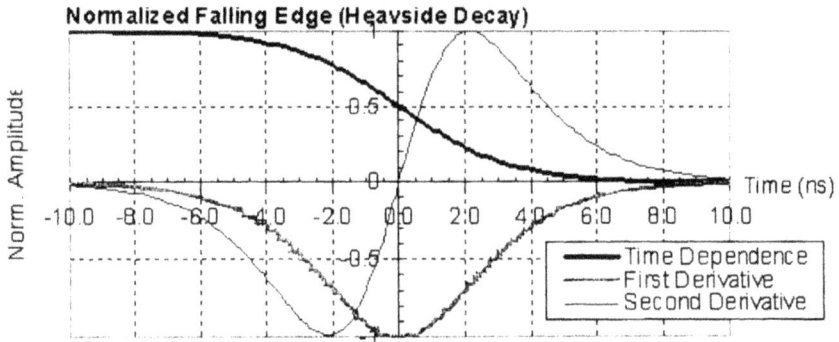

Figure 5.25 Discharging Heaviside step excitation and time derivatives (After [23]).

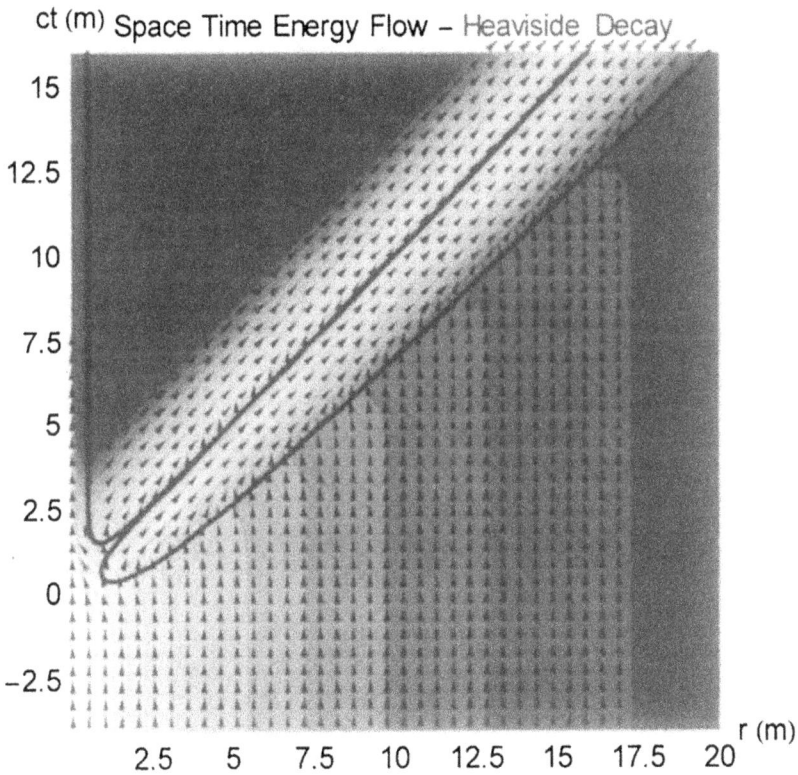

Figure 5.26 A space-time energy-flow diagram for a discharging dipole (After [23]).

Transient Impulse

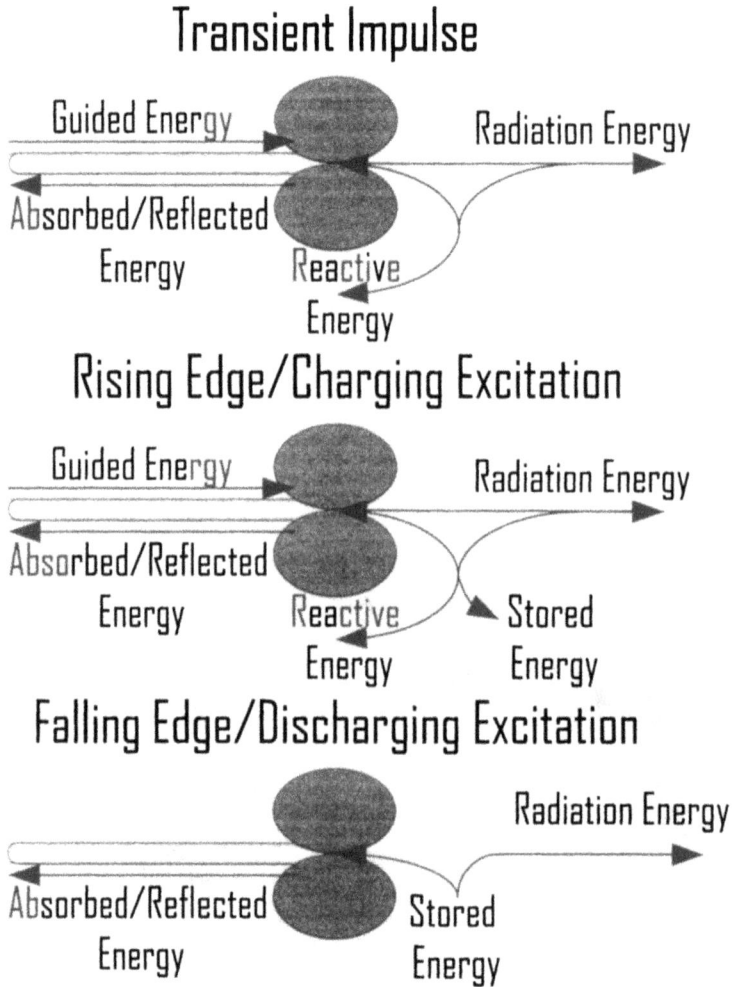

Guided Energy   Radiation Energy

Absorbed/Reflected
Energy   Reactive
Energy

Rising Edge/Charging Excitation

Guided Energy   Radiation Energy

Absorbed/Reflected
Energy   Reactive   Stored
Energy   Energy

Falling Edge/Discharging Excitation

Radiation Energy

Absorbed/Reflected   Stored
Energy   Energy

Figure 5.27   Simplified energetics of typical time domain excitations.

5.3.4.4 Summary of Time Domain Excitations

Radiation is a process of electromagnetic energy conversion. The three preceding examples illustrate the fundamental physics of energy flow around a small antenna. Figure 5.27 presents the simplified energetics of typical time domain excitations.

In the case of a transient impulse, guided energy on a transmission line is incident on an antenna. Some of this energy radiates directly away. Other energy is stored in the reactive fields around the dipole. Some of this reactive energy does subsequently radiate away also. The remaining energy is absorbed by the dipole to give rise to the reflected, or mismatch, energy.

In the case of a rising edge, guided energy on a transmission line is incident on an antenna. Some of this energy radiates directly away. Other energy is stored in the reactive fields around the dipole. Some of this reactive energy does subsequently radiate away also. Some of this reactive energy is left behind in the static field surrounding the antenna. The remaining energy is absorbed by the dipole to give rise to the reflected, or mismatch, energy.

In the case of a falling edge, there is no guided energy per se. Instead the transmission line and antenna support stored energy. As the edge falls, some of the stored energy is absorbed by antenna and taken away by the transmission line. The remaining stored energy decouples and radiates away.

From the energy conversion point of view, the goal of UWB antenna design is to minimize the reactive energy absorbed by the antenna and maximize the conversion of guided or stored energy into radiation energy. The following sections will consider how to accomplish these goals.

> *In other words, a thin antenna has a high Q value, a thick antenna relatively small Q value, so that a thick antenna is not nearly so selective as a thin antenna.*
>
> John C. Slater, 1942

5.4 OPTIMAL ELEMENT DESIGN

An antenna converts bound reactive energy to unbound radiation energy. If the reactive energy remains stored around the antenna, it is reabsorbed by the antenna, giving rise to mismatch reflection losses. Stored reactive energy gives rise to a reactive component in an antenna's impedance. Additionally, the greater the proportion of stored energy, the narrower the bandwidth. The relation between stored reactive energy and bandwidth is encompassed in the concept of Q or "quality factor." Section 5.5.3 addresses this subject at greater length. In summary, however, lower reactive energy means wider bandwidth and better matching. The art of designing a UWB antenna is, in this sense, the art of designing an antenna element with minimal stored reactive energy.

Thin-wire elements or abrupt discontinuities yield large current and field concentrations in their immediate vicinity. The same phenomenon explains why spheres are typically used as high-voltage elements. A sharp or pointed conductor will have very high field strengths in the immediate vicinity of the point. Ionization and breakdown occur near these discontinuities where the field concentration is strongest. In the context of an antenna, a sharp edge or a discontinuity tends to be a locus for undesired concentrations of reactive energy.

The key lesson of Chapter 3 is that a well-designed UWB antenna should have a smoothly varying impedance profile to assure a good broadband match. Empirically, Chapter 3 also demonstrates that a smoothly varying impedance profile generally (although not always) follows from a smoothly varying antenna geometry. The results of Chapter 5 now unveil the physical reasoning behind the observations of Chapter 3: thin-wire elements and abrupt discontinuities concentrate currents and reactive energy, leading not only to a reactive impedance but also to narrower bandwidth.

This section will discuss a basic principle of UWB antenna element design: fatter is better. A fat element may be a bulbous element occupying much volume, but even thin planar cross-sections of bulbous elements offer good performance. This section will further discuss optimal shapes for small-element dipole and loop antennas.

5.4.1 Fatter Is Better

The concept that fatter is better when it comes to broadband antenna design was understood as early as the 1940s by J. C. Slater, R. W. P. King and others [32, 33]. Many UWB antenna designs of that period, like Lindenblad's TV antenna and Kraus's volcano smoke antenna emphasized smoothly varying bulbous shapes that act to minimize stored reactive energy. Thus, these bulbous shapes tend to have relatively broadband performance and a relatively stable match.

The idea that a fat antenna is broadband because it minimizes stored reactive energy was also an element in Harold Wheeler's thinking. In the late 1940s, Wheeler identified the concept of the radian sphere, a spherical boundary centered on a small dipole antenna with radius $r = \lambda/(2\pi)$ [34]. This is the radial distance at which the radiation component of the field and the near, or reactive, field components are equal in peak magnitude. "There is not a definite boundary," Wheeler acknowledged, "but rather a transition, since the terms associated with the near field predominate inside and those associated with the far field predominate outside [35]." Further, Wheeler hypothesized, "If a small antenna is restricted in its maximum dimension but not in its occupied volume, the radiation power factor is increased by utilizing as much as possible of the volume of a sphere whose diameter is equal to this dimension" [36]. Extending this idea to a radian-sphere-sized antenna, Wheeler concluded, "An idealized small spherical antenna is found to have a radiation power factor equal to the ratio of its volume over that of the radian-sphere" [37]. Alternatively, this can be interpreted to mean that the more efficiently an antenna occupies the volume of the radian sphere (call it "volumetric efficiency"), the higher the radiation efficiency of the antenna. Elsewhere, Wheeler provided an excellent and highly readable summary of his ideas on small antennas [38].

The idea that radiation efficiency follows from volumetric efficiency relies on two false premises. First, Wheeler's thinking assumed that reactive energy is uniformly distributed inside the radian sphere. Second, Wheeler assumed that the only way to exclude reactive energy is to expand the antenna to occupy more volume. In fact, a properly designed planar antenna can largely short out or block reactive fields within its boundary sphere while occupying only an infinitesimal fraction of its radian-sphere volume.

One might also anticipate, then, that the sharp edge of a planar antenna would make it unsuitable for use as a UWB antenna. This intuition is incorrect, however. Although currents do tend to concentrate on the edges of a planar antenna element (as described in Section 4.3.1), a planar element still has ample surface area across which currents may spread (as shown in Figure 4.15), thus avoiding the current concentration of a thin-wire element. In fact, planar cross-sections of bulbous elements offer performance substantially equivalent to the corresponding solid bulbous element. Bow tie antennas are comparable to biconicals, and circular element planar dipoles are comparable to spherical dipoles, for instance. So long as an element is "thick" in at least two dimensions, it can offer broadband performance.

5.4.2 Optimal Dipole Shape

What, then, is the optimal design for an electric dipole antenna? A bulbous-shaped planar element is a good starting point, but there are many candidates including circles, ellipses, and ovals. An analysis of energy flow provides a basis for selecting what might be an optimal design.

This chapter presented an analysis of energy flow around a small electric dipole. A small electric dipole pumps energy into the region immediately surrounding the dipole. The energy concentrated along the axis or poles of the dipole is stored reactive energy that the dipole eventually reabsorbs. The energy concentrated around the waist, or equator, in the azimuthal plane of the dipole is reactive energy that decouples and radiates away. Figure 5.28 summarizes these results and presents a qualitative sketch of the energy-flow streamlines around a dipole.

There are two basic principles at work in selecting an optimal element.

• Conformity to streamlines: An optimal element is conformal to the energy-flow streamlines of a desired field configuration.

• Minimization of stored reactive energy: An optimal element is designed so as to short out, or block, the undesired stored reactive energy, while supporting the desired reactive energy that decouples and converts to radiation.

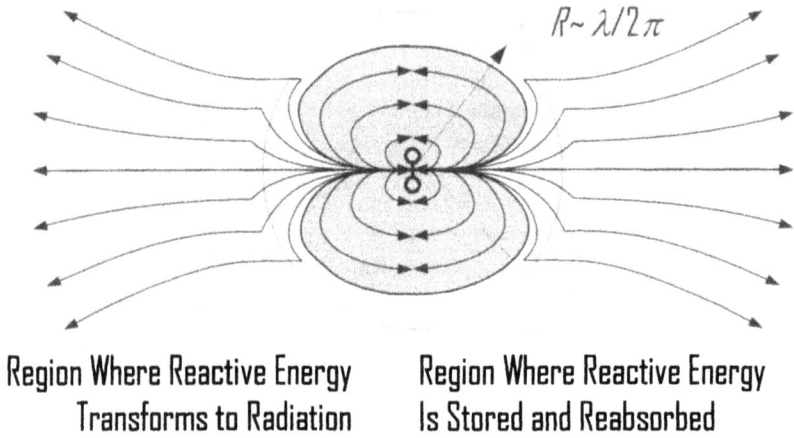

Region Where Reactive Energy   Region Where Reactive Energy
Transforms to Radiation   Is Stored and Reabsorbed

Figure 5.28   Qualitative sketch of energy-flow streamlines around an ideal electric dipole.

In the case of an electric dipole element, the solution is an approximately elliptical shape that shorts out stored reactive energy and conforms to the energy-flow streamlines. The performance of planar elliptical antennas is further examined in Chapter 6.

5.4.3 Optimal Loop Shape

By duality, the field structure of a magnetic dipole is similar to that of an electric dipole: only the electric and magnetic fields swap their arrangements. Thus the energy flow around a magnetic dipole is identical to that around an electric dipole given the same excitation. Where an electric dipole principally stores electric field energy, a magnetic dipole stores magnetic field energy. An effective small-element loop is thus a loop constructed so as to block the undesired stored magnetic field energy along the axis of the loop. Thus, the aim of a small loop is to distribute currents away from the loop axis.

At first glance, a standard planar loop made fat to broaden bandwidth appears ideal. Fed at the periphery, such an antenna has no current circulation near the axis and, thus, minimal stored energy. A standard planar loop is an efficient UWB antenna, well matched over a significant bandwidth. Unfortunately, a standard planar loop is highly dispersive, as seen in Section 2.2.3. Figure 5.29(a) shows a standard planar loop antenna.

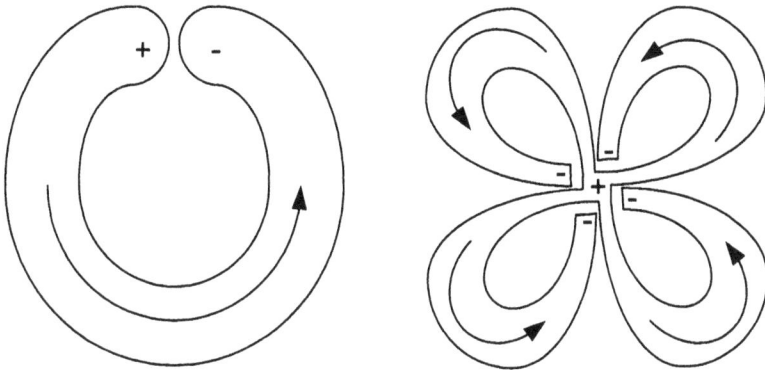

Figure 5.29   (a) A standard planar loop antenna, and (b) a four-lobed planar clover-leaf loop antenna.

A more synchronous distribution of current around the loop is essential. The solution is a clover-leaf-type structure, suggested in the UWB context by Harmuth [39]. Figure 5.29(b) shows a four-lobed UWB clover-leaf loop antenna. This structure avoids loop currents near the antenna axis. In fact, the counterflowing currents near the antenna feed tend to cancel each other out. As the currents diverge near the loop periphery, they yield an overall loop distribution of current. More or fewer lobes may be used to achieve a desired minimization of dispersion. Harmuth suggests shielding is required to minimize radiation from the far side of a clover-leaf loop. In practice however, since the delay of signals from one side of the loop to the other varies little, the superposition of near and far signals create a relatively broadband and nondispersive radiation pattern [40]. Small-element UWB magnetic antennas are addressed further in Chapter 6.

5.5 FUNDAMENTAL LIMITS ON ANTENNA SIZE

The energetics of antennas also enables an analysis of antenna efficiency. This section applies energy analysis to the problem of establishing fundamental limits on antenna size and performance in the UWB context. Fundamental limits are an area of long-standing interest in the narrowband context, so this section begins by reviewing the classical "Chu-Harrington" limit introduced by L.J. Chu and refined by Roger F. Harrington. This section further reviews the revised version of this limit offered by James S. McLean. These theoretical deviations all employ a concept known as the quality factor or Q of an antenna.

Kenneth S. Johnson originated the Q concept in 1914, although he referred to it as K until 1920 [41]. Johnson first applied the concept in the context of inductors, taking Q as the ratio of reactance to effective resistance of a coil. Usual treatments of Q involve approximations valid only in the narrowband limit. This section examines whether and how Q can be applied in the UWB limit. Then, this section reexamines the insights of Chu, Harrington, and McLean and explores their application to UWB antennas. Finally, this section proposes a variety of time domain UWB limits on antenna size and frequency response.

5.5.1 The Chu-Harrington Limit

Some of the earliest insights on antenna efficiency were provided by Chu [42]. Chu modeled an antenna as an arbitrary source distribution contained within a boundary sphere (see Figure 5.30). Outside this spherical shell, he assumed that the fields could be represented by a superposition of orthogonal spherical modes, each with their own wave impedance. Chu was able to construct a circuit with the same impedance and analyze the resonant behavior. He derived an expression for the quality factor Q of his idealized antenna. The quality factor is also the inverse fractional bandwidth of the antenna.

Chu further demonstrated that for a fixed-size boundary sphere with radius R, the lowest Q, thus the broadest bandwidth is achieved by an antenna structure that generates the fields of an infinitesimal point dipole outside the sphere [43]. Following Chu's approach, Harrington derived an expression for Q as a function of R [44]. This expression, the Chu-Harrington limit, is generally regarded as a fundamental physical limitation on antenna performance [45]. More recently, an error in Chu's original formulation was corrected by McLean to yield the result in (5.51) [46]. For the lowest order TM_r mode (i.e., that corresponding to the fields of a point dipole), Chu's expression becomes

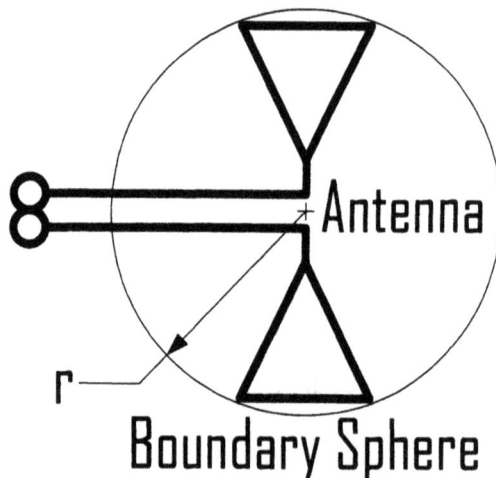

Figure 5.30 The boundary sphere around an antenna (After [42]).

$$Q = \frac{1 + 2(kR)^2}{(kR)^3\left[1 + (kR)^2\right]} = \frac{f_C}{BW} \tag{5.51}$$

In this expression, R is the radius of the boundary sphere, and k is the wave number ($k = 2\pi/\lambda_C$). The wave number is related to the wavelength at the center, or resonant, frequency, a fact which becomes increasingly important to remember in UWB applications, where a broad range of wavelengths are possible with an individual antenna. As noted in (5.51), the quality factor, Q, is also related to the bandwidth, BW, and the center frequency, f_C. The idea of Q as inverse fractional bandwidth is valid (with some qualifications) in the UWB limit, as will be discussed in Section 5.5.3.

The Chu-Harrington limit relies on the fact that whatever structure may lie within the boundary sphere, outside the boundary sphere of an optimal small antenna, the fields must be those of an ideal point dipole.

5.5.2 McLean Limit

Recently, McLean proposed an alternate limit to antenna performance [46]. McLean starts from the customary definition of radiation Q in terms of stored energy and radiated power

$$Q = \begin{cases} \dfrac{2\omega\langle U_E\rangle_{reac}}{\langle P_{rad}\rangle} & \langle U_E\rangle_{reac} > \langle U_M\rangle_{reac} \\[3mm] \dfrac{2\omega\langle U_M\rangle_{reac}}{\langle P_{rad}\rangle} & \langle U_M\rangle_{reac} > \langle U_E\rangle_{reac} \end{cases} \tag{5.52}$$

In (5.52), $<U_E>_{reac}$ is the time-average reactive, or stored, electric energy, and $<U_M>_{reac}$ is the time-average reactive, or stored, magnetic energy.

The time-average reactive electric energy is given by (A.14) and the time average radiated power is given by (A.17). Details of the calculation are in the appendix. Substituting these results into (5.52) yields McLean's expression for radiation Q of an electrically small dipole

$$Q = \frac{2\omega\langle U_E\rangle_{reac}}{\langle P_{rad}\rangle} = 2\omega\frac{p_0^2 k^3}{24\pi\varepsilon_0}\left(\frac{1}{k^3 R^3} + \frac{1}{kR}\right)\frac{12\pi\varepsilon_0}{p_0^2 \omega\, k^3}$$
$$= \frac{1}{k^3 R^3} + \frac{1}{kR} \tag{5.53}$$

Both the Chu-Harrington and McLean expressions are relations involving the quality factor, or Q, of an antenna. Thus, a deeper examination of Q is in order.

5.5.3 Is There a Q in UWB?

The concept of Q originated in the context of narrowband resonant circuits. In extrapolating Q to the UWB limit, certain narrowband approximations break down. This section will reexamine Q from the UWB perspective, beginning with an analysis of a damped harmonic RLC circuit.

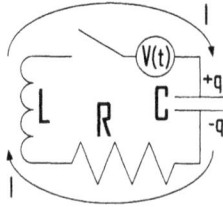

Figure 5.31 A series RLC circuit.

Consider a series resistor, inductor, capacitor (RLC) circuit as in Figure 5.31. Applying the Kirchoff Voltage Law around the resulting loop yields the differential equation

$$L\frac{dI}{dt} + RI + \frac{q}{C} = V(t)$$

(5.54)

Since current is the time derivative of charge ($I = dq/dt$), (5.54) may be rewritten as a second-order differential equation for charge with respect to time

$$\frac{d^2q}{dt^2} + \frac{R}{L}\frac{dq}{dt} + \frac{q}{LC} = \frac{1}{L}V(t)$$

(5.55)

An electric dipole may be thought of as an infinitesimal capacitor with a dipole moment $p = qd$, where the charge separation distance (d) is constant, and the charge magnitude (q) is given by (5.55). The unforced solution has $V(t) = 0$. Assuming a solution of the form $q = e^{-\alpha t}$, this yields the indicial relation

$$\alpha^2 + \frac{R}{L}\alpha + \frac{1}{LC} = 0$$

(5.56)

with solution

$$\alpha = -\frac{R}{2L} \pm \sqrt{\left(\frac{R}{2L}\right)^2 - \frac{1}{LC}} \qquad (5.57)$$

Thus,

$$q = C_0 e^{-\frac{R}{2L}t} e^{j\omega_n t} = C_0 e^{-\frac{\omega_0}{2Q}t} e^{j\omega_n t} \qquad (5.58)$$

where the natural frequency is

$$\omega_n = \sqrt{\frac{1}{LC} - \left(\frac{R}{2L}\right)^2} = \sqrt{\omega_0^2 \left(1 - \frac{1}{4Q^2}\right)} \qquad (5.59)$$

and where the quality factor Q is the ratio of the inductive reactance at resonance to the resistance

$$Q = \frac{\omega_0 L}{R} = \frac{L}{R} \frac{1}{\sqrt{LC}} = \frac{1}{R} \sqrt{\frac{L}{C}} \qquad (5.60)$$

This solution of 5.59 is analogous to that of an underdamped mechanical oscillator and assumes $1/(LC) > (R/(2L))^2$. There are also additional solutions analogous to critically damped and overdamped harmonic oscillations. In the limit, where losses are negligible, $R \rightarrow 0$, and the natural frequency becomes identical to the resonant frequency

$$\lim_{R \rightarrow 0} \omega_n = \omega_0 = \frac{1}{\sqrt{LC}} \qquad (5.61)$$

The natural frequency describes the actual harmonic behavior of the damped oscillation. As damping increases, the natural frequency diverges from the resonant frequency becoming smaller and smaller—increasing the interval between zero crossings. In the limit where $1/(LC) = (R/(2L))^2$ and the oscillation becomes critically damped, the natural frequency goes to zero. The resonant frequency ω_0 remains the peak frequency of the frequency domain response, even as the natural frequency ω_n diverges. Figure 5.32 shows the time domain behavior of damped harmonic oscillations with resonant frequency $\omega_0 = 1$ and various values of Q. The Q factor is also approximately equal to the number of periods in the damped oscillation.

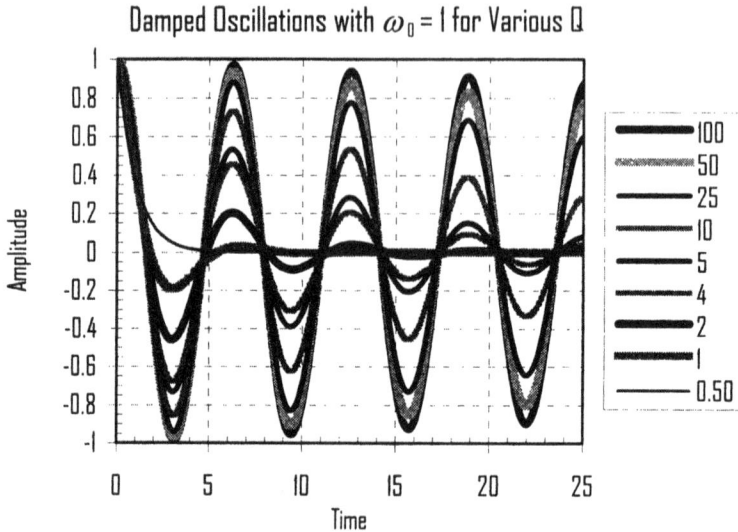

Figure 5.32 Time domain behavior of damped harmonic oscillations with $\omega_0 = 1$ for various Q.

Now consider the steady-state response for $V(t) \neq 0$. In particular, suppose $V(t) = |V_0|\, e^{j\omega t}$ and $I(t) = |I_0|e^{j\omega t}$. Note that the magnitude of the current I_0 is in general a complex quantity ($I_0 \equiv I_0\, e^{j\varphi}$) not necessarily in phase with the voltage. Since $q(t) = I_0\, e^{j\omega t}/(j\omega)$, substituting into (5.55) yields

$$\left(-\omega^2 L + j\omega R + \frac{1}{C}\right)I_0\, \frac{e^{j\omega t}}{j\omega} = V_0 e^{j\omega t} \tag{5.62}$$

Solving for impedance yields

$$Z = \frac{V_0}{I_0} = R + j\left(\omega L - \frac{1}{\omega C}\right) \tag{5.63}$$

The complex term is the reactance ($X = \omega L - 1/(\omega C)$), which comprises an inductive reactance, $X_L = \omega L$, and a capacitive reactance, $X_C = -1/(\omega C)$. "Resonance" is the point where the reactance goes to zero

$$\omega_0 = \frac{1}{\sqrt{LC}} \tag{5.64}$$

The magnitude of the impedance is

$$|Z| = \sqrt{R^2 + \left(\omega L - \frac{1}{\omega C}\right)^2} = R\sqrt{1 + \left(\frac{\omega L}{R} - \frac{1}{\omega RC}\right)^2}$$

$$= R\sqrt{1 + \left(\frac{Q\omega}{\omega_0} - \frac{Q}{\omega\omega_0 LC}\right)^2} \qquad (5.65)$$

$$= R\sqrt{1 + \left(\frac{Q\omega}{\omega_0}\right)^2\left(1 - \frac{\omega_0^2}{\omega^2}\right)^2}$$

Figure 5.33 shows the normalized admittance in decibels plotted versus angular frequency for resonant frequency $\omega_0 = 1$ and various values of Q. As is obvious from a cursory examination of (5.65), the peak of the impedance curve lies at $\omega = 1$, independent of Q. Also, there is a distinct asymmetry in the impedance response that becomes increasingly evident for $Q < 5$. To better understand this asymmetry, one can evaluate the half-power, or -3dB, points.

The half-power points correspond to a reduction of the current, and thus the impedance, by a factor of $\sqrt{2}$. This condition follows when the second term under the radical sign in (5.65) goes to 1. Solving yields

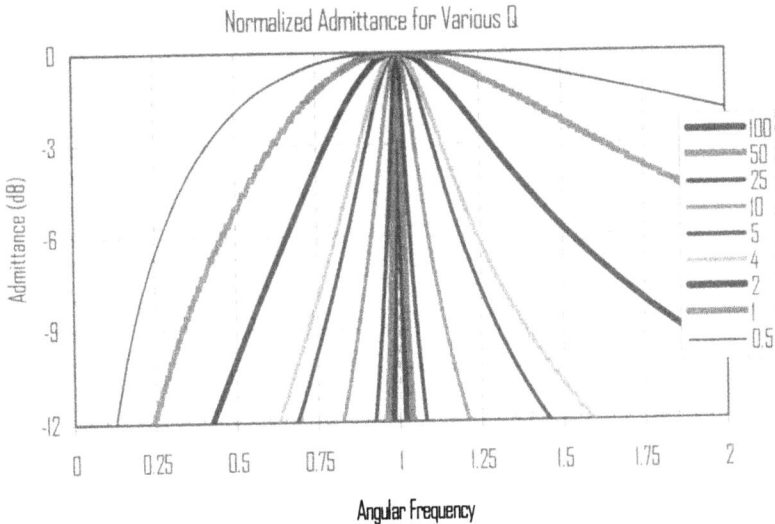

Figure 5.33 Frequency domain normalized admittance ($20\log[Z(\omega_0)/Z(\omega)]$) of damped harmonic oscillations with $\omega_0 = 1$ for various Q.

$$\left(\frac{Q\omega}{\omega_0}\right)^2 \left(1 - \frac{\omega_0^2}{\omega^2}\right)^2 = 1 \rightarrow \omega^2 \pm \frac{\omega_0}{Q}\omega - \omega_0^2 = 0 \rightarrow$$

$$\omega = \frac{\omega_0}{2Q}\sqrt{1+4Q^2} \pm \frac{\omega_0}{2Q} \qquad (5.66)$$

$$= \omega_A \pm \frac{2\pi BW}{2}$$

where two of the four roots are discarded as unphysical. The "arithmetic center" frequency (ω_A) is the arithmetic average of the upper frequency $\omega_H = \omega_C + \pi BW$ and the lower frequency $\omega_L = \omega_C - \pi BW$. Thus, the arithmetic center frequency follows from examination of (5.66)

$$\omega_A = \frac{\omega_0}{2Q}\sqrt{1+4Q^2} \qquad (5.67)$$

The relation between Q and the bandwidth also follows from (5.66)

$$Q = \frac{\omega_0}{2\pi BW} = \frac{f_0}{BW} \qquad (5.68)$$

Q is thus the inverse fractional bandwidth, even in the UWB limit, provided the resonant frequency ($f_0 = 2\pi\omega_0$) is taken as the center frequency ($f_C = f_0$). Although the resonant frequency is not the arithmetic average of the upper and lower frequencies, it is the peak of the frequency response. Further, taking the geometric average of the upper and lower frequencies yields

$$\sqrt{\omega_L\omega_H} = \sqrt{\left(\frac{\omega_0}{2Q}\right)^2 \left(\sqrt{1+4Q^2}+1\right)\left(\sqrt{1+4Q^2}-1\right)}$$

$$= \sqrt{\left(\frac{\omega_0}{2Q}\right)^2 \left(1+4Q^2-1\right)} \qquad (5.69)$$

$$= \omega_0$$

The resonant frequency is the "geometric center frequency," the geometric average of the upper and lower frequencies. The resonant frequency is not only the peak frequency but also the true center frequency. The result of (5.69) suggests that frequency should be thought of logarithmically in the UWB limit. At the very least, the geometric average should be preferred to the arithmetic average in the UWB context. Indeed, the asymmetry in the impedance plot of Figure 5.33

disappears when evaluating frequency response logarithmically. Figure 5.34 shows the behavior of the natural frequency (ω_n), the lower frequency (ω_L), the resonant, or "geometric center," frequency (ω_0), the "arithmetic center" frequency (ω_C), and the upper frequency (ω_H) for various values of Q.

In summary, the quality factor may be applied in the UWB limit with a few caveats. First, the upper and lower frequencies are defined by the half-power, or –3dB, points. Using other definitions of bandwidth will not yield valid results for Q. Second, the center frequency is the geometric average of the upper and lower frequency. Using the arithmetic average of the upper and lower frequencies also yields an invalid result for Q. Remembering these two simple rules allows for application of the Q concept in the UWB context.

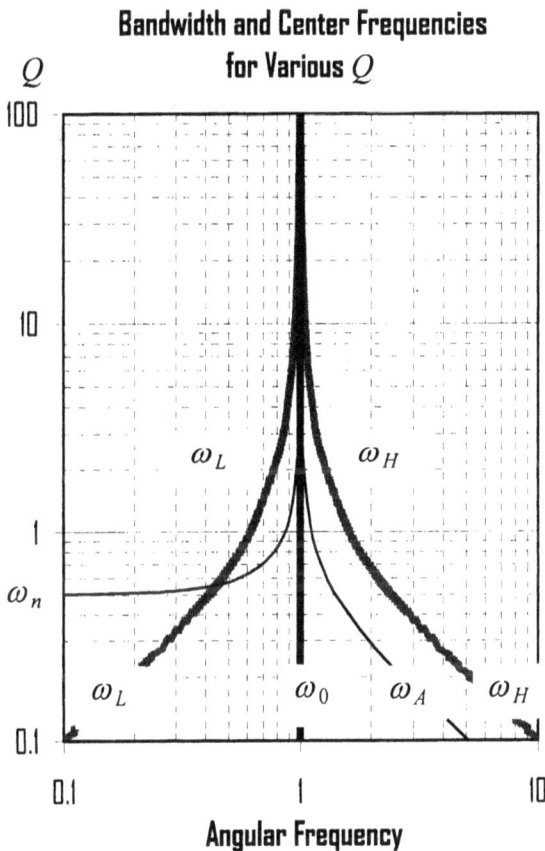

Figure 5.34 Behavior of natural frequency (ω_n), lower frequency (ω_L), resonant and "geometric center" frequency (ω_0), "arithmetic center" frequency (ω_A), and upper frequency (ω_H) for various Q.

5.5.4 Q-Based Antenna Limits in UWB Practice

The Chu-Harrington and McLean limits converge to the same answer in the limit of high-Q, narrowband antennas. These limits diverge, however, in the limit of low-Q broadband antennas. Q-based limits to antenna performance may be extended to the ultra-wideband context subject to the two basic observations of the last section: first, bandwidth is delineated by the half-power or –3 dB points and second, the center frequency (f_C) is the geometric average of the upper (f_H) and lower (f_L) frequencies. Subject to these conditions, the Q is the inverse fractional bandwidth

$$Q = \frac{f_C}{BW} = \frac{\sqrt{f_H f_L}}{f_H - f_L} \tag{5.70}$$

Traditional applications of Q-based limits focus on performance at the center of the bandwidth. Usually these applications measure the dimensions of the antenna boundary sphere in radian units using the product of the wave number (k) and the radial dimension (R). In this system, a half-wave dipole fits within a quarter-wavelength radius boundary sphere and is a $kR = \pi/2$ antenna. Radian measure is much beloved by theorists, but in actual engineering practice, expressing antenna dimensions in terms of wavelengths is generally more useful. For instance, if R is the radius of an antenna's boundary sphere, then $R_{\lambda C} = R/\lambda_C$ is the boundary radius in terms of the wavelength (λ_C) at the center frequency (f_C). A half wave dipole has a boundary sphere of radius $R_{\lambda C} = 0.25$. Using the wavelength rather than the radian measure, the Chu-Harrington limit of (5.68) becomes

$$Q = \frac{1 + 2(kR)^2}{(kR)^3 \left[1 + (kR)^2\right]} = \frac{1 + 2(2\pi R_{\lambda C})^2}{(2\pi R_{\lambda C})^3 \left[1 + (2\pi R_{\lambda C})^2\right]} = \frac{f_C}{BW} = \frac{\sqrt{f_H f_L}}{f_H - f_L} \tag{5.71}$$

The McLean limit of (5.69) may be cast in a similar form

$$Q = \frac{1}{k^3 R^3} + \frac{1}{kR} = \frac{1}{(2\pi R_{\lambda C})^3} + \frac{1}{2\pi R_{\lambda C}} = \frac{f_C}{BW} = \frac{\sqrt{f_H f_L}}{f_H - f_L} \tag{5.72}$$

The expressions above, both "radian measure" and "wavelength measure," all relate Q to antenna scale as referenced to the center frequency of an antenna's response. In the narrowband limit, the wavelength varies little across an antenna's operational band. There being so little distinction between the lower, center, and upper frequencies and their corresponding wavelengths, referencing to the center frequency is as good a choice as any in this case.

The story is different in the UWB limit, however. By definition, a UWB antenna is characterized by very large deviations in frequency and wavelength from one end of the band to the other. Worse, all of the analyses of fundamental antenna limits start from the assumption that an antenna is operating in a dipole mode. Most UWB antennas will transition from a dipole mode to higher-order modes at the upper end of the operating band. For instance, a typical quarter-wave scale antenna with a good 3:1 bandwidth dipole mode will be transitioning to a three quarter-wave mode at the upper end of the band. Thus, a naïve attempt to evaluate bandwidth will often miss the effective upper frequency of the dipole mode because the antenna still presents good gain and impedance characteristics through the transition from dipole to higher-order modes.

The lower end of the operating band is usually far clearer and more distinct than the upper end. The dipole mode is, after all, the fundamental and lowest-order antenna mode. To avoid unjustified bandwidth inflation due to upper-frequency exaggeration, one should evaluate antenna limits with respect to the lower frequency. This practice is particularly critical in the limit as $Q \rightarrow 1$ or less.

The wavelength at the low frequency end of an antenna's operating band (λ_L) may be related to the wavelength at the center of the band (λ_C) and the Q

$$\lambda_L = \frac{c}{f_L} = \frac{c}{f_C}\frac{\sqrt{1+4Q^2}+1}{2Q} = \lambda_C\frac{\sqrt{1+4Q^2}+1}{2Q} \qquad (5.73)$$

Then $R_{\lambda L} = R/\lambda_L$ is the boundary radius in terms of the wavelength (λ_L) at the lower frequency (f_L). A similar treatment may be applied to the wavelength at the high-frequency end of an antenna's operating band

$$\lambda_H = \frac{c}{f_H} = \frac{c}{f_C}\frac{\sqrt{1+4Q^2}-1}{2Q} = \lambda_C\frac{\sqrt{1+4Q^2}-1}{2Q} \qquad (5.74)$$

Here, $R_{\lambda H} = R/\lambda_H$ is the boundary radius in terms of the wavelength (λ_H) at the upper frequency (f_H).

Traditionally, when the Chu-Harrington or McLean limits are plotted showing a curve of antenna Q versus boundary-sphere radius, only the center frequency result is plotted. Figures 5.33 and 5.34 depict these limits, showing not only the size of the boundary spheres as referenced to the center frequency's wavelength, but also with respect to the wavelengths corresponding to the upper and lower frequencies' wavelengths.

One feature immediately leaps out of the result for the Chu-Harrington limit shown in Figure 5.35. If the Chu-Harrington result is to be accepted, the lower-frequency dimension of an antenna does not increase monotonically with increasing bandwidth and decreasing Q.

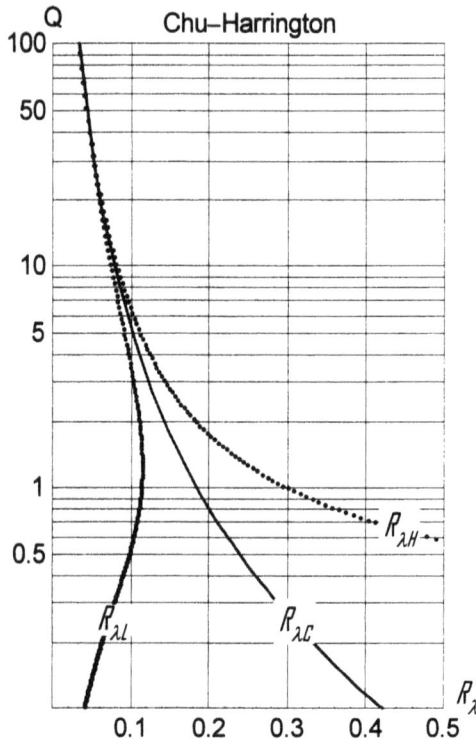

Figure 5.35 The Chu-Harrington result for boundary-sphere radius in units of wavelength (R_λ). The $R_{\lambda L}$, $R_{\lambda C}$, and $R_{\lambda H}$ denote the boundary-sphere radius as referenced in units of the low-frequency wavelength, center-frequency wavelength, and high-frequency wavelength, respectively (After [47]).

For instance, the Chu-Harrington result suggests one could build a $Q = 0.5$ antenna within a $R_{\lambda L} = 0.10$ radius boundary sphere. By incorporating a low pass filter to truncate the antenna's high end response, one could reduce the upper frequency, reduce the bandwidth, and increase Q without changing the result that the antenna fits within a $R_{\lambda L} = 0.10$ radius boundary sphere. Suppose the low pass filter decreases bandwidth so as to increase Q to $Q = 1$. But the Chu-Harrington result claims that a $Q = 1$, $R_{\lambda L} = 0.10$ antenna is smaller than the $R_{\lambda L} = 0.12$ ideal limit for $Q = 1$. This contradiction highlights the unphysical nature of any proposed antenna limit in which the size of the lower-frequency dimension of an antenna does not increase monotonically with increasing bandwidth and decreasing Q [47].

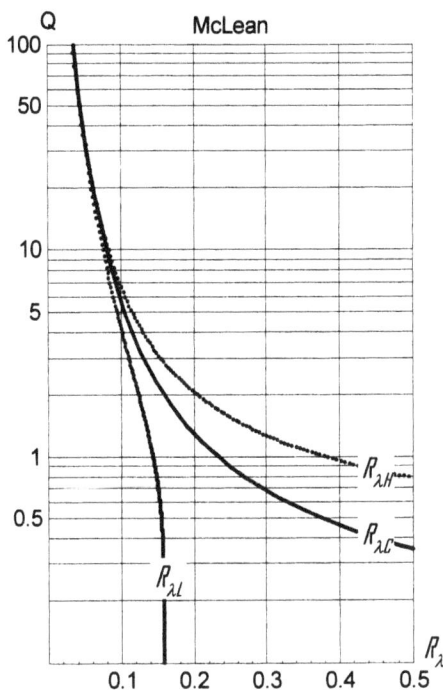

Figure 5.36 The McLean result for boundary-sphere radius in units of wavelength (R_λ). Note that $R_{\lambda L}$, $R_{\lambda C}$, and $R_{\lambda H}$ denote the boundary-sphere radius as referenced in units of the low-frequency wavelength, center-frequency wavelength, and high-frequency wavelength, respectively (After [47]).

Conversely, the McLean limit, shown in Figure 5.36, does involve a monotonic increase of the size of an antenna referenced to the lower frequency wavelength with increasing bandwidth and decreasing Q. In fact, the McLean limit monotonically converges to the radian sphere, $R_{\lambda L} = 1/(2\pi) = 0.1592$, as $Q \to 0$. For a 3:1 bandwidth antenna where $Q = f_C/BW = 0.577$, the McLean limit predicts that a boundary sphere radius of about $R_{\lambda L} = 0.14$ is ideal.

These Q-based antenna limits were never originally intended to be applied in the UWB limit. As Chu noted, "It is understood that the physical interpretation of Q as so computed becomes rather vague whenever the value of Q is low" [48]. McLean's limit represents an attempt to extrapolate what is essentially a narrowband analysis of efficiency to the UWB limit. The time domain energy-flow concepts of this chapter enable a variety of alternate antenna limits worthy of examination. The following section will present energy-flow-based limits to antenna performance.

> *While the oscillator is at work the energy oscillates in and out through the spherical surfaces surrounding the origin. But the energy which goes out during each period of oscillation is greater than that which returns, and is greater by the same amount for all surfaces. This excess represents the loss of energy due to radiation during each period of oscillation.*

Heinrich Hertz, 1893

5.5.5 Energy-Flow-Based Limits to Antenna Performance

This chapter has analyzed and documented the often considerable ebbs and flows of reactive energy around a dipole source. The energy distribution about a source "cleaves" at a causal surface. Energy outside decouples and radiates away. Energy inside remains bound and may be reabsorbed by the antenna, or alternatively stored about the antenna to decouple at a later time. Hertz realized that that the difference between the outflowing and inflowing energy during each period was exactly the energy lost to radiation [49]. This section will examine the ebb and flow of electromagnetic energy about a dipole source to establish limits to antenna performance.

Examine the oscillation of a dipole in closer detail. Through a given spherical boundary shell around a dipole, a certain amount of energy is pushed out (U_{out}), and a lesser amount is pulled back in (U_{in}). The average, or net, energy flow at any given distance from a point dipole is just the radiated energy ($U_{rad} = U_{out} - U_{in}$). This total radiated energy is independent of the radius of the shell. Close in to an ideal point dipole, the quantity of energy pushed out is much bigger than the radiated energy ($U_{out} \gg U_{rad}$); sufficiently far away from the dipole, virtually all the energy pushed out through the shell is radiated away ($U_{out} \cong U_{rad}$).

In the case of a narrowband antenna, when a signal is applied, reactive energy builds up around the antenna. Large quantities of reactive energy oscillate with each cycle of an antenna once steady state is achieved. But because this initial investment in reactive field energy is "amortized" over many cycles of the oscillation, the quantity of reactive energy does not enter into a calculation of efficiency. The reactive energy loitering in the near field does not directly impact the efficiency (unless it is prone to dissipation in conduction or dielectric losses). In the narrowband case, reactive energy is a bystander whose presence serves to determine Q, the sharpness of the resonance, and, thus, the bandwidth. The more reactive energy, the bigger Q and the smaller the bandwidth.

In the case of a UWB antenna, the applied signal is typically a short-duration impulse or excitation. Reactive energy builds up around the antenna, and then is either radiated or reabsorbed. The reabsorbed reactive energy (U_{in}) manifests itself in the form of a reflected signal bouncing back from the antenna. In the case of a nonresistively loaded UWB antenna, this reflected reactive energy is usually the most significant loss term in the overall energy budget.

What is needed, though, is an approach to antenna efficiency that is well suited to application in the UWB case. Therefore, we can define the antenna efficiency (η) as the ratio of the radiation energy (U_{rad}) to the total energy (U_{out}) that had to be pushed through the antenna's boundary shell

$$\eta = \frac{U_{rad}}{U_{out}} = \frac{\int\limits_{-\infty}^{+\infty} P(r)\,dt}{\sum\limits_{i=1}^{n} \int\limits_{\tau_i^-}^{\tau_i^+} P(r)\,dt} \qquad (5.75)$$

Calculation of the radiated energy is straightforward. Calculation of the total out-flowing energy is tricky because it requires knowledge of the start time τ^- and the end time τ^+ for each of the n periods during which there is a net outward flow of power through a shell of radius r around the dipole. This is a calculation best performed numerically for a given excitation. The following section discusses the results of an analytic treatment of the energy-flow efficiency limit for a harmonic dipole.

5.5.5.1 Energy-Flow Limit for a Harmonic Dipole

The energy-flow-based efficiency limit described in (5.75) may be subject to exact analytical treatment where the space-time energy flow around a point dipole has been characterized. The analytical treatment may be quite long and involved, making this a calculation best suited for a numerical analysis. In the time-harmonic case, however, a detailed calculation (available elsewhere [50]) yields an exact analytic expression for efficiency of a dipole as a function of the boundary sphere radius referenced to wavelength (R_λ)

$$\eta(R_\lambda) =$$

$$\begin{cases} \dfrac{8\pi^4 R_\lambda^3}{1 + 4\pi^3 R_\lambda^3 \left(\cos^{-1}\dfrac{4\pi^2 R_\lambda^2 - 1}{4\pi^2 R_\lambda^2 + 1} + \cos^{-1}\dfrac{1 - 12\pi^2 R_\lambda^2 + 16\pi^4 R_\lambda^4}{1 - 4\pi^2 R_\lambda^2 + 16\pi^4 R_\lambda^4} \right)} & R_\lambda \le \dfrac{1}{2\pi} \\[3em] \dfrac{8\pi^4 R_\lambda^3}{1 + 4\pi^3 R_\lambda^3 \left(\cos^{-1}\dfrac{4\pi^2 R_\lambda^2 - 1}{4\pi^2 R_\lambda^2 + 1} - \cos^{-1}\dfrac{1 - 12\pi^2 R_\lambda^2 + 16\pi^4 R_\lambda^4}{1 - 4\pi^2 R_\lambda^2 + 16\pi^4 R_\lambda^4} \right)} & R_\lambda \ge \dfrac{1}{2\pi} \end{cases} \qquad (5.76)$$

Evaluating this expression for efficiency yields the result that for $R_\lambda = 0.07$, only 25% of the energy pushed through the boundary sphere (of radius 0.07λ) ends up being radiated away. The remaining 75% is reabsorbed by the dipole or stored in the dipole's near fields to escape in a later cycle. For $R_\lambda = 0.10$, half of the energy emerging from the boundary sphere decouples and radiates away. The remaining 50% retreats back within the boundary sphere. At $R_\lambda = 0.30$, fully 75% of the energy pushed through the boundary sphere escapes, and 25% remains coupled. Figure 5.37 presents a qualitative diagram of these results.

Figure 5.38 presents a graph of antenna efficiency η versus antenna boundary-sphere radius as referenced to wavelength (R_λ) for an ideal harmonic dipole. Figure 5.39 presents an analogous plot showing the variation of efficiency with respect to frequency as referenced to the frequency at the low end of the band (f_L). This is the frequency where the boundary-sphere radius is precisely $R = 0.0964\lambda$ and $\eta = 0.50$ (in other words, the frequency of the half-power point).

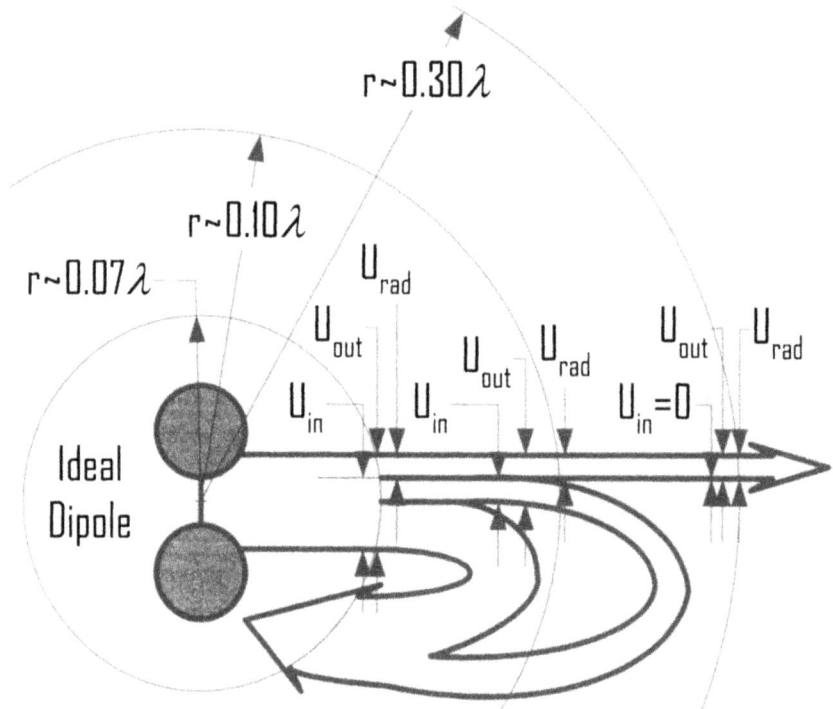

Figure 5.37 Conceptual diagram of the net radial energy flow around an ideal harmonic dipole. Only about 25% of the energy pushed through a 0.07λ boundary shell radiates. About 50% of the energy pushed through a 0.10λ boundary shell radiates. By the time the boundary shell extends out to 0.30λ, virtually all the energy pushed through the boundary shell radiates.

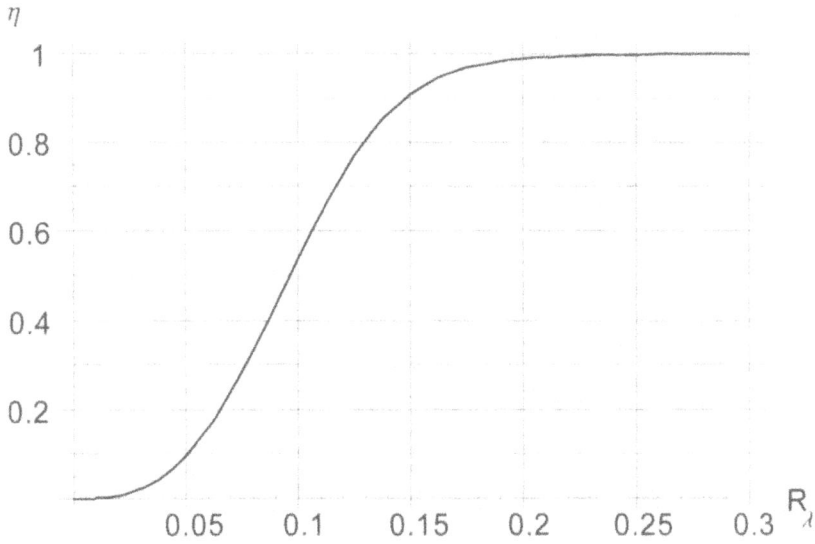

Figure 5.38 Diagram of antenna efficiency (η) versus antenna boundary-sphere radius (R_λ) as referenced to wavelength for an ideal harmonic dipole.

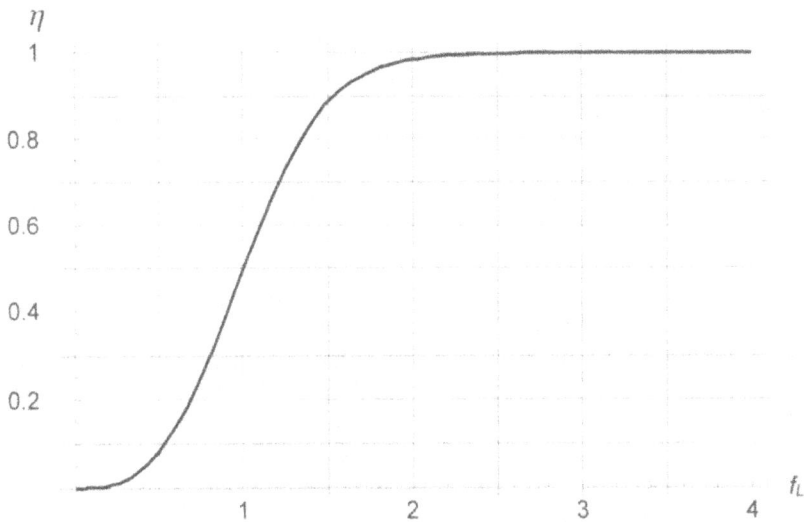

Figure 5.39 Efficiency (η) as a function of frequency referenced to the half-power frequency f_L.

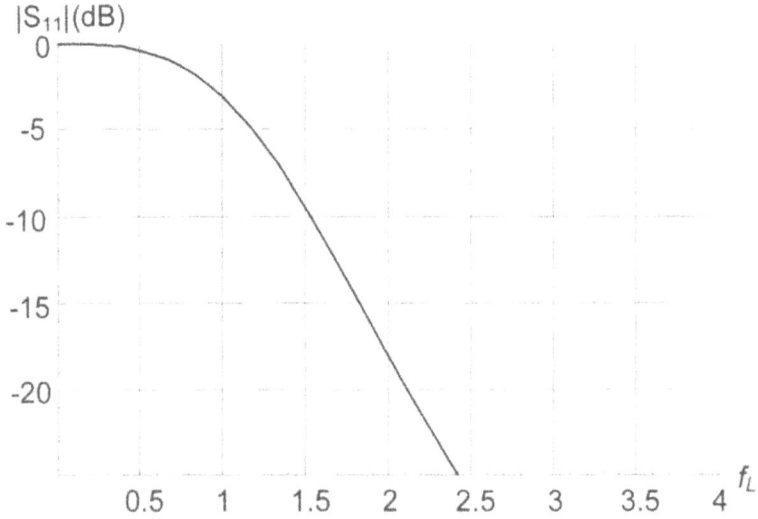

Figure 5.40 Return loss $|S_{11}|$ in decibels as a function of frequency referenced to the half-power frequency f_L.

Most well-designed UWB antennas have negligible ohmic, or dielectric, losses, so the predominant loss term is the return (or mismatch) loss. In this case, the efficiency may be related to the return loss ($|S_{11}|$)

$$\eta = 1 - |S_{11}|^2 \tag{5.77}$$

The *VSWR* may be similarly written in terms of efficiency

$$VSWR = \frac{1+|S_{11}|}{1-|S_{11}|} = \frac{1+\sqrt{1-\eta}}{1-\sqrt{1-\eta}} \tag{5.78}$$

Figure 5.40 plots the return loss as a function of the frequency (normalized to f_L). The –3-dB point technically defines the low end of the band, but practically speaking, at least a –10-dB S_{11} would be preferred. This requires an antenna approximately 50% larger than the –3-dB point indicates. Figure 5.41 plots the *VSWR* as a function of the frequency (again normalized to f_L). Note that *VSWR* = 5.83, where the antenna reflects half the applied power at the low end of the band. The VSWR also achieves a more respectable value (*VSWR* ≤ 2:1) at a frequency about 50% higher than the low end of the band corresponding to $R_{\lambda L} = 0.0964$. Unless a poorly matched antenna can be tolerated in a particular overall system design, a size approximately 50% larger than the $R_{\lambda L} = 0.0964$ limit derived here is required. Of course, a variety of design trade-offs are available.

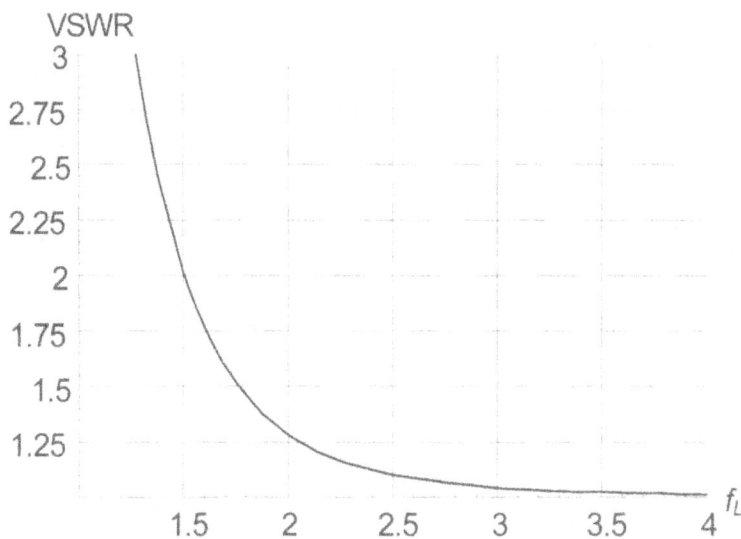

Figure 5.41 *VSWR* as a function of frequency referenced to the half-power frequency f_L.

5.5.5.2 Efficiency of a Transient Dipole

The limits and results defined in the previous section are best suited for use in the UWB context. In the narrowband limit, the energy absorbed by the dipole is stored in a matching network. With an efficient matching network, this stored energy is not lost but rather made available to be radiated in later cycles. Q-based limits (like McLean's) really set limits on the efficiency of a small antenna in combination with an associated matching network. Thus, with an appropriate efficient matching network, an antenna even smaller than $R_{\lambda L} = 0.0964$ is possible in the narrowband context. Interestingly, a $R_{\lambda L} = 0.0964$ is substantially smaller than predicted by McLean's limit for $Q < 5$, which encompasses the entire UWB regime. One goal of this section will be to verify the $R_{\lambda L} = 0.0964$ limit by considering a transient excitation.

Another goal is to evaluate the commensurability of time domain and frequency domain efficiency results. As shown in Section 5.5.5.1, inefficiencies due to mismatch are necessarily higher at lower frequencies. To call a 1–11GHz UWB antenna 90% efficient is misleading if the antenna is actually 0% efficient from 1GHz to 2GHz and 100% efficient above 2GHz. Although radiation of transients is inherently a time domain process, for practical applications it is useful to assess the efficiency of a transient radiation process in the frequency domain.

Accordingly, consider the transient excitation of (5.34). This excitation describes the fundamental dipole mode on a conducting sphere charged up to a particular static voltage and then discharged across the equator. As discussed in Section 5.3.2, the energy outside the sphere decouples and radiates away. If we assume that the original static field is created by charging the dipole adiabatically (i.e., so slowly that negligible radiation occurs), then no energy is lost in the charging process. If we assume that no energy is stored within the sphere, then the energy outside the sphere decouples and radiates away with 100% efficiency. This model is a crude approximation in that it considers only the fundamental dipole mode and not the higher-order modes on the sphere. Nevertheless, this example captures the essential low-frequency behavior of interest in the present context.

The dipole-mode radiated signal is proportional to the second time derivative

$$\ddot{T}(t) = \frac{c^2}{3R_S^2} e^{-\frac{ct}{2R_S}} \left[-3\cos\left(\frac{\sqrt{3}c}{2R_S}t\right) + \sqrt{3}\sin\left(\frac{\sqrt{3}c}{2R_S}t\right) \right] \tag{5.79}$$

If we assume $R_S = 1.0$cm, then the time dependence of the radiated signal is shown in Figure 5.42(a). Figure 5.42(b) presents the corresponding spectral power density. The spectral power density follows from the Fourier Integral Transform of the damped harmonic excitation

$$F(\omega) = -\frac{3jc^2\omega}{3c^2 + 3jcR_S\omega - 3R_S^2\omega^2} \tag{5.80}$$

This spectral density has half-power points at $f_L = 2.96$ GHz and $f_H = 7.70$ GHz. Thus, the center frequency is $f_C = 4.77$ GHz, and the bandwidth is $BW = 4.74$ GHz. The corresponding quality factor is $Q = f_C/BW = 1.01$. The wavelength at the lower end of the band is $\lambda_L = 10.1$cm so $R_{\lambda L} = 1.0$cm/10.1cm $= 0.099$. This spherical antenna is smaller than would be allowed by the McLean limit yet falls just within what would be allowed by the $R_{\lambda L} = 0.0964$ limit.

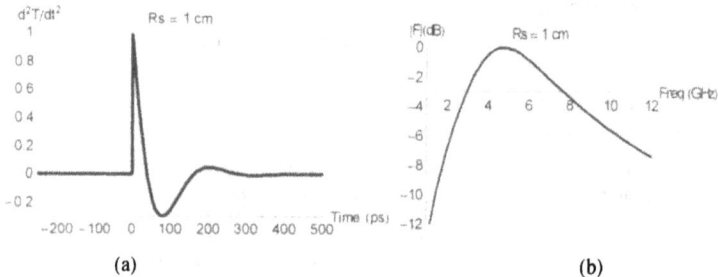

(a)   (b)

Figure 5.42 (a) Damped harmonic time domain response, and (b) spectral response of a damped harmonic spherical dipole mode where $R_S = 1.0$cm.

5.6 ANTENNAS AS ENERGY CONVERTERS

This chapter has introduced and explored a novel way to think of antennas: treating an antenna as a device that converts bound, or guided, electromagnetic energy into radiation energy. Treating antennas as energy converters is particularly useful in the UWB context because energy considerations are crucial to proper element design and to antenna efficiency.

Long ago, electromagnetic energy was thought to be an inherent property of charges. Energy was thought to transfer from one point to another via a mysterious "action-at-a-distance." Thanks to the discoveries of Kelvin, Maxwell, Poynting, and Heaviside, the underlying theory of electromagnetic energy transfer may be understood. This theory was confirmed by Hertz and by more than a century of engineering practice. The confusing and flawed model of radiation as a kinked field line may be replaced with a more elaborate and physically correct model. A better model of radiation takes into account not only the acceleration of charges but also the applied fields that give rise to their acceleration. The energy flows in electromagnetic systems may be tracked and understood using a variety of methods, including the method of causal surfaces. These are the surfaces through which there is no net flow of electromagnetic energy.

The theory of electromagnetic energy may be applied to a variety of interesting physical examples, including harmonic and transient radiation processes. These examples highlight several very important antenna design principles. First, reactive energy stored around an antenna is undesirable. This reactive energy narrows bandwidth and yields severe return losses or mismatch reflections. Second, reactive energy flows through and occupies particular zones or regions around an antenna. These two physical principles yield two basic antenna design rules.

First, fatter is better for a UWB antenna. A fat element tends to reduce the reactive electromagnetic energy. This improves bandwidth and matching. Second, by understanding the geometry of energy flow, one may design elements that are conformal to natural energy flows, while blocking, or shorting out, undesired reactive energy. A properly designed planar element can accomplish both these missions as effectively as a more difficult to build 3-D volumetric or surface-of-revolution-type element.

Energy considerations also allow one to determine fundamental limits to antenna size and bandwidth. Traditional analysis starts from narrowband resonance concepts and a parameter known as the quality factor, or Q. These limits may be extrapolated into the UWB limit with a couple of caveats. First, bandwidth is determined from the half-power, or –3dB, points. Second, the center frequency is the geometric average (not the arithmetic average) of the high and low frequencies. The traditional Chu-Harrington limit is demonstrably incorrect in the UWB context, but the more recent McLean limit is well behaved when extrapolated to ultra-wide bandwidths. Still these limits follow from an inherently narrowband approach to antenna behavior.

Time domain energetics allow a new definition of ideal antenna efficiency. According to this new definition, the ideal efficiency of an antenna is the fraction of energy pushed through a particular boundary sphere that successfully decouples and radiates away. In the case of an ideal harmonic dipole, a boundary sphere of radius $R_{\lambda L} = 0.0964$ defines the surface through which half of the emerging energy decouples and radiates away. This limit is consistent with the fundamental dipole mode of a sphere, which in principle can achieve a 100% efficient radiation of energy where the sphere has radius $R = 0.099\lambda$ at the low frequency end of the excitation spectral response.

Any designer of antennas benefits from a fundamental understanding of electromagnetic radiation. This kind of fundamental understanding is important not only for implementing tried-and-true antenna designs but also for advancing the state of the antenna art.

Endnotes

[1]   Heaviside, Oliver, *Electromagnetic Theory*, Vol. 3, London: "The Electrician" Printing and Publishing Company, Limited, 1912, p. 55. Heaviside's remarkable three volume work on electromagnetic theory was recently reprinted in facsimile edition (2003) and is now available from Elibron Classics at www.elibron.com. Originally published 1893.

[2]   Thompson, J. J., *Energy and Matter*, New York: Charles Scribner and Sons, 1904, p. 56.

[3]   Some of the material in this section was originally presented by the author in H. Schantz, "Localization of Electromagnetic Energy," *Ultra-Wideband, Short Pulse Electromagnetics* Vol. 5, New York: Kluwer Academic/Plenum Publishers, 2002, pp. 89–96. Reprinted by the kind permission of Kluwer Academic/Plenum Publishers.

[4]   von Helmholtz, Hermann, "The Conservation of Force: A Physical Memoir" (1847), in *Selected Writings of Hermann von Helmholtz*, Russel Kahl, ed., Middletown, CT: Wesleyan University Press, 1971, pp. 3–55.

[5]   Whittaker, Sir Edmund, *A History of the Theories of Aether and Electricity*, Vol. 1, New York: Harper and Brothers, 1951, p. 222. This is the classic work for students of electromagnetic history.

[6]   Maxwell, James Clerk, *A Treatise on Electricity and Magnetism*, Vol. 2, 3rd ed., Oxford: Clarendon Press, 1892, pp. 270–274. See Section 631 in particular.

[7]   Maxwell, James Clerk, "On Action-at-a-Distance," Proceedings of the Royal Institution of Great Britain, Vol. 7, 1873–1875, pp. 48–49. This paper is reprinted in *The Scientific Papers of James Clerk Maxwell*, Vol. 2, W. D. Nivens, ed., New York: Dover Publications, pp.311–323. This remarkable paper traces the history of theories of physical interaction from Newton, through Faraday, to Maxwell's own discoveries and thinking.

[8]   Poynting, John Henry, "On the Transfer of Energy in the Electromagnetic Field," Philosophical Transactions, Royal Society, London, Vol. 175 Part II, 1885, pp. 334–361.

[9]   Heaviside, Oliver, *Op. Cit.*, pp. 73–74.

[10] Hertz, Heinrich, *Electric Waves*, London: MacMillan and Co., 1893. See the insightful preface by Lord Kelvin for the philosophical implications of Hertz's discoveries.

[11] Hertz, *Op. Cit.*, p. 146.

[12] King, Ronold W. P., *Fundamental Electromagnetic Theory*, 2nd ed., New York: Dover, 1963, pp. 191–192.

[13] Jeans, Sir James, *The Mathematical Theory of Electricity and Magnetism*, Cambridge: University Press, Cambridge, 1933, p. 518.

[14] King, Ronold W. P., *Op. Cit.*, p. 180.

[15] Feynman, Richard, *The Feynman Lectures in Physics*, Vol. 2, Reading, MA: Addison-Wesley Pub. Co., 1964, pp. II-27–8.

[16] Pugh, E. M. and G. E. Pugh, "Physical significance of the Poynting vector in static fields," American Journal of Physics, Vol. 35, 1967, pp. 153–156.

[17] Lombardi, Gabriel G., "Feynman's disk paradox," American Journal of Physics, Vol. 51, 1983, pp. 213–214.

[18] Beth, Richard A., "Mechanical Detection and Measurement of the Angular Momentum of Light," Physical Review, Vol. 50, 1936, pp. 115–125.

[19] Nichols, E. E. and G. F. Hull, "The Pressure Due to Radiation," Physical Review, Vol. 17, 1903, pp. 26–50, 91–104.

[20] Backhaus, Udo and Klaus Schäfer, "On the uniqueness of the vector for energy flow density in electromagnetic fields," American Journal of Physics, Vol. 54, 1986, pp. 279–280.

[21] Birkeland, K., "Über die Strahlung electromagnetischer Energie im Raume." Ann. Phys., Vol. 52, 1894, pp. 357–380.

[22] Schantz, Hans, "The Energy Flow and Frequency Spectrum Around Electric and Magnetic Dipoles," Ph. D. dissertation, University of Texas, Austin, Austin, TX, 1995, p. 41.

[23] Schantz, Hans, " Electromagnetic energy around Hertzian dipoles," IEEE Antenna and Propagation Magazine, April 2001, pp. 50–62.

[24] Thompson, J. J., "On Electrical Oscillations and the effects produced by the motion of an Electrified Sphere," Proceedings of the London Mathematical Society, April 8, 1884, pp. 197–218.

[25] Sommerfeld, Arnold, *Electrodynamics*, publisher unavailable, 1952, pp. 154–155.

[26] Larmor, J., "On the Theory of the Magnetic Influence on Spectra; and on the Radiation from Moving Ions," Philos. Mag., Series 5, Vol. 44, 1897, pp. 503–512.

[27] Hertz, *Op. Cit.*, pp. 146–147.

[28] Hertz, *Op. Cit.*, p. 152.

[29] Wheeler, Harold A., "Fundamental Limitations of Small Antennas," Proceedings of the IRE, Vol. 35, 1947, pp. 1479–1484.

[30] Wheeler, Harold A., "The Radiansphere Around a Small Antenna," Proceedings of the IRE, Vol. 47, 1959, pp. 1325–1331.

[31] Capps, Charles, "Near Field or Far Field?" EDN Magazine, August 16, 2001, pp. 95–102. Available online at: www.edn.com/contents/images/150828.pdf.

[32] Slater, J. C., *Microwave Transmission,* New York: McGraw Hill, 1942, pp. 230–232

[33] King, R. W. P., et al., *Transmission Lines, Antennas, and Waveguides*, New York: McGraw-Hill, 1945, p. 107.

[34] Wheeler, Harold A., "The Radiansphere Around a Small Antenna," Proceedings of the IRE, Vol. 44, August 1959, pp. 1325–1331.

[35] Ibid. p. 1330.

[36] Ibid. p. 1330.

[37] Ibid. p. 1331.

[38] Wheeler, Harold A., "Small antennas," IEEE Transactions on Antennas and Propagation, Vol. AP–23, No. 4, July 1975, pp. 462–469.

[39] Harmuth, Henning, *Antennas and Waveguides for Nonsinusoidal Waves*, New York, Academic Press, 1984, pp. 98–99.

[40] Schantz, Hans, "Broadband electric-magnetic antenna apparatus and system," U.S. Patent Pending.

[41] Green, Estill I., "The Story of Q," American Scientist, Vol. 43, October 1955, pp. 584–594. This excellent and insightful article is must reading for students of resonance phenomena. Kenneth S. Johnson (1885–1956), the originator of *Q*, was a Western Electric (later Bell Labs) engineer regarded as one of the leading experts on wire transmission. A prolific inventor (with 53 patents to his credit) he also designed the 1937 telephone which saw wide use in the United States.

[42] Chu, L. J., "Physical Limitations of Omni-Directional Antennas," Journal of Applied Physics, Vol. 19, December 1948, pp. 1163–1175. Chu was by no means the first to consider this question. See Chu's paper for a summary of earlier investigations.

[43] Chu, L. J., *Op. Cit.*, p. 1171–1172.

[44] Harrington, Roger F., "Effect of antenna size on gain, bandwidth, and efficiency," J. Res. Nat. Bur. Stand., Vol. 64D, January–February 1960, pp. 1–12.

[45] Hansen, Robert C. "Fundamental Limitations in Antennas," Proceedings of the IEEE, Vol. 69, No. 2, February 1981, pp. 170–182.

[46] McLean, James S., "A re-examination of the fundamental limits on the radiation *Q* of electrically small antennas," IEEE Transactions on Antennas and Propagation, Vol. 44, No. 5, May 1996, pp. 672–675.

[47] Schantz, Hans "Introduction to ultra-wideband antennas," Plenary Talk, 2003 IEEE Conference on Ultra Wideband Systems and Technologies. Reston, Virginia. November 16-19, 2003, pp. 1-9.

[48] Chu, L. J., *Op. Cit.*, p. 1165.

[49] Hertz, *Op. Cit.*, p. 149.

[50] Schantz, Hans, "Harmonic dipole energy flow and antenna efficiency," 2005 IEEE Antennas and Propagation Society Symposium. Washington, D. C., July 4-7, 2005.

Chapter 6

A Taxonomy of UWB Antennas

> *Generally [UWB element antennas] are characterized by linear polarization, low directivity and relatively limited bandwidth unless either end loading or distributed loading techniques are employed, in which case bandwidth is increased at the expense of radiation efficiency.*

IEEE Spread Spectrum Conference, 1996

> *The main limiting factor of a UWB system is the antenna, which acts as a filter and limits the transmission bandwidth.*

Government Report, 2000

> *Traditionally, antenna's bandwidth has been the main bottleneck of a ultra-wide-band wireless system.*

IEEE International Symposium on Antennas and Propagation, 2004

There is a widespread and continuing, erroneous impression that UWB antennas represent a fundamental impediment to the performance of a UWB radio system. In fact, there are a wide variety of UWB antenna designs with excellent 3:1 or better bandwidth performance well suited for modern 3.1–10.6GHz systems. Unless system architects have unreasonable expectations for antenna size and form factor, the designs presented throughout this book, and particularly in this chapter, will not disappoint.

The aim of this chapter is to provide a thorough survey of the UWB antenna kingdom. This taxonomy begins by considering some of the important classes of UWB antennas including frequency-independent antennas, small-element antennas (both electric and magnetic), electrically small antennas, UWB antenna arrays, horn antennas, and reflector antennas. Within each class, this taxonomy discusses the important families of antennas and presents representative examples of particular antenna species. Given the number of designs considered in this chapter, discussion is necessarily brief. The reader is welcome and encouraged to peruse the references for additional information on the antennas surveyed.

6.1 FREQUENCY-INDEPENDENT ANTENNAS

Frequency-independent antennas form a relatively mature, well-understood, and appreciated family of UWB antennas. These antennas are discussed in their historical contexts in Section 1.1.3. Frequency-independent antennas rely on an angle invariant scaling from a small-scale portion, which defines the high-frequency limit, to a large-scale portion, which defines the low-frequency limit. This property of frequency-independent antennas has significant advantages as well as disadvantages. The principal advantage is that bandwidth is limited only by the range of scaling physically achievable in antenna construction. Thus, a decade or more of bandwidth is possible. The principal disadvantage is that this same distributed scaling gives rise to the dispersion documented in Section 2.2.1.

Some of the more common frequency-independent antennas include spirals, conical spirals, and log periodics. There is a vast and detailed body of literature on the subject of frequency-independent antennas that need not be reexamined here. One of the best references remains Victor Rumsey's classic book on the antennas he pioneered, now regrettably out of print [1]. Frequency-independent antennas are also a topic in most contemporary antenna textbooks.

One more-recent development in frequency-independent antennas is the "fractal" antenna like the Minkowski Island loop antenna of Figure 6.1. Fractal antennas are, in a sense, a generalization of the standard self-similar frequency-independent antenna. Some fractal geometries involve a looser sort of self-similarity called "self-affinity" in which a scaled replication of a particular geometric relation is skewed, distorted, or otherwise altered from one scale to another within the structure. Fractal antennas have attracted attention more for their beauty than for their behavior.

The performance of fractal antennas has generally been disappointing [2]. Thin-wire fractal antennas offer improved performance compared to conventional straight-wire designs but are not superior to a variety of nonfractal, non-Euclidean designs. In the context of planar elements, geometric features at scales much smaller than $\lambda/10$ are unlikely to have much of an effect on the overall performance of an antenna. Hertz's principle requires that currents concentrate on and near the edges of an element, so the large-scale structure and boundary of an element is likely to remain the dominant factor governing antenna performance.

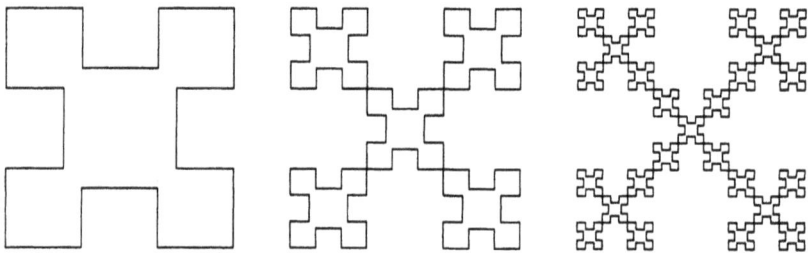

Figure 6.1   Minkowski Island fractal loop antennas.

6.2 SMALL-ELEMENT ELECTRIC ANTENNAS

Small-element electric antennas are physical realizations of an ideal Hertzian dipole. These are some of the most important UWB antennas because they combine the compact size and omnidirectional pattern useful in a wide variety of consumer electronic and other applications. There are a variety of families of small-element electric antennas. These families include conical element antennas, planar conical antennas, bulbous antennas, and planar bulbous antennas. Electric antennas may be thought of as voltage driven and have predominantly electric near fields. This section will address some general principles of small-element electric antenna design.

6.2.1 Conical Antennas

Conical antennas are among the oldest of UWB small-element electric antennas having been pioneered by Lodge in the 1890s. The conical antenna family includes biconical dipoles, "monocone" monopoles, and discone antennas. The electromagnetics of conical elements were examined by Schelkunoff and others in the 1930s. Now, biconicals are a standard topic in most antenna textbooks [3]. Also, UWB application of these antennas is an area of current interest [4]. This section will address biconical, monocone, and discone antennas, and discuss how to construct conical elements.

6.2.1.1 Biconical Antennas

In the limit of an infinitely long biconical element, the impedance becomes

$$Z = \frac{Z_S}{\pi} \ln \cot \frac{\theta_{hc}}{2} \qquad (6.1)$$

where $Z_S = 376.7\Omega$ is the impedance of free-space and θ_{hc} is the half-angle of the bicone. A half-angle in the neighborhood of 67° yields a 50Ω match. Infinitely long biconical elements are difficult to build. The challenge for an antenna designer is how to terminate the ends of the biconical elements. One approach is to make the biconical relatively long with respect to frequencies. Derek McNamara and his colleagues found that a one-wavelength-diameter, $\theta_{hc} = 60°$ biconical offered excellent matching over a 6:1 range of frequencies [5]. Figure 6.2(a) presents a sketch of this antenna.

An abrupt termination will yield an undesirable reflection. Thus, a gradual taper is preferred. Figure 6.2(b) shows a teardrop biconical proposed by Sergei Schelkunoff and Harald Friis [6]. Carl Baum also studied a similar configuration

[7, 8]. The gradual taper and rounded termination on the ends of the biconical elements reduce reflections and improve matching. The low-frequency limit is governed by the overall length of the antenna. The high-frequency limit is governed by the mechanical precision of the feed. Glenn Wolenec constructed the particular teardrop biconical in Figure 6.2(b) using quarter-inch hardware cloth mounted on half-inch PVC pipe. The conical sections used a planar template with a planar pattern arc angle $\theta_P = 90°$. This resulted in a conical half angle $\theta_P \cong 15°$. Further construction details for conical elements may be found in Section 6.2.1.3. Using a 4:1 TV transformer, the impedance of this antenna was 65Ω as measured by a 50Ω network analyzer. Thus the actual antenna impedance was around 260Ω. Applying (6.1) yields a predicted impedance of 247Ω, for an error of about 5%. The spherical endcaps are about 20cm (8in) in diameter and the total length of each element was about 50cm (20in). This antenna exhibits a -10dB S_{11} above about 200MHz and is limited in impedance bandwidth only by the transformer which began degrading above 1GHz.

Just as two conical elements may be combined to create a biconical dipole antenna, so also may a single conical element be driven against a ground plane to form a monocone monopole antenna. Marconi introduced the monocone in thin-wire form around 1900 (see Section 1.1.1.4), and Carter proposed a taper fed version in the late 1930s (see Section 3.3.2). In the limit where the ground plane becomes small, the monocone becomes the "discone" antenna of the next section.

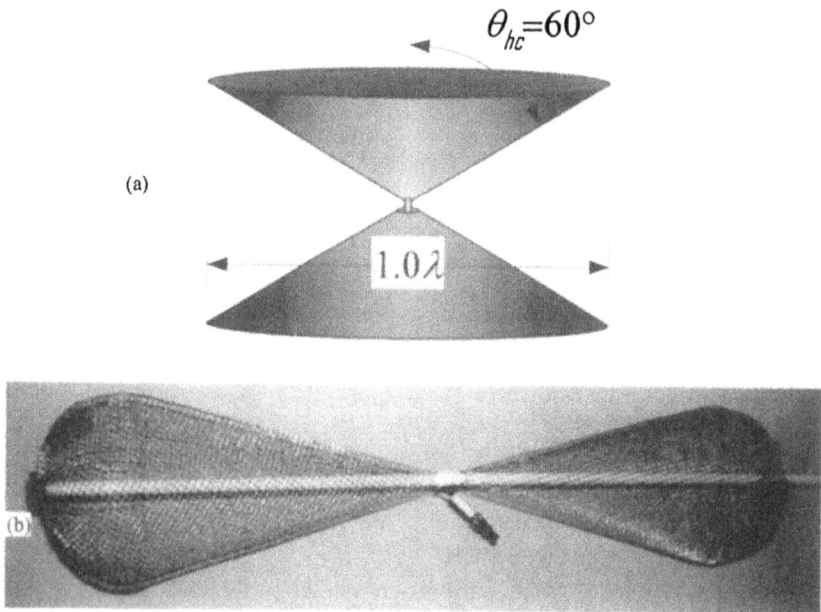

Figure 6.2   (a) UWB biconical of McNamara et al. (After [5]), and (b) a teardrop biconical antenna (Courtesy: Glenn Wolenec).

6.2.1.2 Discone Antennas

Figure 6.3(a) presents a diagram of a discone antenna. Figure 6.3(b) is a UWB discone antenna employed by Innovative Wireless Technologies (IWT). As discussed in Section 1.1.2, Kandoian first disclosed these antennas in 1945 [9, 10]. Discones are characterized by a dipolelike pattern for approximately one octave and a multioctave impedance bandwidth. Thus they are not well suited for applications demanding a uniform omni pattern over a 3:1 range of frequencies.

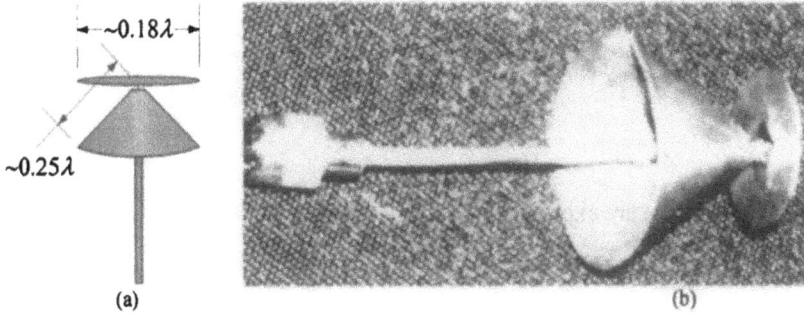

(a) (b)

Figure 6.3 (a) Diagram of a discone antenna with approximate dimensions, and (b) IWT UWB discone antenna (Courtesy: Innovative Wireless Technologies, Forest, Virginia).

Somewhat better performance may be obtained by taking Kandoian's basic discone architecture and rolling the edges to create more gradual discontinuities. Figure 6.4 shows the rolled-edge discone devised by Keith A. Snyder and Gary L. Peisley [11]. This antenna operates over a four-octave bandwidth achieving a VSWR of 2:1 or better over much of that bandwidth. Bulbous-type antennas like this one are addressed in much greater detail in Section 6.2.3.

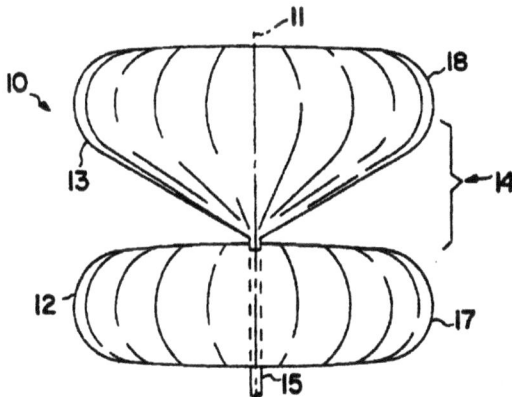

Figure 6.4 The rolled-edge discone of Snyder and Peisley [11].

6.2.1.3 Construction of Conical Elements

This section discusses the construction of conical elements from planar templates. Figure 6.5 shows how to design a planar template to yield a conical element with a desired height (h) and a desired cone half-angle (θ_{hc}). The radius of the planar template is the length (L) of the side of conical element:

$$L = \frac{h}{\cos\theta_{hc}} \tag{6.2}$$

The planar pattern arc angle is

$$\theta_p = 2\pi\sin\theta_{hc} \tag{6.3}$$

where both angles are expressed in radians, and the conical element radius is

$$R = h\tan\theta_{hc} \tag{6.4}$$

Conical elements may be constructed from sheet metal or foil. Hardware cloth, chicken wire, or other metallic meshes are well suited for construction of larger conical elements where the weight of a continuous element might be considerable. Particular care must be paid to the feed region. Small elements might be soldered directly to the metallic ground sheath of a coaxial feed line and to the center conductor. Larger elements will require additional mechanical support.

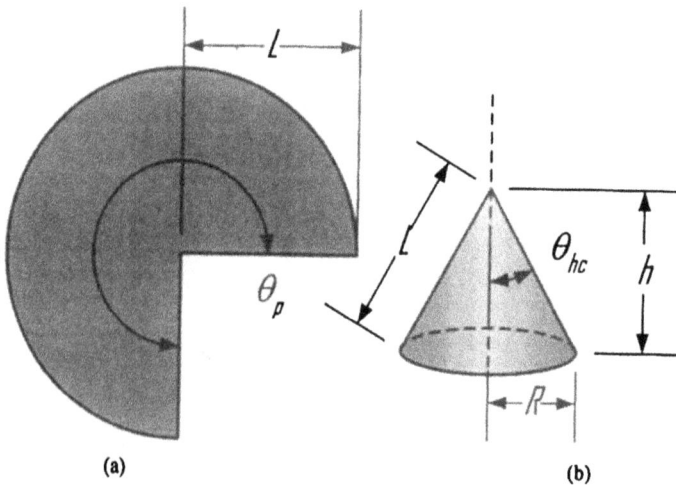

(a)   (b)

Figure 6.5   (a) Planar template to create (b) a conical element.

6.2.2 Planar Conical Antennas

Planar conical antennas have many of the same advantages as conical antennas in a more compact and easier-to-manufacture form factor. Examples of planar conical antennas include bow tie antennas and diamond dipoles. This section discusses each of these two species of planar conical antennas.

6.2.2.1 Bow Tie Antennas

Bow tie antennas were also introduced by Lodge in the 1890s. Brown and Woodward further popularized bow tie antennas in the 1950s [12]. A bow tie antenna is a planar cross-section of a biconical antenna. Figure 6.6(a) shows a sketch of three circular arc bow tie antennas with different feed angles that fit within a boundary-circle radius of 4cm [13]. Figure 6.6(b) is a photo of these antennas.

Figure 6.7 shows the gain of these three antennas as a function of frequency. These bow tie antennas have constant gain over a 5:1 bandwidth. Gain is relatively insensitive to apex angle. Figure 6.8 shows the impedance of these three antennas. Conventional wisdom is that the apex angle is critical for determining the element impedance. My experience, however, has been that impedance depends principally upon the fine-detail structure of the feed region. In any event, the significant variations these antennas show in their impedance means that they are difficult to match.

(a)

(b)

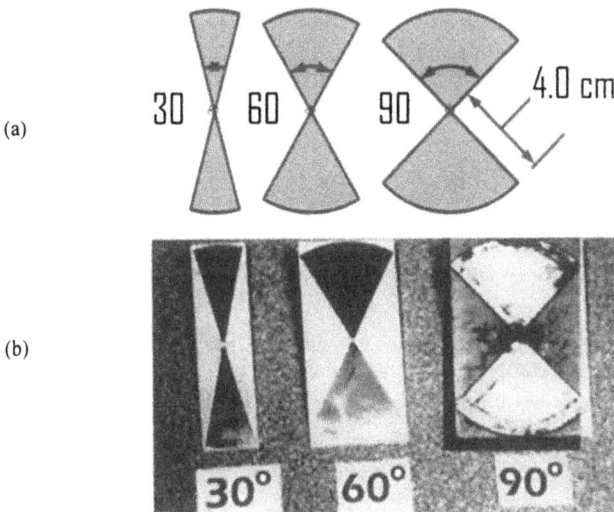

Figure 6.6   (a) Diagram of three bow tie antennas with various apex angles ([13]; © 2004 by CMP Media, LLC), and (b) photograph of three bow tie antennas with various apex angles.

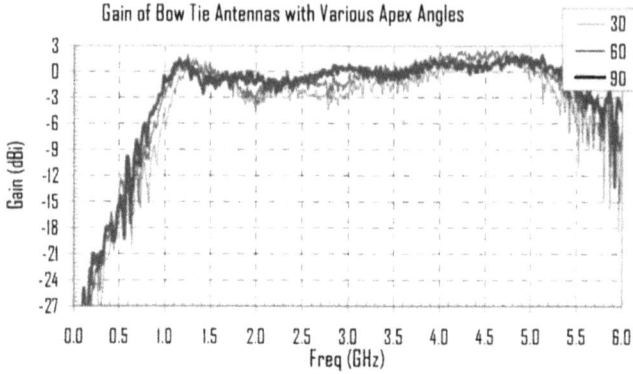

Figure 6.7   Gain is relatively insensitive to the bow tie apex angle. Uniform performance across a
5:1 band is possible ([13]; © 2004, CMP Media, LLC).

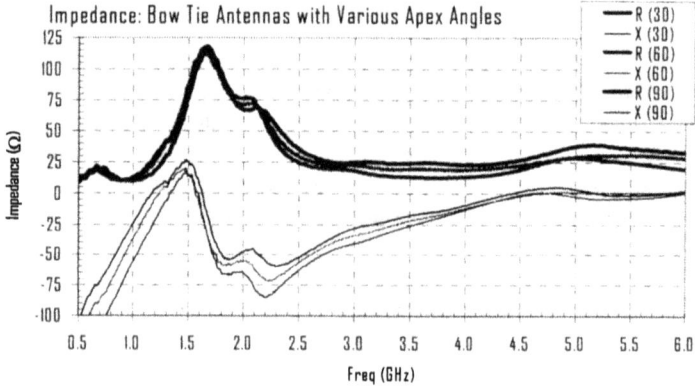

Figure 6.8   Impedance, real (*R*) and imaginary (*X*) of bow tie antennas with various apex angles
([13]; © 2004, CMP Media, LLC).

6.2.2.2 Diamond Dipole Antennas

Masters introduced the inverse bow tie, or diamond dipole, antenna in 1947 [14].
This antenna was rediscovered and popularized by Larry Fullerton [15, 16].
Figure 6.9(a) shows a diamond dipole formerly used in UWB radio systems by the
Time Domain Corporation. These antennas have isosceles triangular elements
whose height and base are scaled to be $\lambda/4$ at the center frequency of interest.
Figure 6.10 shows the gain of a typical 2GHz Diamond Dipole antenna. This
diamond dipole has a peak gain of around 1dBi with about a 2:1 quasi-Gaussian
tapered impedance bandwidth.

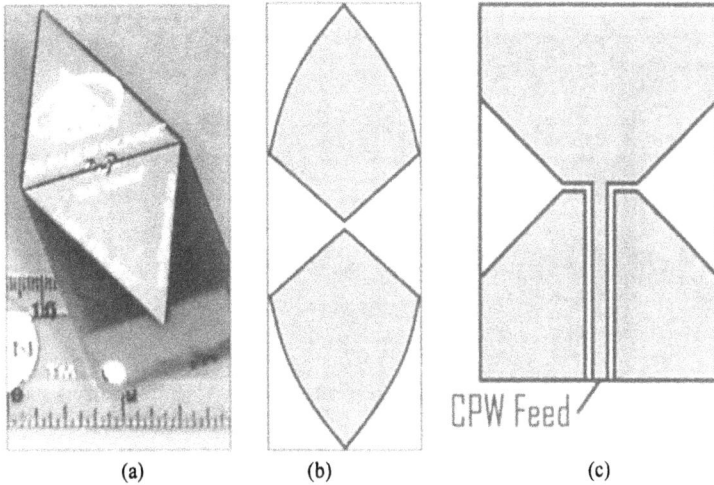

(a)   (b)   (c)

Figure 6.9   (a) The diamond dipole (After [14-16]), (b) the Bishop's Hat antenna of Smith, Starkie, and Lang (After [19, 20]), and (c) the hexagonal dipole of Kwon and Kim (After [21]).

This quasi-Gaussian spectral response means that the Diamond Dipole radiates a uniformly shaped impulse response reminiscent of a Gaussian doublet: the third derivative of a Gaussian, or $T_3(t)$, in the nomenclature of Section 4.1.3.1. Unfortunately, this quasi-Gaussian response is created by reflecting back significant energy.

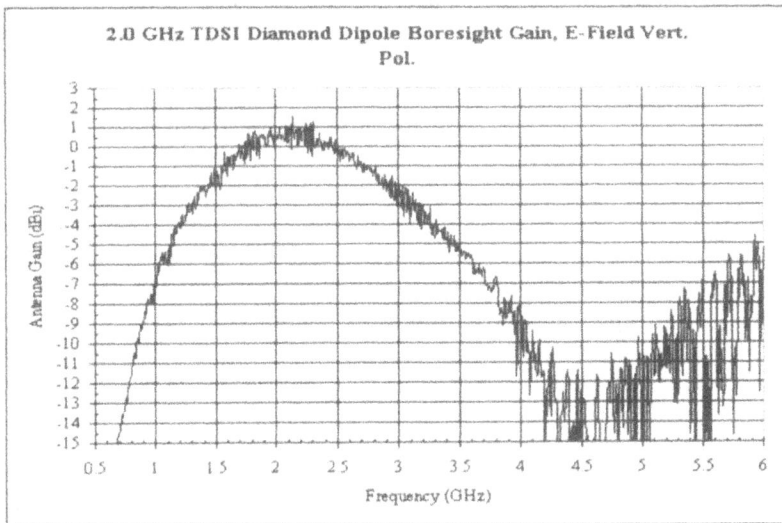

Figure 6.10   Gain of a matched pair of 2GHz diamond dipoles ([15]; © 2001, IEEE).

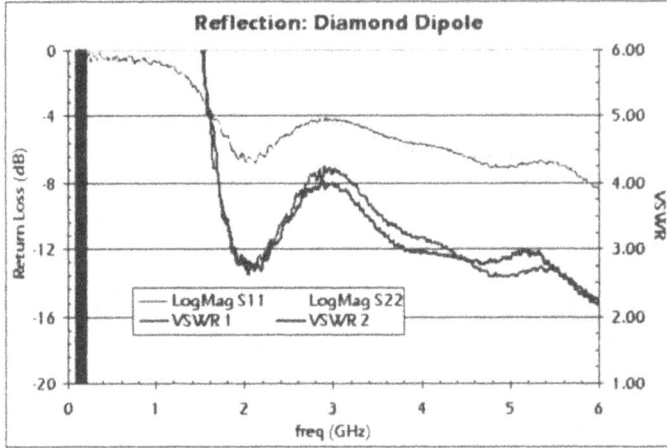

Figure 6.11 Reflection from a matched pair of 2GHz diamond dipoles ([15]; © 2001, IEEE).

The 50Ω connector at the feed point couples into a 100Ω slot line between the two elements. Two 100Ω slots in parallel yield a 50Ω load, so the feed region is well matched to the 50Ω connector. The abrupt discontinuities at the end of the slot lines on either side of the antenna are the geometric features most responsible for the significant reflections from this antenna. Figure 6.11 shows the return loss and VSWR versus frequency. The best-case return loss around the center frequency of 2.0GHz is about –6.4dB, and the VSWR is about 2.75:1. The diamond dipole's poor matching makes it a less desirable candidate for UWB applications, unless one requires a symmetric Gaussian waveform and can tolerate significant reflections.

6.2.2.3 Other Angular Antennas

Despite their long pedigree and difficulty to match, bow ties have seen use in modern UWB applications [17, 18]. One recent hybrid of the bow tie and diamond dipole antenna is Les Smith, Tim Starkie, and Jack Lang's "bishop's hat" antenna [19, 20]. The bishop's hat antenna is a bow tie antenna terminated in a curved taper culminating in a point (or conversely, a diamond dipole with a tapered slot). This antenna exhibits a return loss of no worse than –8dB referenced to 100Ω over about a 4:1 bandwidth. Figure 6.9(b) shows the bishop's hat antenna employed by Artimi, Ltd. Figure 6.9(c) shows a hexagonal coplanar waveguide (CPW)-fed dipole proposed by Do-Hoon Kwon and Yongjin Kim [21]. The slot line feed region of this dipole allows for a better-controlled feed impedance than a standard bow tie element. However, the angular discontinuities are likely to serve as sources of reflection.

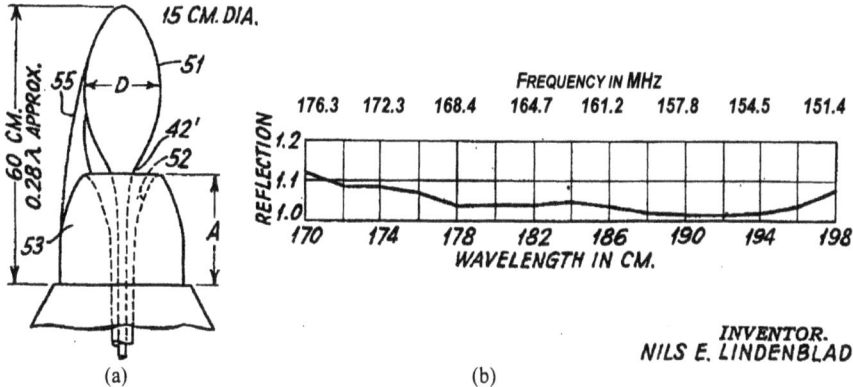

Figure 6.12 (a) Lindenblad's element, and (b) VSWR of Lindenblad's element as a function of wavelength and frequency [22].

6.2.3 Bulbous Antennas

The archetypical and original bulbous antenna was Lindenblad's televison-broadcast antenna described in Section 1.1.2 [22, 23]. Figure 6.12(a) shows a diagram of Lindenblad's element in profile, and Figure 6.12(b) shows the reflection or VSWR across approximately a 15% fractional bandwidth. This element could readily accommodate four 6MHz television channels with a VSWR of 1.1 or better. With an antenna about 0.32λ long in the operating band, elements are approximately 0.16λ, smaller than the 0.25λ length expected for a resonant thin-wire quarter-wave dipole. This resonant, or center, frequency depends not so much on the height of the antenna element from the feed but rather on the circumferential path from the feed around to the far side of the element. Thus, a bulbous element not only has a wide bandwidth due to the exclusion of reactive energy but also tends to be more compact than a comparable thin narrowband element at the same frequency.

Lindenblad's element is an example of a surface-of-revolution antenna. Such an antenna takes a particular curve or taper and rotates it about an axis to enclose a solid volume. Examples of surface-of-revolution antennas include Kraus's volcano smoke antenna (see Section 1.1.2) and the tapered biconicals of Schelkunoff and Friis and of Baum (see Section 6.2.1.1). These surface-of-revolution antennas require intricate lathing, elaborate machining, or other difficult assembly processes to implement. Stöhr took an important step toward simplifying implementation of surface-of-revolution antennas with his discovery of spherical dipoles and monopoles (see also Section 1.1.3) [24].

Stöhr discovered that an elaborately tapered antenna element was not critical to achieving good results. Figure 6.13(a) shows Stöhr's spherical monopole design in cross-section. Figure 6.13(b) depicts Stöhr's spherical dipole. In fact, spherical UWB antennas are capable of good broadband matching (better than 2:1 VSWR) and dipole pattern across a 3:1 bandwidth.

In practical implementations, one can drill and tap a sphere for use as an antenna element. I have found that chrome-plated, steel Chinese meditation spheres offer an inexpensive and readily available source of UWB antenna elements. Figure 6.13(c) shows my spherical monopole antenna, and Figure 6.13(d) shows details of this antenna's feed region. There are also commercial sources of metallic spheres for purists who prefer copper, brass, or aluminum elements [25]. A coaxial feed line may comprise a copper tube and a threaded screw center conductor. One may then optimize antenna matching by screwing the spherical element to adjust the gap at the feed region.

Figure 6.14 presents measured reflection data from a (different) spherical monopole with a diameter of 1.5in (3.8cm). The lower-frequency limit for $S_{11} <$ –10dB is about 1.6GHz. This corresponds to a spherical monopole diameter $d = 0.20\lambda$. The lower frequency limit for $S_{11} < $ –3dB is about 1.0GHz. This corresponds to a spherical monopole diameter $d = 0.13\lambda$. Note that this is smaller than McLean's limit suggests should be possible for $Q < 2.5$ but lies comfortably within the $R = 0.0964\lambda$ limit proposed in Section 5.5.5.1.

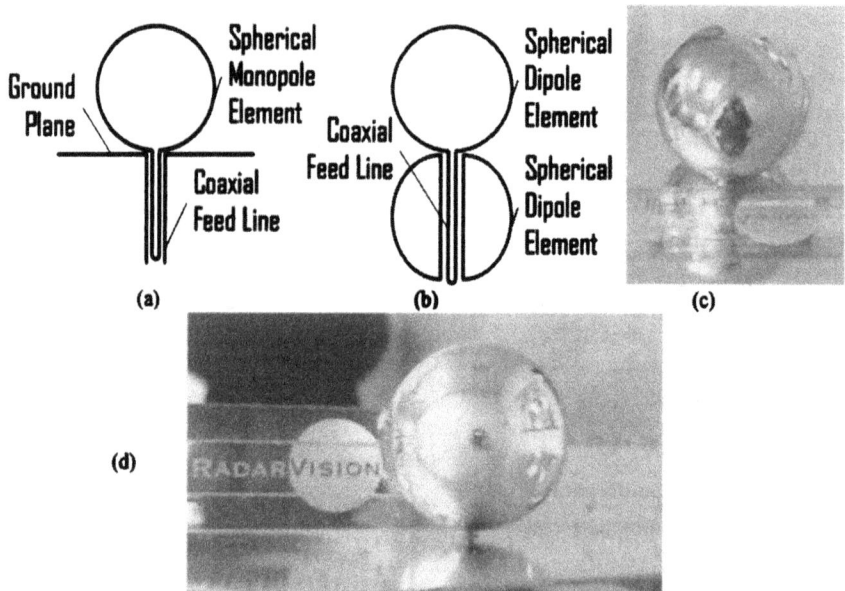

Figure 6.13 (a) Cross-section of Stöhr's spherical monopole (After [24]), (b) cross-section of Stöhr's spherical dipole antennas (After [24]), (c) detail of my Chinese meditation spherical monopole, and (d) Chinese meditation spherical monopole showing detail of feed region (Courtesy: Time Domain Corporation, Huntsville, Alabama).

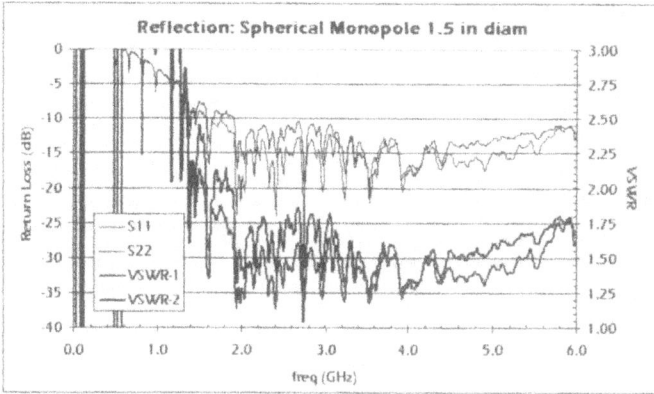

Figure 6.14 Reflection and VSWR for a pair of 1.5in (3.8cm)-diameter spherical monopoles (Courtesy: Time Domain Corporation, Huntsville, Alabama).

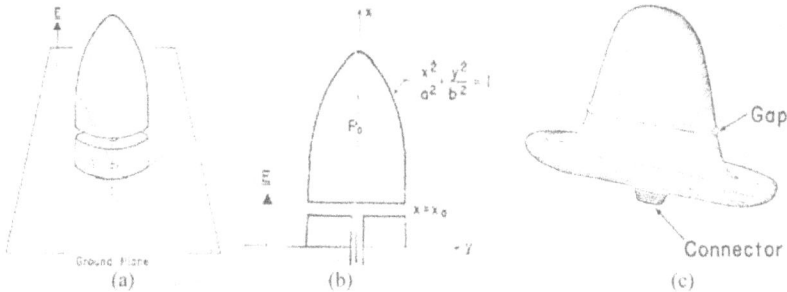

Figure 6.15 (a) Lee's ellipsoidal element, (b) cross-section of Lee's ellipsoidal element, and (c) streamlined version of Lee's ellipsoidal element for aviation applications [26].

Stöhr also proposed ellipsoidal shaped elements in addition to spherical elements. These ellipsoidal elements work particularly well when reduced to planar implementations. Planar elliptical elements are the subject of Section 6.4.3.2.

In practice, surface-of-revolution and other 3-D solid antennas are preferred only where other considerations justify the added bulk and difficulty to manufacture. For instance, K. S. H. Lee examined ellipsoidal antennas [26]. These elements, shown in Figure 6.15(a) and in cross-section in Figure 6.15(b), are well suited for use as broadband field probes. Lee's antenna also accommodates aerodynamic streamlining for use in aviation applications, as shown in Figure 6.15(c).

For most applications, however, planar antennas offer nearly comparable performance with significantly reduced manufacturing cost and bulk. The following section will show how planar cross-sections of bulbous elements make excellent low-cost UWB antennas for a wide variety of applications.

6.2.4 Planar Bulbous Antennas

The planar bulbous antenna family includes a wide variety of antennas well suited for low-cost consumer UWB applications. Virtually any geometric shape imaginable may be used as a planar element. Section 6.2.2.1 discusses "pointy-feed" planar antennas, like the bow tie, bishop's hat, and hexagonal antennas. "Blunt feed" planar antennas date back to the square plate elements first used in Marconi's original radio (see Figure 1.7). Rounded planar elements are a more recent development, however. This section will discuss semicircular antennas, circular antennas, elliptical antennas, and a variety of other planar bulbous antennas.

6.2.4.1 Semicircular Antennas

Kraus described a "dish" dipole in his 1988 antenna textbook [27]. Figure 6.16(a) shows a dish dipole. This antenna comprises two bowls or hemispheres driven against each other in a dipole configuration. Lalezari, Gilbert, and Rogers introduced one of the earliest semicircular planar antennas (see Figure 1.23(a)) [28]. In the low frequency, long wavelength limit, their broadband notch antenna behaves as a dipole. In the high frequency, short wavelength limit, the curved slot behaves as a dual horn antenna with lobes of radiation aimed along the opposing axes of the slot.

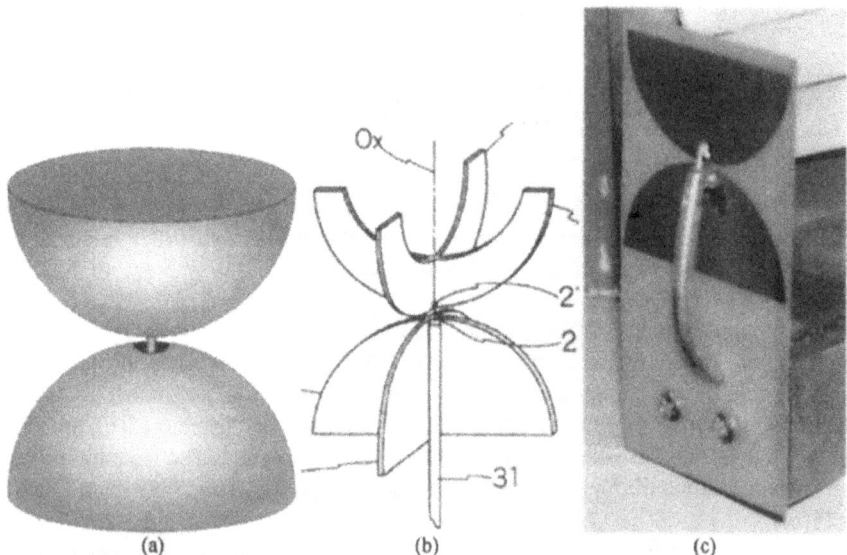

Figure 6.16   (a) A dish dipole (After [27]), (b) semicircular dipole antenna of Ihara et al [29], and (c) semicircular dipole with coaxial feed line coupled to UWB RF device (Courtesy: Centre for Wireless Communications, Oulu, Finland).

According to Hertz's principle, currents concentrate near the edges of a planar element. Taisuke Ihara, Koichi Tsunekawa, and Makoto Kijima took advantage of this fact in designing a broadband antenna using a semicircular radiator [29]. This antenna uses semicircular arc elements and achieves multioctave performance with a VSWR better than 2:1. Figure 6.16(b) shows the semicircular antenna of Ihara and his collaborators, which illustrates two important design principles. First, a surface-of-revolution element may be approximated by a multiple-plate structure. Second, a plate element may be replaced by an outline. Here, the semicircular arc outlines stand in for semicircular plates. This design option follows directly from Hertz's principle. Thus, an outline of a planar element has performance comparable to the corresponding element. Similarly, a multiple orthogonal plate antenna has performance comparable to the corresponding surface of revolution. Ihara et al.'s semicircular antenna, shown in Figure 6.16(b), offers performance comparable to that of the dish dipole of Figure 6.16(a).

Semicircular elements behave almost as well as the circular elements discussed in Section 6.2.4.2. Thus, the principal advantage of semicircular elements is that their truncated form factor delivers adequate performance from a relatively compact size. Semicircular dipoles can yield acceptable matching, yet are limited by the discontinuity in the circular taper element at the corner. These elements tend not to be as well matched as comparable circular or elliptical elements. Finally, there is the difficulty of feeding these elements. A transmission line must be coupled to the feed point between the two elements. Figure 6.16(c) shows a semicircular antenna constructed by Electrobit, Ltd. and used for research purposes by the Centre for Wireless Communications in Oulu, Finland. A coaxial cable soldered across the feed is an adequate coupling mechanism for small volume prototype antennas. Commercial antennas require less expensive coupling mechanisms such as printed circuit board transmission lines.

In the early 1990s, the Time Domain Corporation, in conjunction with Sarnoff Labs, undertook a study of semicircular antennas and constructed a pair of prototype antennas. Figure 6.17 shows these antennas, which were constructed on FR-4 substrate with a parabolic slot taper and a bazooka balun feed. One antenna included foil tape and wire stapling to connect electrically elements on both sides of the substrate. The other antenna did not include this feature. Figure 6.18 shows that these antennas have respectable matching: approximately a 2:1 VSWR from about 1.2GHz to nearly 5.0GHz. The plain dipole has better matching than the stapled dipole. The matching response shows significant variations in matching, probably due to the bazooka balun. Figure 6.19 shows the boresight and edge-on gain for these antennas. For low frequencies, the gain is similar in both orientations. This indicates a more-or-less omnidirectional pattern. For higher frequencies, the edge-on gain is 6dB or more greater than the boresight gain. This shows that the antenna has transitioned from a dipole behavior to a dual-horn behavior.

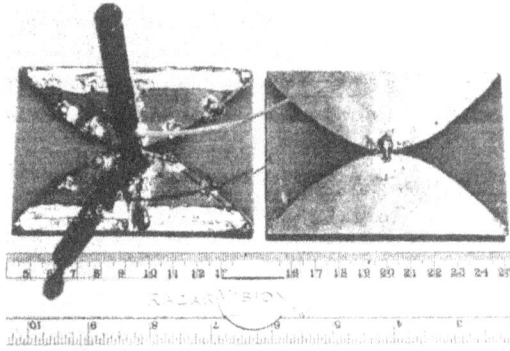

Figure 6.17 A pair of semicircular dipoles with a parabolic taper (Courtesy: Time Domain Corporation, Huntsville, Alabama).

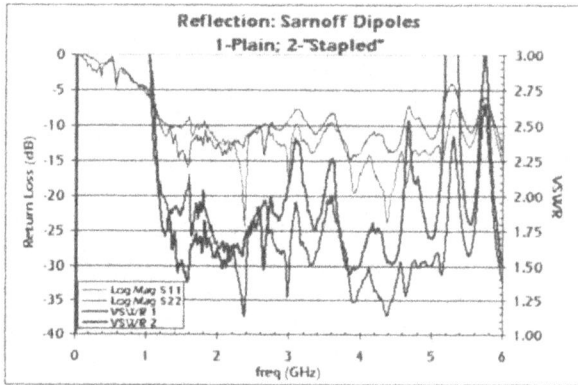

Figure 6.18 Reflection and matching of a pair of semicircular dipoles with a parabolic taper (Courtesy: Time Domain Corporation, Huntsville, Alabama).

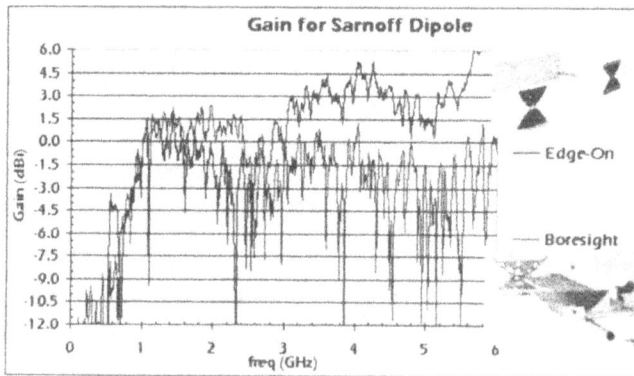

Figure 6.19 Boresight and edge-on gain of a pair of semicircular dipoles with a parabolic taper (Courtesy: Time Domain Corporation, Huntsville, AL).

There have been a great many variations on the semicircular element idea. P. V. Anob, K. P. Ray, and Girish Kumar constructed a hybrid of Lamberty's square plate monopole and a semicircular monopole, shown in Figure 6.20(a) [30]. Anob et al. observed that the semicircular base of their proposed element not only exhibited a wide impedance bandwidth but also showed an omnidirectional pattern on azimuth. Seong-Youp Suh, Warren L. Stutzman, and William Davis proposed the semicircular wedge monopole with a triangular termination shown in Figure 6.20(b) [31]. Taeyoung Yang and William Davis of Virginia Polytechnical Institute introduced a semicircular monopole with a CPW feed and a coplanar counterpoise replacing the traditional ground plane [32]. Figure 6.20(c) shows this antenna. This CPW feed mechanism makes it much easier to connect the antenna to a low-cost consumer-electronic device. All these elements offer excellent matching and nondispersive omnidirectional UWB performance.

Semicircular elements have also been the foundation of attempts to create electrically small UWB antennas for commercial applications. Do-Hoon Kwon, Yongjin Kim, Minoru Hasegawa, and Takao Shimamori proposed the small, dielectrically loaded, semicircular monopole with a coplanar counterpoise shown in Figure 6.20(d) [33]. This 10mm × 5mm × 1mm antenna exhibits a VSWR < 2.3:1 across the 3.1–10.6GHz band but has gain as low as –9dBi in band.

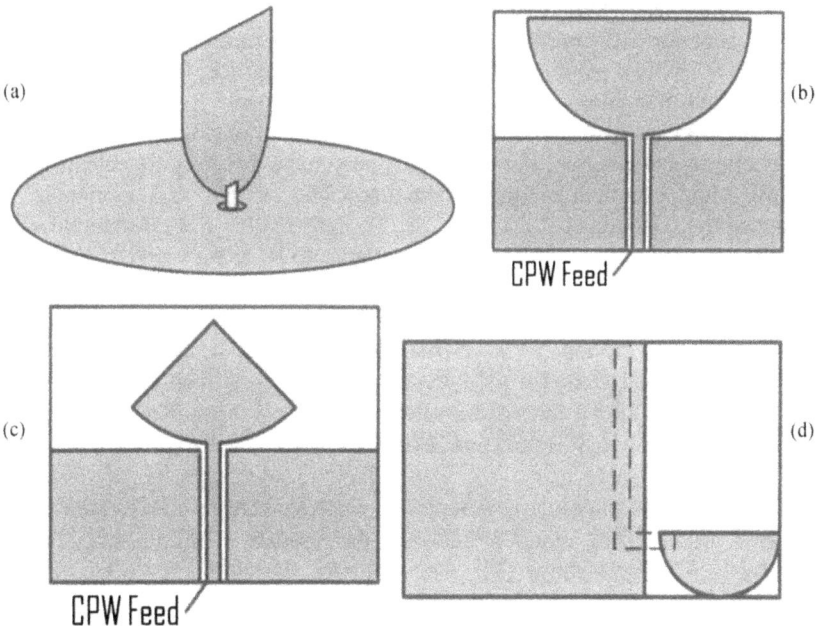

Figure 6.20 (a) Circular tapered square plate monopole (After [30]), (b) planar semicircular monopole with CPW feed (After [31]), (c) semicircular wedge monopole with CPW feed (After [32]), and (d) semicircular monopole with strip line feed (After [33]).

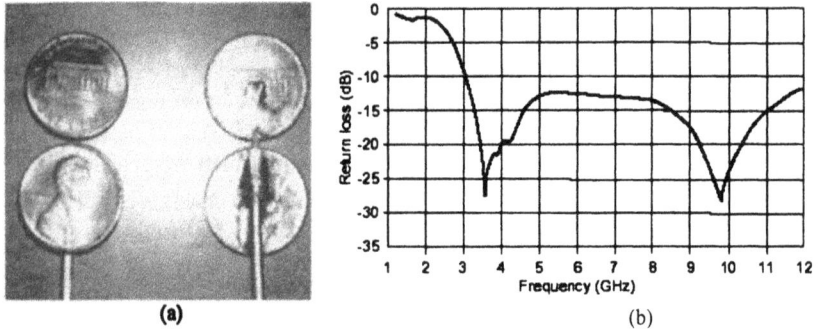

Figure 6.21 (a) A two penny dipole (Courtesy Next-RF, Inc., Huntsville, Alabama), and (b) return loss of a two penny dipole (Courtesy: Time Derivative, Inc., Coral Springs, Florida).

6.2.4.2 Circular and Elliptical Antennas

Mike Thomas and Ronald I. Wolfson pioneered the circular planar dipole in the early 1990s [34]. This simple-to-implement antenna has remarkably good performance. Coincidentally, a U.S. one-cent piece (penny) is ideally sized to yield a −10dB bandwidth from 3.1–10.6GHz. Thus, a two cent price point is the target to beat for a commercial UWB antenna (feed line not included!). Figure 6.21(a) shows a "two penny dipole" constructed by Next-RF, Inc. [35], and Figure 6.21(b) shows matching data extrapolated from a lower frequency implementation. Engineers at Artimi, Ltd. have also investigated this design [19,20].

As one might imagine, if circular elements make good dipole antennas, they will also yield excellent monopole antennas. The circular disk monopole was introduced by S. Honda et al. in 1992 [36]. Soon thereafter, P. P. Hammoud and F. Colomel investigated matching these antennas [37]. Circular disk monopoles remain a high-profile UWB antenna candidate [38]. Figure 6.22(a) shows a disk monopole.

Elliptical disk monopoles were first investigated by N. P. Agrawall et al. in 1998 [39]. Elliptical elements offer even better matching than circular elements due to their more oblong form factor and more gradual taper. These elements are also a high profile UWB antenna candidate [40]. Figure 6.22(b) shows an elliptical disk monopole.

One recent and particularly noteworthy improvement in disk elements is the multiband multiple-ring monopole antenna proposed by C. T. P. Song, Peter S. Hall, and H. Ghafouri-Shiraz [41]. This antenna, shown in Figure 6.22(c), uses nested rings to extend the bandwidth of an elliptical disk beyond the 3:1 bandwidth associated with a simple elliptical structure. If the rings are sized and spaced properly, as each outer ring reaches its high-frequency limit, the corresponding inner ring takes over, maintaining a more-or-less uniform pattern and behavior across as many 3:1 bands as there are rings.

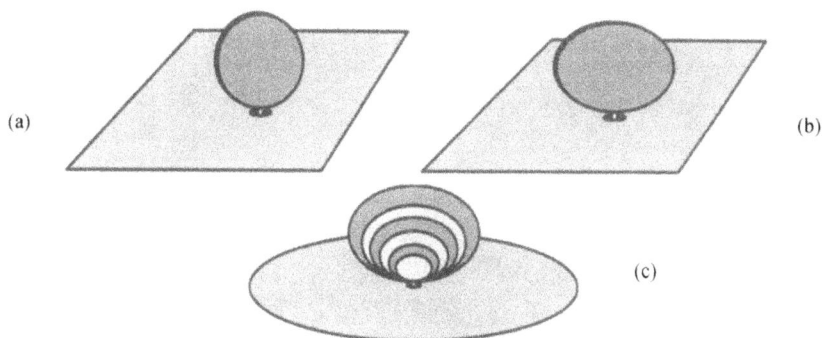

Figure 6.22 (a) The circular disk monopole of Honda et al. (After [36]), (b) Agrawall et al.'s elliptical disk monopole (After [39]), and (c) Song et al.'s multiband multiple ring monopole (After [41]).

I undertook a systematic investigation of circular and elliptical planar dipoles [42]. Figure 6.23 shows a test matrix of antennas implemented on 60mil Rogers RO-4003 substrate. The axial ratio of these antennas varied from 1:1 (circular elements), to a modest eccentricity (1.25:1), to more eccentricity (1.50:1), and finally to highly eccentric elliptical elements (1.75:1). A sub-miniature type A (SMA) connector coupled each antenna to a transmission line in the manner shown in Figure 3.12.

Figure 6.24 shows the return loss, or S_{11}, for the four planar elliptical antennas with varying axial ratio or eccentricity. All four antennas have a return loss -10dB or better for minor axis $l_\lambda \geq 0.20\lambda$. This is somewhat smaller than the 0.25λ usually thought necessary for efficient radiation. In fact, the half power point for these antennas falls at about $l_\lambda \approx 0.14\lambda$. This lies close to McLean's limit for an antenna with $Q \approx 1$. Match improves with increasing eccentricity. The 1.00:1 elliptical (circular) elements yield an S_{11} of around -12dB; for the 1.25:1 elliptical elements, the S_{11} is about -15dB; the 1.50:1 elliptical element's reflection is about -20dB; and the 1.75:1 elliptical elements exhibit an S_{11} around -30dB. These results are further supported by finite difference time domain (FDTD) simulation [43].

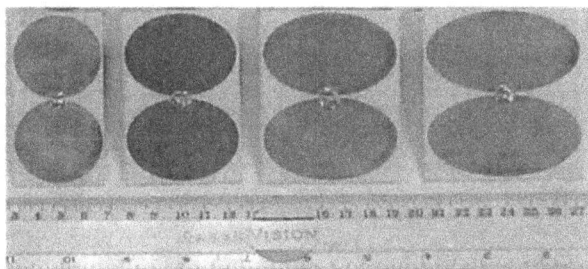

Figure 6.23 Matched pairs of elliptical dipoles with varying axial ratios ([42]; © 2002 IEEE).

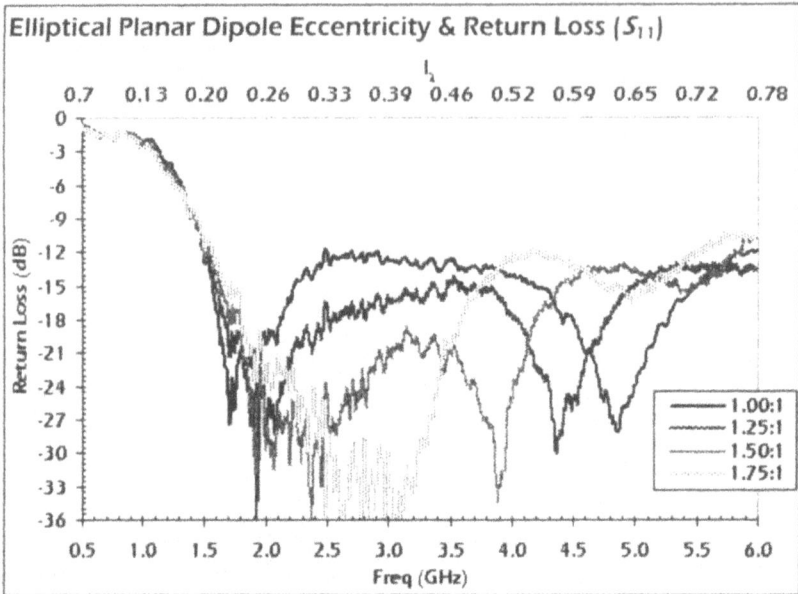

Figure 6.24 Return loss for elliptical dipoles with varying axial ratios plotted against frequency and against minor axis length l_λ in units of wavelength. The more eccentric the elements, the better matched the antenna ([42]; © 2002 IEEE).

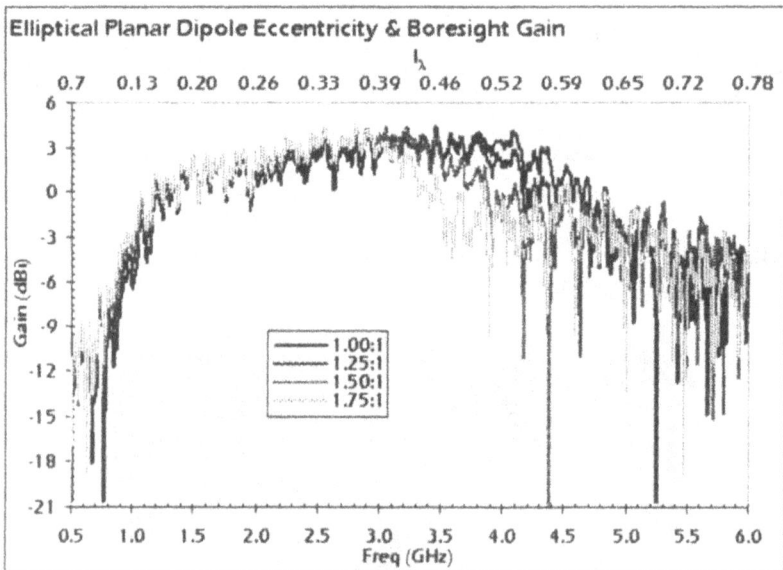

Figure 6.25 Boresight gain for elliptical dipoles with varying axial ratios plotted against frequency and against minor axis length l_λ in units of wavelength ([42]; © 2002 IEEE).

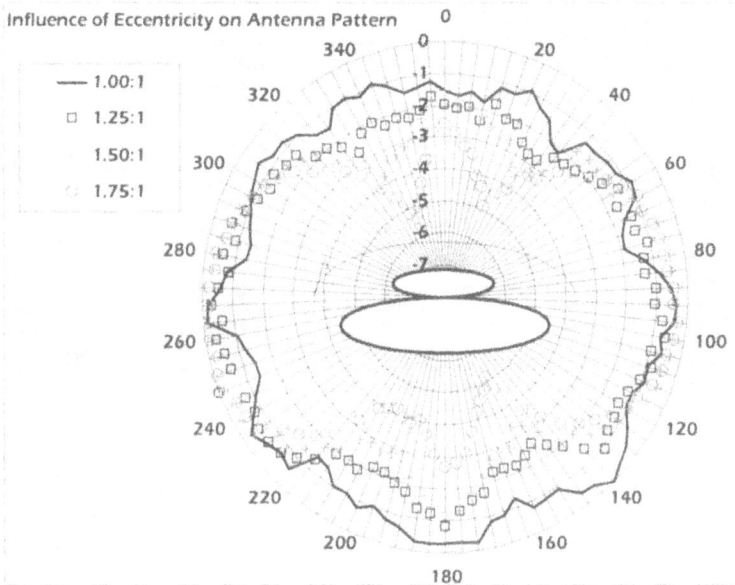

Figure 6.26 Azimuthal (*H*-plane) peak power patterns for elliptical dipoles with varying axial ratios. The more eccentric the elements, the less omnidirectional the pattern (After [42]).

In their lowest-order mode, these antennas behave like dipoles. This dipole behavior dominates for a roughly 1:3 span in frequency, making these antennas well suited for application to the 3.1–10.6GHz regime. Figure 6.25 shows boresight gain. Boresight gain is nominally around 2.0dBi for these antennas, as would be expected for dipoles. Figure 6.26 shows the azimuthal (H-plane) peak power patterns for these antennas. The more eccentric the elements, the less omnidirectional the pattern. Elements with axial ratios of 1.00:1 and 1.25:1 have less than 3dB variation in the azimuthal plane. Elements with axial ratios 1.50:1 and 1.75:1 have more than 3dB variation.

These peak power patterns are collected by using a Picosecond Pulse Labs 4050B pulser to radiate impulses from the antenna under test [44]. A Farr Research Model TEM-2-50 horn antenna collects the signals [45]. A digitizing oscilloscope collects the waveforms, and a computer records the peak voltage received as the antenna under test rotates through 360°. A peak power pattern is an excellent way to acquire pattern information from an antenna whose pattern varies little across the bandwidth of the excitation.

Elliptical dipole selection involves a trade-off between matching and pattern. For applications where better matching is more important than more uniform pattern, eccentric elements are preferred. For applications where more uniform pattern is the priority, less eccentric elements get the nod. For most applications, the optimum lies somewhere between 1.25:1 and 1.50:1.

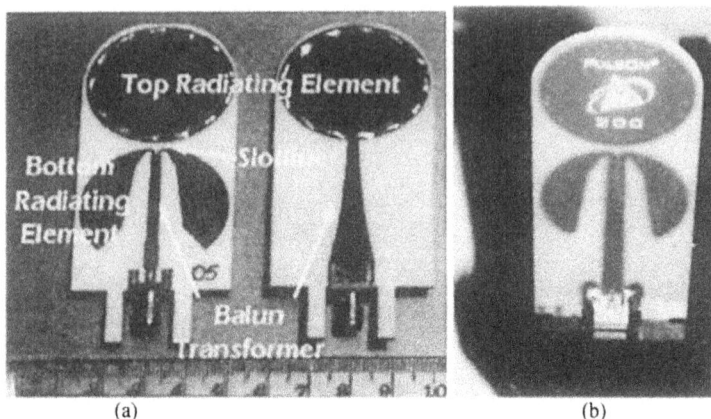

Figure 6.27   (a) Bottom-fed planar elliptical dipole antenna, or "catfish" dipole ([46]; © 2003 IEEE). The figure shows the "front" of the antenna to the left and the "back" of the antenna to the right. Both sides of the top radiating element in this prototype are electrically connected by "vias" stitched with thin wire. (b) Production version optimized for 3–10GHz performance (Courtesy: Time Domain Corporation, Huntsville, Alabama).

The disadvantage of these planar elliptical elements is that they are center fed. A transmission line must couple to the gap between the elements. Necessarily this transmission line lies in the heart of the reactive fields surrounding the feed and is thus particularly vulnerable to undesired sheath coupling. This coupling is liable to distort the antenna pattern directly due to blockage and indirectly due to undesired cable currents.

A solution to this problem is to incorporate a PCB transmission-line feed into the antenna structure itself. After experimenting with the CPW and strip line transmission lines shown in Figure 3.11, I settled on a strip line transmission line. Although CPW feeding yielded acceptable results, strip line feeding resulted in more uniform patterns and better matching during testing of prototypes. The result was the bottom-fed planar elliptical dipole of Figure 6.27(a) [46, 47, 48]. Being a bottom-feeder, this antenna was originally dubbed a "catfish dipole."

Figure 6.27(a) depicts a prototype bottom-fed UWB planar dipole antenna designed to radiate and receive UWB impulses with frequency content from about 2.5GHz to about 6GHz. Figure 6.27(b) shows a production version of this same antenna. A coaxial feed line delivers a UWB impulse to a balun transformer at the base of the antenna. The tapered balun transformer allows the inherently unbalanced coaxial line (+ signal and ground) to be connected to an inherently balanced dipole structure (+ signal and – signal). This helps avoid spurious currents on the sheath of the coaxial feed line that could cause distortions in the antenna pattern and undesired variations in overall system performance. The balun transformer divides the bottom radiating element into two sections as it connects to a slot line between the top and bottom radiating elements. The balun transformer allows the 50Ω impedance of the coaxial feedline to be matched precisely to the somewhat higher impedance of the slot line. This slot line guides energy between the top and bottom radiating elements, decoupling it cleanly from the antenna so that it can radiate away with minimal reflection.

The process works in reverse when the antenna is receiving. The radiating elements collect energy, and the slot line guides it into the balun transformer. The balun transformer couples this energy to the coaxial line. The coaxial line then guides the energy into the RF front end, where the energy is made available to the receiver. Ideally, antennas like this could be integrated onto the same board as a complete UWB radio.

The bottom-fed planar elliptical dipole antenna offers an excellent match to 50Ω. The VSWR is 1.5:1 or better from just below 3.0GHz to nearly 5.5GHz. Reflections from the antenna are down about −14dB across this same band; thus, the antenna accepts 96% of the applied power over these frequencies. Figure 6.28 shows this matching as a function of frequency from 0–7GHz.

The prototype antenna has a nominal gain of about +3dBi in the direction normal to either face of the antenna. Edge-on gain is on the order of about 0dBi. These gain numbers are representative of antenna response from 2.5–6.0GHz. Gain was determined from the through response (S_{12}) of a matched pair of antennas using a 10MHz–20GHz Rohde and Schwartz Vector Network Analyzer Model ZVM. Figure 6.29 shows gain for a variety of orientations from 0–7GHz. Note that even when the nulls of the antenna were aligned, a gain of about −6dBi was still obtained. This is likely due to indirect propagation paths between the antennas under test and represents a worst-case scenario.

The bottom-fed planar elliptical dipole antenna has a dipole like pattern: omni (to within 3dB). Peak gain for the antenna is about +3dBi perpendicular to the face of the antenna, front and back. Gain is on the order of +0dBi edge-on. The antenna has the usual dipole nulls along the axis of the antenna top and bottom. Thus, this antenna provides good coverage in the plane normal to the axis of the antenna but may have difficulty achieving optimal range directly above or below the antenna. In most cases, an indirect path is likely to exist, however. In typical operation, the antenna is oriented with the long axis positioned vertically. This places the dipole nulls along the vertical axis and provides the best response in the equatorial, azimuthal, or horizontal plane.

Ansoft HFSS (high frequency structure simulator) software was used to model the prototype antenna and calculate its radiation pattern. Figure 6.30 provides a typical result. This view shows the antenna edge-on with the front face of the antenna oriented to the right, and the back face of the antenna oriented to the left.

Bottom-fed planar elliptical dipoles are well matched and radiation efficient. They are omnidirectional and are thus well suited for ad hoc networks with arbitrary azimuthal orientations. Furthermore, these antennas are electrically small and inexpensive without compromising performance. Thus, bottom-fed planar elliptical dipoles are well suited for commercial applications.

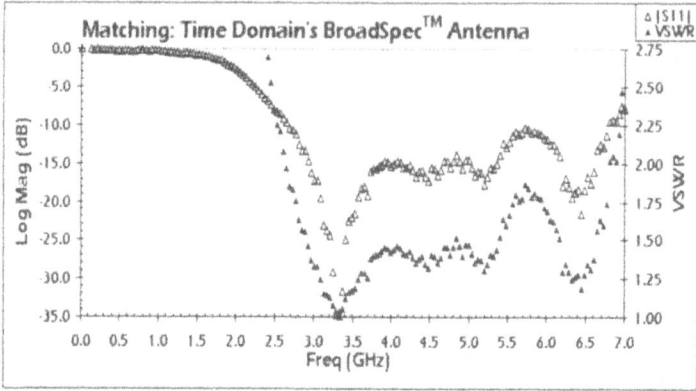

Figure 6.28 Matching for a bottom-fed planar elliptical dipole ([46]; © 2003 IEEE).

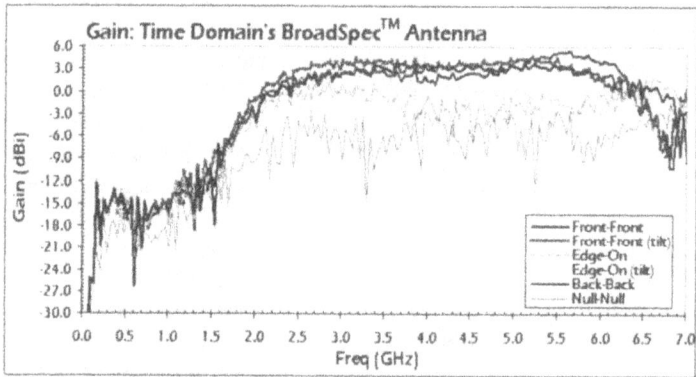

Figure 6.29 Gain for a bottom-fed planar elliptical dipole ([46]; © 2003 IEEE).

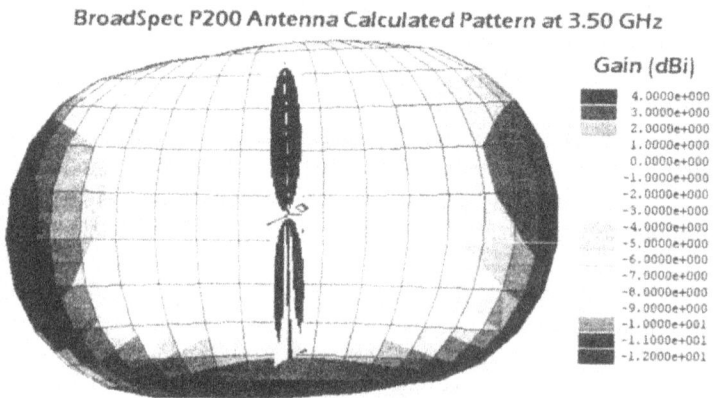

Figure 6.30 Typical pattern for a bottom-fed planar elliptical dipole ([46]; © 2003 IEEE).

6.2.4.3 A Panoply of Planar Bulbous Antennas

Semicircles, circles, and ellipses hardly exhaust the possibilities for planar bulbous antennas. The "fat" planar elliptical dipole of Figure 6.31(a) has major axes aligned so as to be substantially parallel. This family of antennas combines good all-around performance with a relatively compact form factor. The particular fat planar elliptical dipole shown in Figure 6.31(a) has a –3-dB gain bandwidth from 1.6GHz to 5.1GHz and a Q of 0.82. This antenna fits within a boundary sphere $R_{\lambda L} = 0.14$ in units of wavelength at the low-frequency end of the operating band. This is a good performance benchmark; however, there are many alternate planar bulbous designs that offer comparable performance. Figure 6.32 shows the boresight gain of the fat planar elliptical antenna and selected other planar bulbous designs.

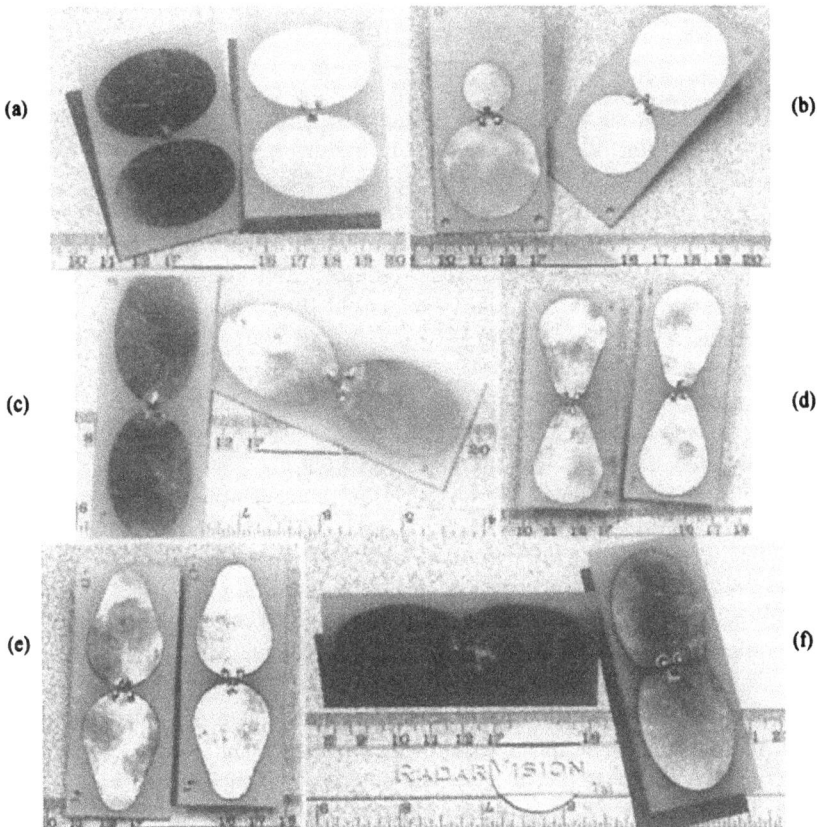

Figure 6.31 (a) "Fat" planar elliptical dipoles, (b) asymmetric planar circle dipoles, (c) "skinny" planar elliptical dipoles, (d) "Lilliputian" planar ovoidal dipoles, (e) "Blefuscudian" planar ovoidal dipoles, and (f) truncated skinny planar elliptical dipoles (Courtesy: Time Domain Corporation, Huntsville, Alabama).

Gain Versus Frequency: Planar Bulbous Dipole Antennas

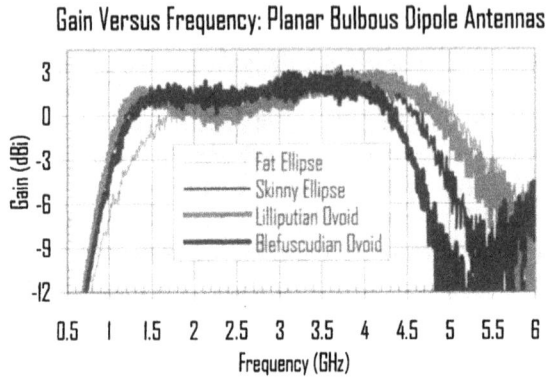

Figure 6.32 Boresight gain for selected planar bulbous antennas (Courtesy: Time Domain Corporation, Huntsville, Alabama).

Figure 6.31(b) shows a pair of asymmetric circle dipoles. Varying element sizes in this fashion allows the range of frequency operation to be extended beyond the 3:1 range typical for symmetric planar bulbous antennas.

Figure 6.31(c) presents a pair of "skinny" elliptical dipoles. In contrast to the fat elliptical dipoles of Figure 6.31(a), these skinny elliptical dipoles are aligned so that their major axes are substantially colinear. This antenna has a −3-dB gain bandwidth from 1.2GHz to 4.5GHz and a Q of 0.70. This antenna fits within a boundary sphere $R_{\lambda L} = 0.16$ in units of wavelength at the low-frequency end of the operating band.

The planar ovoidal dipoles of Figure 6.30(d) are fed from the small ends of the ovals. I dubbed these "Lilliputian ovoids" after the little people from Jonathan Swift's *Gulliver's Travels* who ate their eggs from the small end. The Lilliputian ovoid antenna of Figure 6.31(d) has a −3-dB gain bandwidth from 1.1GHz to 4.9GHz and a Q of 0.61. This antenna fits within a boundary sphere $R_{\lambda L} = 0.14$ in units of wavelength at the low-frequency end of the operating band.

Figure 6.31(e) displays a pair of planar ovoidal dipoles fed from the large ends of the ovals. I dubbed these "Blefuscudian ovoids" after the hereditary rivals of the Lilliputians who ate their eggs from the large end. The Blefuscudian ovoid antenna of Figure 6.30(e) has a −3-dB gain bandwidth from 1.3GHz to 4.4GHz and a Q of 0.77. This antenna fits within a boundary sphere $R_{\lambda L} = 0.16$ in units of wavelength at the low frequency end of the operating band. Seong-Yup Suh and Warren Stutzman investigated a similar antenna, which they dubbed the "low profile dipole planar inverted cone" antenna [49, 50]. Figure 6.33(a) shows a sketch of Suh's and Stutzman's design. Planar bulbous elements may also be employed as monopole elements in conjunction with a coplanar counterpoise. Figure 6.33(b) shows such a design invented by Ernst Zollinger [51].

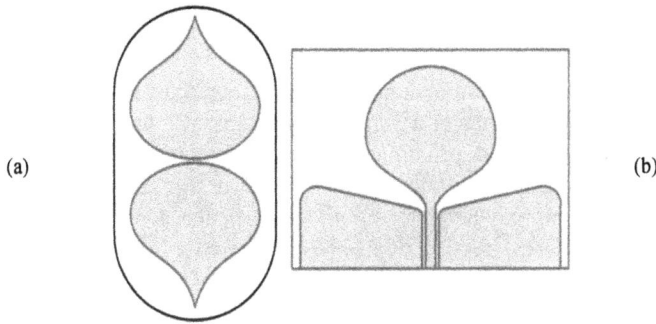

Figure 6.33 (a) Suh's and Stutzman's "low profile dipole planar inverted cone" antenna (After [49]), and (b) Zollinger's planar broadband antenna (After Ref. 51).

The skinny elliptical dipoles of Figure 6.31(f) have elements truncated at their foci. Thus, the feed point couples to a constant width slot line reminiscent of a diamond dipole. This design achieves a more compact size than the standard skinny planar elliptical dipoles of Figure 6.31(c) at the cost of increased return loss, worse matching and narrower bandwidth.

Table 6.1 summarizes the performance of select planar bulbous antennas. The 3.1–10.6GHz commercial UWB antenna calls for an antenna with a 3.42:1 frequency span, or in other words, a Q of about 0.76. The antennas summarized in Table 6.1 are all appropriate for use in the commercial UWB band. The fat elliptical dipole and the Lilliputian ovoid dipole are particularly compact choices. The Lilliputian ovoid dipole exhibits about a 3dB variation in gain across the operational band. The fat elliptical dipole, on the other hand, tends to have a more uniform gain response in band. Thus, for most applications, fat elliptical dipoles (described at length in Section 6.2.4.2) are preferred.

Table 6.1
Performance of Selected Planar Bulbous Dipoles

	Fat Ellipse	Skinny Ellipse	Lilliputian Ovoid	Blefuscudian Ovoid
f_L (GHz)	1.6	1.2	1.1	1.3
f_C (GHz)	2.86	2.32	2.32	2.39
f_H (GHz)	5.1	4.5	4.9	4.4
BW (GHz)	3.5	3.3	3.8	3.1
$Q = f_C/BW$	0.82	0.70	0.61	0.77
R (cm)	2.6	4.0	3.7	3.7
λ_L (cm)	18.7	25.0	27.3	23.1
$R_{\lambda L} = R/\lambda_L$	0.14	0.16	0.14	0.16

6.2.5 General Principles of Small-Element Design

There are many more possible small-element antenna designs than could be described within the confines of a single book. Thus, it is important to step back and consider some general principles of small-element design. These general principals allow a designer to take a known good design and derive a family of analogous designs that may be better suited for particular applications.

First, the planar cross-section of a good surface-of-revolution antenna yields a good planar antenna. Naturally, this principle works in reverse: a surface of revolution derived from a good planar antenna yields a good surface-of-revolution antenna. Surface-of-revolution antennas will tend to have more uniform pattern and better matching than comparable planar elements. Planar antennas may have somewhat compromised performance but are much easier to manufacture. Instead of the lathing, casting, stamping, or other precision machining required for a surface-of-revolution antenna, a planar antenna may be implemented using relatively inexpensive PCB techniques. This gives planar antennas a significant advantage over surface-of-revolution antennas for commercial applications. The planar elements may lie on the same or opposite sides of a PCB substrate. Alternatively, a planar element may occupy both sides of a substrate. In this case, using through-hole vias to electrically connect the two portions of an element on opposite sides of the substrate may be a good idea. Figure 6.34 shows a variety of surface-of-revolution antennas and their corresponding planar counterparts.

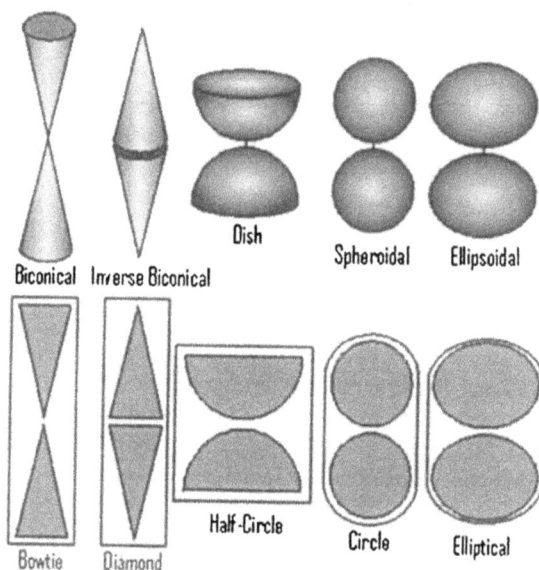

Figure 6.34 Planar cross-sections of good surface-of-revolution antennas make good UWB antennas. Conversely, a surface of revolution of a good planar antenna yields a good UWB antenna.

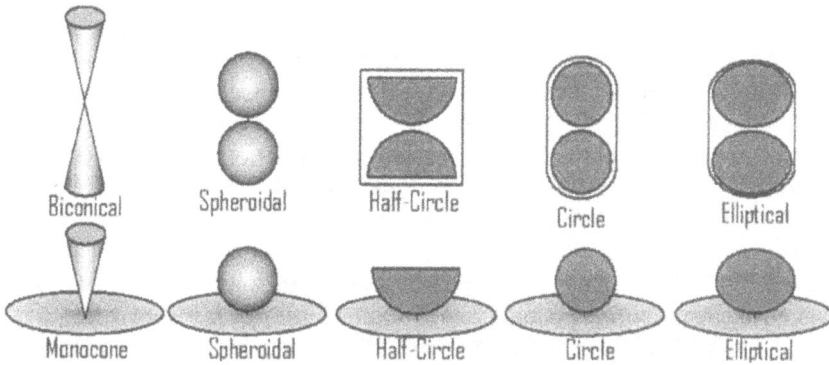

Figure 6.35 A good dipole element makes a good monopole element and vice versa. Unlike narrowband antennas, impedance may be adjusted by sizing the feed gap, so dipole and monopole versions of the same element may have comparable feed impedances.

Second, a good monopole element yields a good dipole element and vice versa. In the narrowband case, a dipole has twice the feed impedance of the corresponding monopole. In the UWB limit, a properly designed monopole antenna can have the same impedance as the corresponding dipole. This is because antenna impedance depends not so much on element geometry as on the feed gap and taper of the slot line between the elements of a dipole or between a monopole element and a ground plane. By adjusting the feed gap, one may vary antenna impedance over a significant range so as to achieve a desired match. Figure 6.35 shows a few dipole antennas and their corresponding monopole antennas.

Third, a surface-of-revolution antenna may be approximated by a wire, plate, or other grid-type structure. Such antenna implementations are an intermediate step between a full surface-of-revolution antenna and a planar cross-section. They may offer performance superior to a planar cross-section antenna yet be easier to construct than a surface-of-revolution antenna. For instance, an eccentric planar elliptical antenna will have good matching and poor omnidirectionality. A multiple-parallel-plate implementation will combine good matching and good omnidirectionality. In some cases, construction of a true surface of revolution is not practical for other reasons. For instance, Marconi's wire monocone antenna would have been cost prohibitive if implemented as a solid surface. In addition, wind loading of a solid surface may drive one toward a wire or grid-type structure.

Parallel wires form a low pass filter with the half-power point where wire spacing is about $\lambda/2\pi$. Thus, more closely spaced wires begin to approximate a continuous conducting sheet, provided wires are aligned in the direction of the desired current flow. Chicken wire or hardware cloth are good, inexpensive materials for large-scale UWB antennas in the VHF or low UHF regime. Fine mesh is required for use at microwave frequencies. Conventional meshes of interwoven wires may not have adequate connections at microwave frequencies. The preferred option is expanded metal mesh. This is mesh cut from a single continuous sheet of conductive foil. Because the strands that form the mesh are from the same continuous conductor, the electrical connections throughout the mesh are superior to those of woven mesh. Figure 6.36(a) shows a few examples of wire and plate approximations to surface-of-revolution antennas. Figure 6.36(b) shows a planar elliptical dipole constructed by Glenn Wolenec using quarter-inch-spaced hardware cloth, attached to a cardboard substrate using duct tape. This antenna fits within a 24in × 36in picture frame and works as a 100–300MHz scanner antenna. Another example is the teardrop biconical of Figure 6.2(b).

Fourth, a planar antenna may be replaced by an outline. This is a consequence of Hertz's principle that currents tend to concentrate on the edges of a planar antenna. Ihara's semicircular arc antenna exploits this principle, but the same concept may be applied to any planar antenna. Enough of the edge must be retained so as to provide a broad path for current. Computational antenna analysis showing the current density on a planar element is useful for showing exactly how wide a strip and which edges should be retained. Figure 6.37 shows a variety of planar outline antennas.

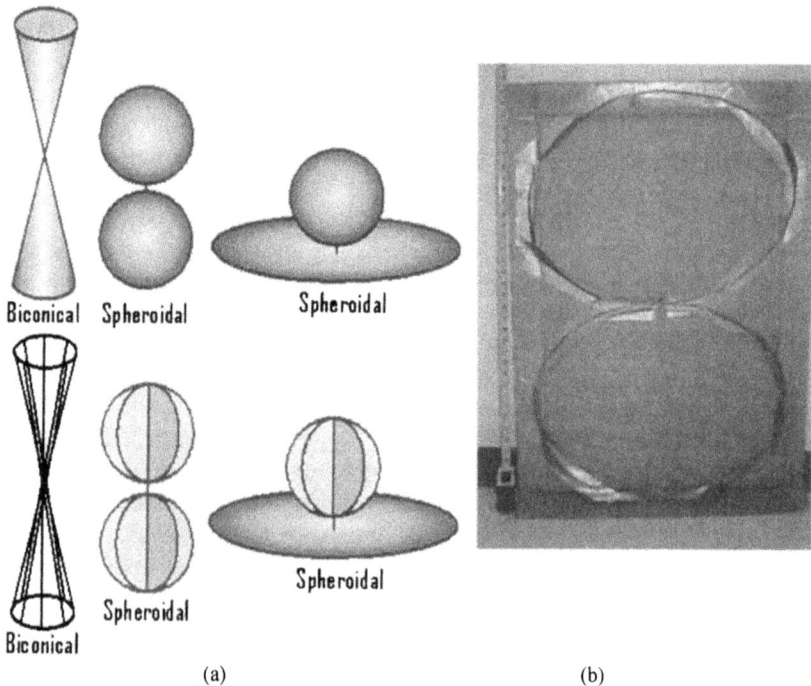

Biconical   Spheroidal   Spheroidal

Spheroidal

Spheroidal

Biconical

(a)   (b)

Figure 6.36 (a) Wire and thin plate approximations to solid or surface-of-revolution antennas make good UWB antennas. (b) A planar elliptical dipole 100–300 MHz scanner antenna is implemented in hardware cloth. A tape measure in inches is to the left of the antenna for scale (Courtesy: Glenn Wolenec).

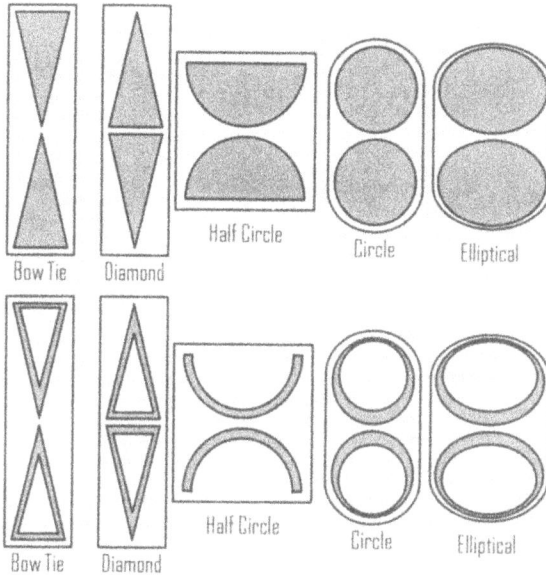

Figure 6.37 A continuous conducting planar element may be replaced with an outline.

Fifth, the ground plane of a monopole may be replaced with a coplanar counterpoise. The coplanar counterpoise may lie on the same side of a substrate as the monopole element in which case a CPW is a convenient feed mechanism. Alternatively, the coplanar counterpoise may lie on the opposite side from the element, in which case a microstrip feed is convenient. Either the monopole element or the counterpoise may comprise metallization on both sides of the substrate. In this case, a CPW with ground plane might be useful. A coplanar counterpoise is particularly valuable in the context of a PCMCIA card or other context in which a coplanar monopole antenna must stick out from a ground plane. Figure 6.38 shows a variety of coplanar counterpoises.

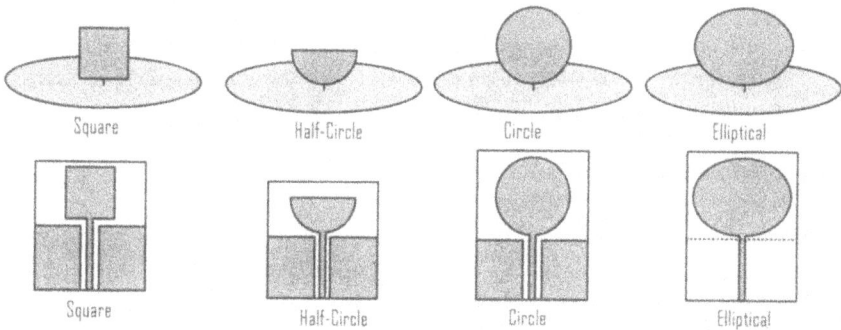

Figure 6.38 Monopole ground planes may be replaced by a coplanar counterpoise.

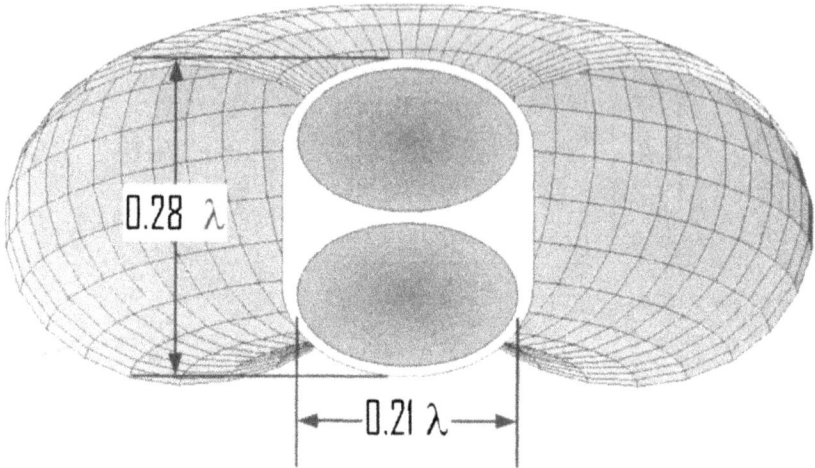

Figure 6.39   The archetypical small element electric antenna is a planar elliptical dipole ([13]; © 2004, CMP Media, LLC).

In summary, five basic principles are particularly useful to designers of small-element antennas:

- **Planar-solid correspondence**: For every good surface-of-revolution antenna, there is a corresponding good planar cross-section antenna.
- **Monopole-dipole correspondence**: With proper antenna design, a UWB dipole and a UWB monopole can both be designed to exhibit the same impedance match.
- **Meshed surface equivalence**: In many cases, these wire, mesh, or grid structures have performance comparable to the corresponding surface-of-revolution antenna but with less material, easier construction, and less wind loading.
- **Outline design equivalence**: In many cases, an outline of a planar antenna element has performance comparable to the original planar element.
- **Coplanar counterpoise ground equivalence**: A coplanar counterpoise is ideal for PCB antenna fabrication and is often preferred to a standard monopole ground plane.

6.2.6 Summary of Small-Element Electric Antennas

Small-element electric antennas combine relatively compact size with the omnidirectional patterns desired for most consumer-electronic applications. Contrary to the conventional wisdom encapsulated in this chapter's epigraphs, it is possible to have an efficient small-element antenna with adequate bandwidth for a commercial UWB system.

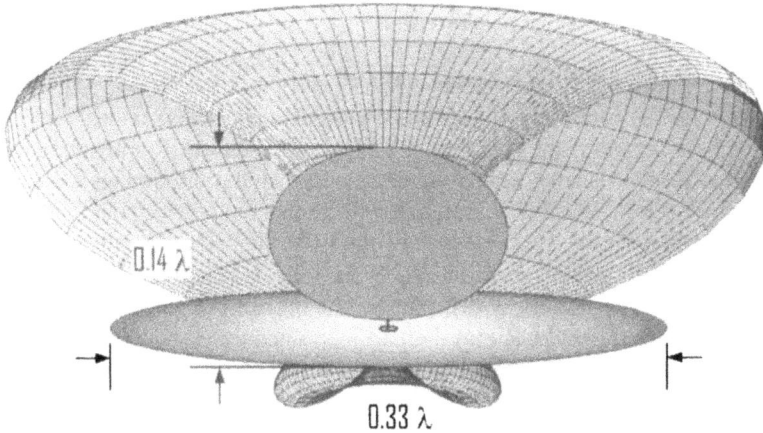

Figure 6.40 A small-element electric monopole antenna ([13]; © 2004, CMP Media, LLC).

Although a variety of alternate designs with comparable performance are available, the archetypal small-element electric antenna is the planar elliptical dipole of Figure 6.39. The elliptical elements of these antennas have a minor axis of about 0.14λ. Thus, a 3.1–10.6GHz version of these antennas may be made about 1.1in × 0.83in (2.8cm × 2.1cm).

Monopole antennas provide an alternate way to attach an antenna to an enclosure. Here again, there are many options, although planar elliptical elements work quite well. Figure 6.40 shows a typical monopole antenna with its pattern. A ground plane at least 0.33λ in diameter is preferred. The enclosure itself often makes a convenient ground plane. A 3.1–10.6GHz version of these antennas may be made about 1.3in × 0.55in (3.3cm × 1.4 cm).

6.3 SMALL-ELEMENT MAGNETIC ANTENNAS

Small-element magnetic antennas are physical realizations of an ideal Hertzian magnetic dipole. These antennas involve one or more current loops. An ideal Hertzian electric dipole provides an omnidirectional vertical polarization pattern, while an ideal Hertzian magnetic dipole yields an omnidirectional horizontal polarization pattern. Magnetic antennas may be thought of as current driven and have predominantly magnetic near fields. Since electric fields tend to couple more strongly to nearby objects, magnetic antennas are better suited for embedded applications. A partial overview of UWB magnetic antennas has been presented elsewhere [52]. There is a variety of families of small-element electric antennas. These families include large current radiators (LCRs), monoloops, loops, and slot antennas. This section will discuss each of these families in turn.

6.3.1 Large Current Radiator Antennas

Harmuth pioneered the LCR antenna in the early 1980s [53, 54, 55]. An LCR antenna is ideally a uniformly distributed current sheet whose return currents are shielded by a ground plane [see Figure 6.41(a)]. Thus, an LCR antenna requires a differential feed. This antenna has some difficulties and shortcomings.

First, devotees of Hertz's Principle will immediately recognize that currents will tend to concentrate on the edges of the LCR element, making it difficult to achieve a uniform current distribution. If the LCR is electrically small, currents will tend to be more uniformly distributed. The problem of current nonuniformity in an electrically larger LCR may be partially ameliorated by dividing the LCR element into a plurality of parallel conducting strips.

The second and principal disadvantage of an LCR is that it tends to be an inefficient antenna. The conducting current sheet necessarily radiates from both sides of the sheet. Half of the energy potentially radiated by an LCR is trapped within the resonant cavity between the LCR element and the ground plane. Thus, an LCR must use ferrite or other lossy coatings to control the resulting narrowband resonances. This makes an LCR a relatively inefficient antenna.

Nevertheless, LCR antennas have seen significant UWB use. Aether Wire and Location, Inc., has used LCR-type antennas in their localizing transceivers [56]. Figure 6.41(b) shows an Aether Wire LCR type antenna and localizing transceiver. Aether Wire exploits the fact that an LCR has a low input impedance, making it suitable to be driven by a complementary metal oxide semiconductor (CMOS) current source. Other investigators have also looked into the properties of these antennas [57].

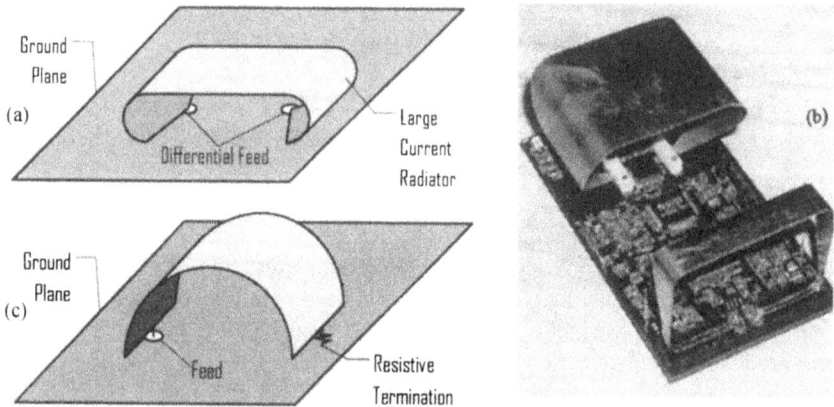

Figure 6.41 (a) A large current radiator or LCR antenna comprises a planar conducting radiating element and a parallel ground or shield plane ([52]; © 2003, IEEE, After [53]). (b) An Aether Wire localizer includes separate LCR type antennas for transmit and receive ([56]; Courtesy: Aether Wire & Location, Inc.). (c) A "balanced-dipole antenna" has a dynamically switchable pattern ([52]; © 2003, IEEE, After [58]).

Everett Farr and his collaborators proposed an interesting variation on the basic architecture of an LCR antenna [58]. Their "balanced-dipole" antenna [shown in Figure 6.41(c)] has the interesting property that it can be fed from either end, while the opposite end is terminated with an appropriate impedance. Thus, the pattern of this antenna may be dynamically switched. The resistive termination also implies that this antenna will be somewhat inefficient.

6.3.2 Monoloop Antennas

The main disadvantage of LCR antennas follows from the fact that they trap energy between their radiating elements and their ground planes. This suggests that it might be fruitful to consider radiating elements oriented in a plane perpendicular to the ground plane. This type of half-loop antenna architecture has been dubbed a "monoloop" [59]. Just as a monopole is a half-dipole driven against a ground plane, a monoloop is a half-loop driven against a ground plane. Similar thin-wire structures have long been used to excite waveguides [60]. Figure 6.42(a) shows such a thin-wire half-current loop. A thin-wire structure like this, however, would not make an efficient broadband radiating antenna. As was the case for electric antennas, "fatter" magnetic elements tend to have wider bandwidths.

One early antenna with the monoloop architecture was Edwin M. Turner and William P. Turner's "scimitar" antenna shown in Figure 6.42(b) [61]. This antenna is characterized by an operating bandwidth in excess of 1:10. Although well matched, the scimitar antenna exhibits some variation in pattern as a function of frequency. Ideally, a UWB antenna should have a stable and consistent pattern as a function of frequency. Similar antennas have been used as broadband field probes [62].

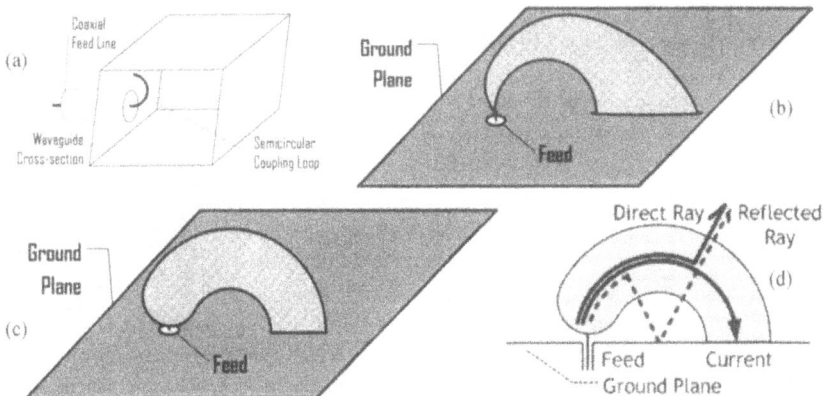

Figure 6.42 (a) A half current loop used to excite a waveguide (After [60]), (b) Turner & Turner's scimitar antenna ([52]; © 2003, IEEE, After [61]), (c) the author's monoloop antenna ([52]; © 2003, IEEE), and (d) a simple current and ray tracing model of a monoloop antenna ([52]; © 2003, IEEE).

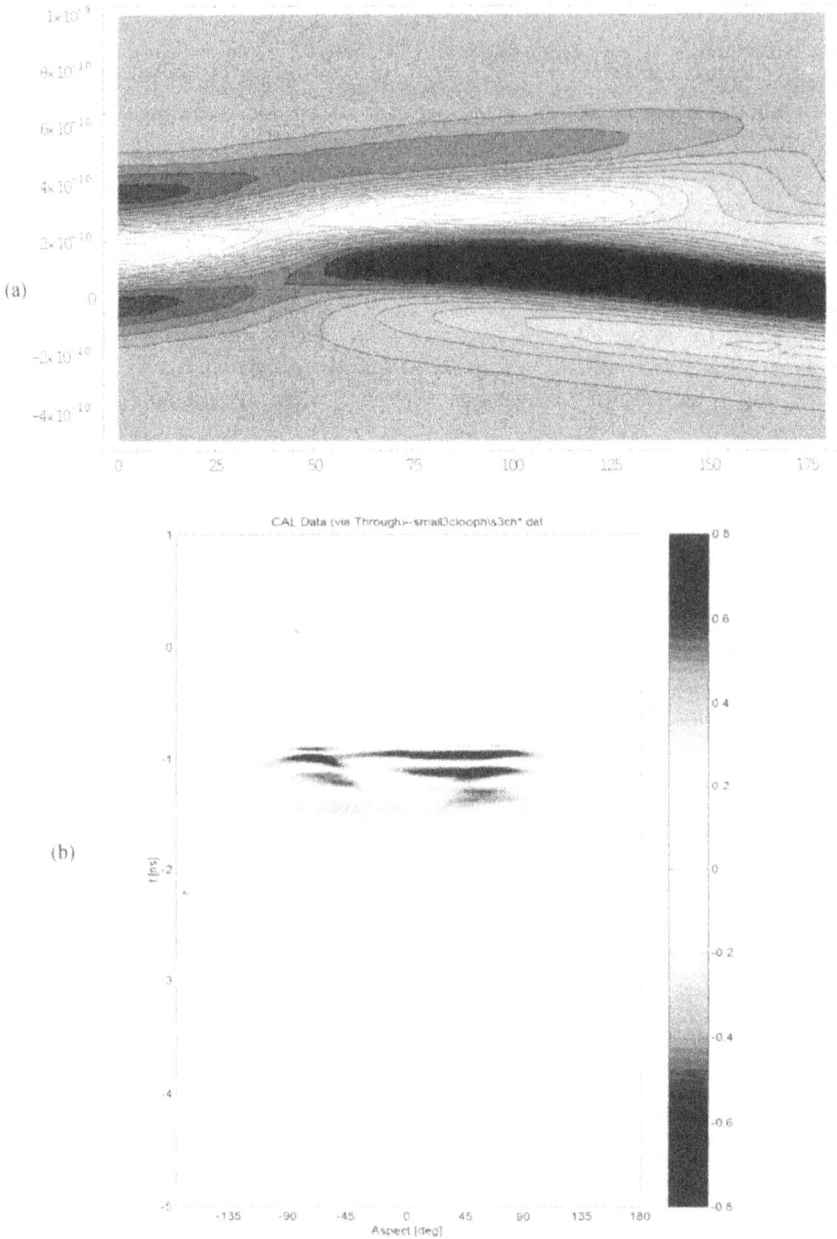

Figure 6.43 (a) A waterfall plot of the time domain radiated waveform from a monoloop antenna. Time lies on the vertical axis, and angle with respect to the ground plane lies on the horizontal axis. (b) The currents and radiation rays from a center-fed monoloop. The direct and reflected rays have a uniform time delay yielding a nondispersive pattern. (Courtesy: Time Domain Corporation, Huntsville, Alabama, and Soumya Nag).

Figure 6.42(c) shows an alternate monoloop [63]. The blunt feed of this antenna tends to offer a lower impedance than that of the scimitar design. Like the scimitar antenna, this monoloop antenna suffers from some variation in pattern as a function of frequency. Also, the monoloop pattern is not uniform in the plane of the monoloop element. The reason for this behavior may be understood by considering the current flow in the monoloop element and the resulting radiation.

Assume each infinitesimal current element along the monoloop is the source of radiation along the radius of curvature at that point. If the monoloop element is considered in cross-section, we see that each infinitesimal current element generates a direct ray of radiation radially outward and a radially inward ray directed toward the ground plane. This radially inward ray ends up being reflected.

The radiation in any given direction is the sum of a direct ray from one part of the monoloop and a reflected ray from a different part of the monoloop element. A quick calculation of the path lengths involved demonstrates an asymmetry of the direct and reflected paths: the relative path difference varies as a function of angle. This variation is the root cause of the nonuniform pattern: a direct impulse waveform combines with a reflected impulse waveform with a relative delay that varies as a function of look angle. Figure 6.42(b) illustrates this behavior.

The model in Figure 6.42(b) may be used to predict the waveform and dispersion as a function of angle. Figure 6.43(a) shows the predicted angle dependence of the radiated waveform in a waterfall contour plot format with time on the vertical axis and angle measured from the ground plane on the horizontal axis. Thus, the normal to the ground plane is centered in the plot at 90°. Figure 6.43(b) shows measured waveform data in a similar format, but presents data for the full 360° (both in front of and behind the ground plane). These waterfall contour plots display time on the vertical axis and angle on the horizontal axis. The shading denotes waveform intensity on a linear scale, with light shading being positive and dark shading being negative. Both plots dramatically illustrate the impact of the direct and reflected signals sweeping past each other in time, yielding a dispersive wave front. Despite these deficiencies, this wide field of view of this antenna makes it well suited for use in a through-wall radar system [64].

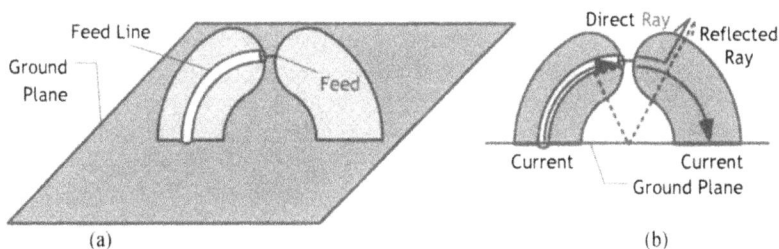

Figure 6.44 (a) A center-fed monoloop antenna, and (b) the currents and radiation rays from a center-fed monoloop. The direct and reflected rays have a uniform time delay yielding a nondispersive pattern ([52]; © 2003, IEEE).

One way to rectify the undesired asymmetry is to feed the monoloop in a symmetric manner. This was the fundamental idea behind the center-fed monoloop [see Figure 6.44(a)]. By moving the feed to the center top of the monoloop antenna, symmetry is restored. The difference between the direct and reflected paths is now constant as the look angle is varied. Thus the pattern and delay spread of a center-fed monoloop is more uniform as a function of look angle. Figure 6.44(b) shows characteristic current and radiation rays for the center-fed monoloop.

6.3.3 Loop Antennas

The principal lesson of Chapter 5 for UWB antenna design was that fatter is better. A thin-wire dipole becomes broadbanded by fattening it up to a biconical or ellipsoidal design. Similarly, a fat loop antenna also exhibits UWB performance.

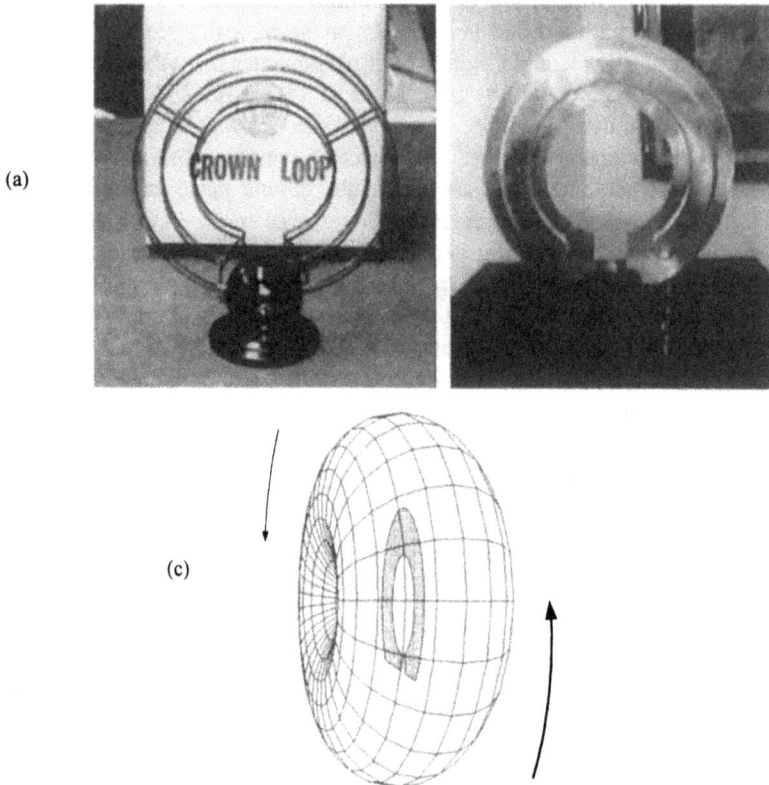

Figure 6.45 (a) A sparse wire approximation to a planar loop antenna, (b) a planar loop antenna, and (c) pattern of an ideal loop antenna.

Figure 6.45(a, b) shows broadband planar loop antennas originally designed for UHF TV reception. The "Crown loop" of Figure 6.45(a) shows how a relatively sparse thin-wire loop may still form an acceptable broadband commercial antenna. The "Spico loop" of Figure 6.45(b) is a planar metallic element. Both these antennas are designed to provide a match to a 300Ω twin-lead. A similar antenna designed for UWB purposes has a rounded feed region and is supported on a PCB substrate [65]. Ideally, a small loop antenna has a dipole doughnut-type pattern in the plane of the loop. Figure 6.45(c) shows an ideal loop pattern. A very small loop can achieve this ideal pattern. However, such a small loop will tend to be inefficient.

A larger loop, one with a diameter on the order of λ/π, is necessary in order to radiate energy efficiently. Just as a $\lambda/2$ dipole is naturally resonant, so also is a λ circumference (λ/π diameter) loop naturally resonant. Figure 2.6(b) shows how such a loop radiates when excited by a UWB impulse. A loop with dimensions that are a significant fraction of a wavelength will experience significant dispersion due to the timing differences from one side to the other. This is similar to the dispersion problems exhibited by monoloop antennas. The solution to this problem is the "clover-leaf" loop antenna originally introduced by Brown in the narrowband context and proposed by Harmuth in the UWB context [66]. The planar clover-leaf design has the dual benefit of tending to block near-field reactive energy while providing a relatively low-dispersion uniform loop excitation.

Figure 6.46(a) shows a prototype 3.0–5.0GHz clover-leaf loop antenna. Figure 6.46(b) shows the gain and the through (S_{21}) phase response of this antenna. Gain is relatively independent of whether the antenna is oriented toward a feed or toward a leaf. The phase response (calibrated to remove the delay effect of the feed line) is fairly linear across a wide bandwidth, indicating nonfrequency-dispersive radiation. Figure 6.46(c) shows the matching of the antenna. This prototype has good matching, with VSWR 1.75:1 or better across the band.

6.3.4 UWB Slot Antennas

Magnetic slot antennas are a final UWB magnetic antenna architecture. A slot antenna is not, strictly speaking, analogous to a Hertzian magnetic dipole or ideal small loop antenna. Instead, a slot antenna involves two counterflowing currents on either side of a loop. The resulting "quadrupole" pattern consists of two lobes of radiation aligned along the normals of the antenna plane. Figure 6.47 shows the radiation pattern for a typical slot. Polarization is orthogonal to the axis of the antenna as denoted by the arrows in Figure 6.47. Slot antennas date back to the 1940s and linearly tapered (or bow tie) slots were described by Schelkunoff and Friis. See Figure 1.20(a, b) and associated discussion for details.

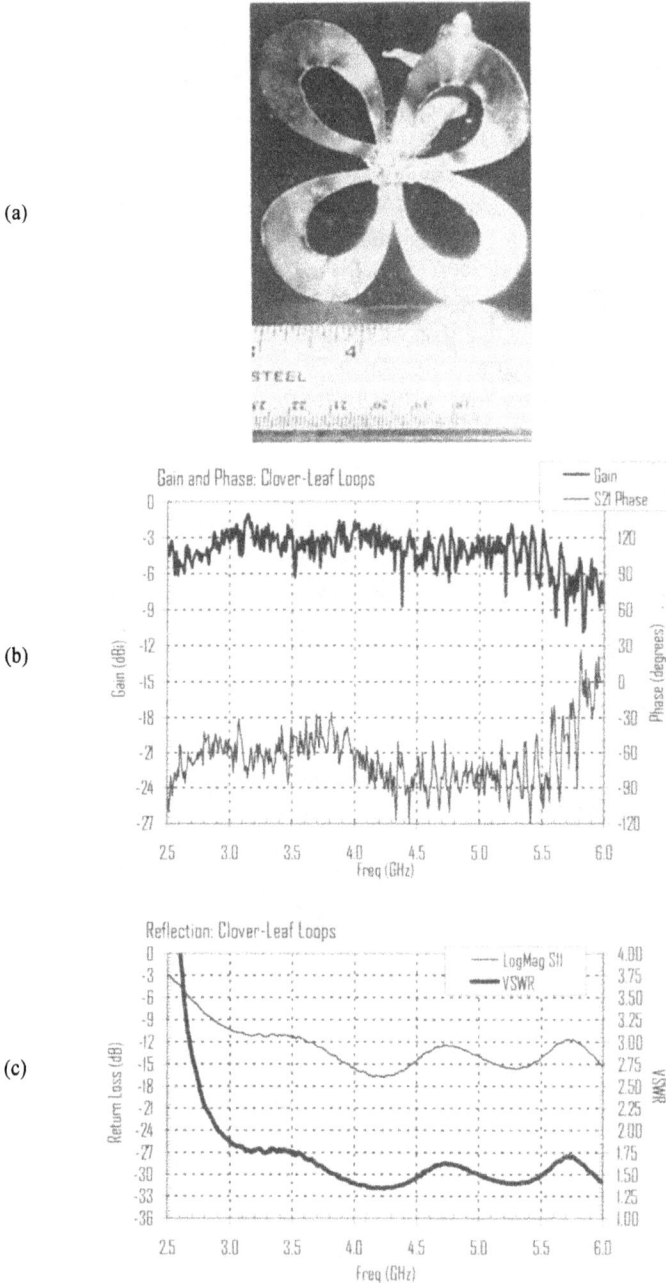

Figure 6.46 (a) A clover leaf loop antenna, (b) gain and phase of clover leaf loop antennas, and (c) matching and reflection of clover leaf loop antennas (Courtesy: Pulse~LINK, San Diego, California).

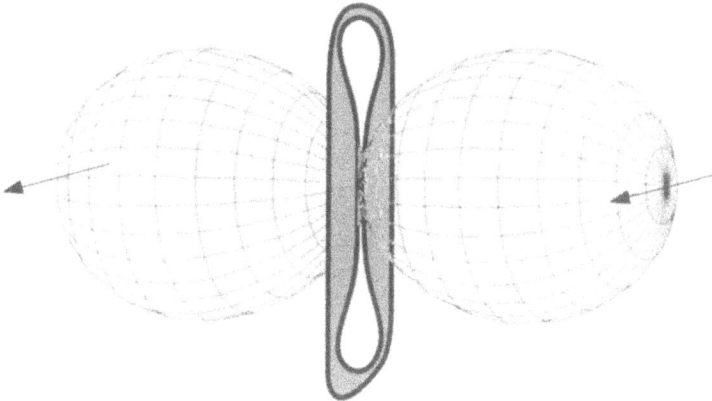

Figure 6.47 A slot antenna has a two-lobed quadrupole pattern aligned along the normals to the plane of the antenna. Polarization is orthogonal to the axis of the slot.

Mark Barnes proposed a continuously tapered magnetic slot antenna [67, 68, 69]. If the taper is properly designed, this antenna is capable of an excellent match to 50Ω. Figure 6.48(a) shows a pair of continuously tapered slot antennas, and Figure 6.48(b) presents a Smith chart showing the match of this antenna. These antennas have a multiple-octave impedance bandwidth and a nominal match better than 1.5:1 over much of the band. Figure 6.48(c) plots the match versus frequency. Peak boresight gain is typically 5–6dBi. The gain curve of Figure 6.48(d) shows that this antenna has about a 3:1, 3-dB gain bandwidth. Figure 6.48(d) also shows that the phase is linear across the operational band of the antenna. This indicates that the antenna is nonfrequency dispersive.

The gain bandwidth extends between about 1.0GHz and 3.0GHz. The impedance bandwidth, however, begins around 1.0GHz and extends past 6.0GHz. This is an unmistakable clue that the antenna enters a higher-order mode past 3.0GHz, probably a four-lobed pattern with a null along the normal of the antenna plane.

Tapered slot antennas are a viable UWB antenna option capable of bandwidths covering the 3.1–10.6GHz band. The principal shortcoming of these antennas is their lack of an omnidirectional pattern. If an application can tolerate the directional pattern, the higher gain of tapered UWB slots makes these antennas an attractive option.

Linear taper slot antennas [like those of Figure 1.20(a)] date back to the 1950s but are still of active interest today. For instance, Andrew Hannigan, Steve Pencock, and Peter Shepard recently investigated a bow tie slot antenna [70]. Figure 6.49 shows this antenna. The antenna comprises a sector bow tie slot with about a 90° angle etched into the front metallization of a PCB. The rear metallization of the PCB provides a microstrip feed line that couples to the slot, terminating in an open stub.

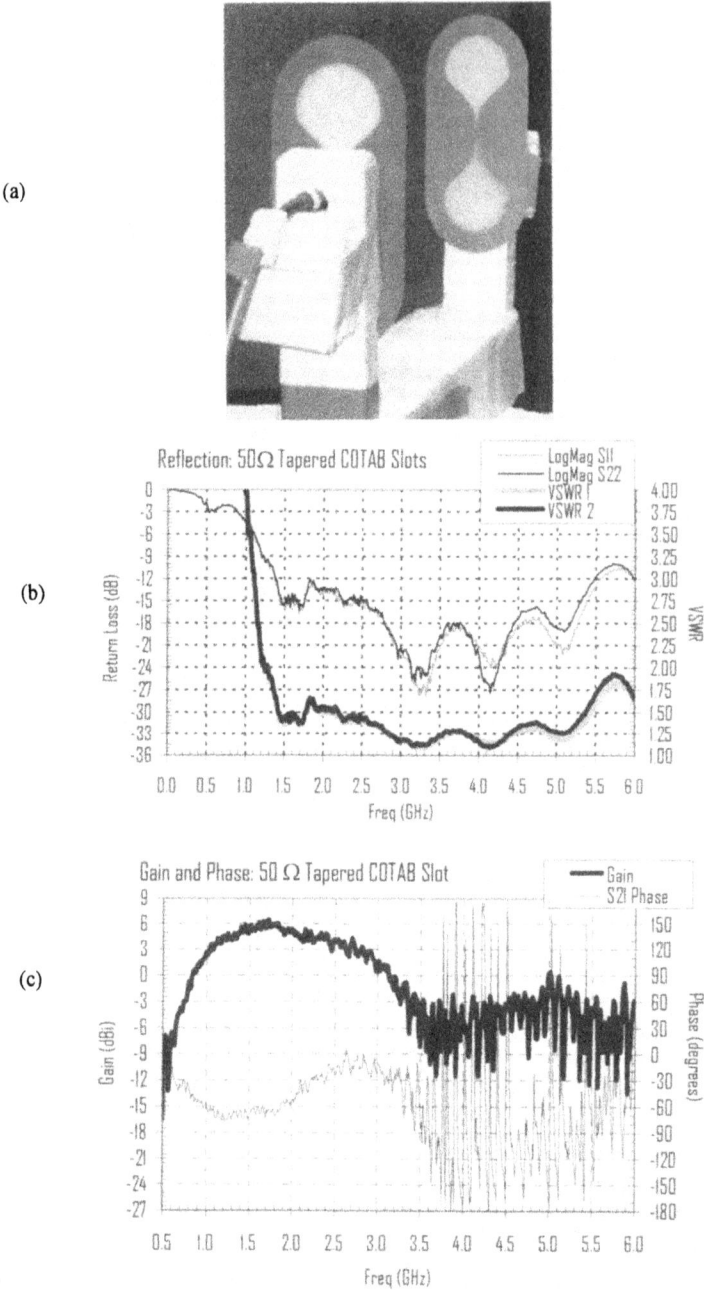

Figure 6.48 (a) A pair of continuously tapered UWB slot antennas, (b) reflection of a continuously tapered UWB slot antenna, and (c) boresight gain and phase of a pair of continuously tapered UWB slot antennas (Courtesy: Time Domain Corporation, Huntsville, Alabama).

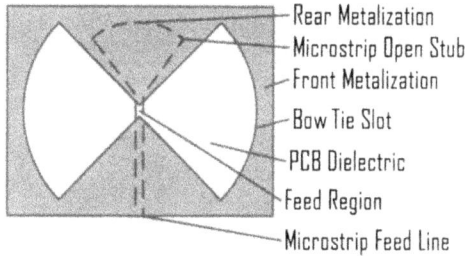

Figure 6.49 The bow tie slot antenna of Hannigan, Pencock, and Shepard (After [70]).

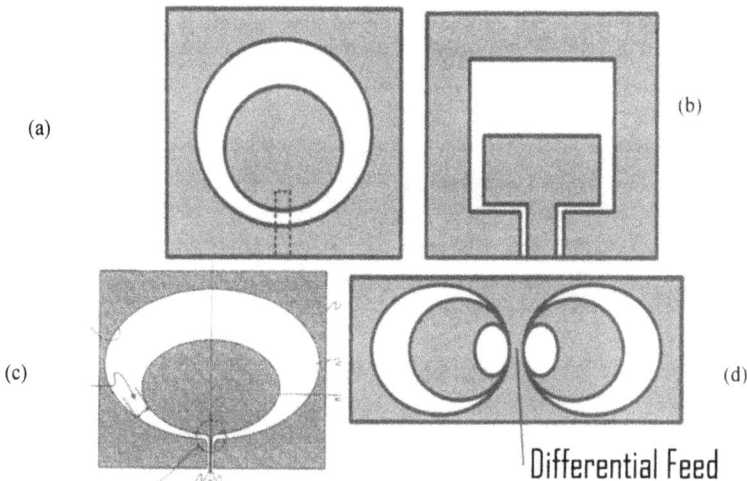

Figure 6.50 (a) Suh and Park's annular slot antenna (After [71]), (b) Chen's CPW fed square slot antenna (After [72]), (c) McCorkle's elliptical annular slot antenna [73], and (d) Powell and Chandrakasan's differential annular slot antenna (After [74]).

Annular slot antennas have also attracted significant interest. Young Hoon Suh and Ikmo Park introduced a broadband eccentric annular slot antenna [71]. This microstrip-fed 30-mm-diameter element [depicted in Figure 6.50(a)] exhibits an impedance bandwidth in excess of an octave. Horng-Dean Chen suggested the square form factor slot of Figure 6.50(b) [72]. This square slot CPW-fed antenna was limited to about a 60% impedance bandwidth, however. Annular slots have also been investigated for commercial applications by John McCorkle [73]. McCorkle's elliptical annular slot antenna, shown in Figure 6.50(c), yields impedance bandwidths in excess of 100%. More recently, Johanna Powell and Anantha Chandrakasan examined both single-ended and differential versions of annular slot antennas [74]. Their differential annular slot is featured in Figure 6.50(d). Coplanar-fed annular slots have also been used for millimeter wave applications [75].

6.3.5 Summary of Small-Element Magnetic Antennas

Small-element magnetic antennas are excellent for embedded applications because their magnetic near fields are less vulnerable to undesired coupling than the near electric fields of small electric field antennas. This section has discussed four general families of small-element magnetic antennas.

LCRs attempt to approximate a uniform current sheet. These antennas tend to be relatively inefficient. A monoloop is a half-loop antenna driven against a ground plane. Introduced by Turner and Turner in the late 1950s, these antennas tend to have a dispersive radiation pattern unless fed in a symmetric fashion. Loop antennas are similarly prone to dispersion, unless fed symmetrically, as in the clover-leaf loop antenna. Slot antennas are a final family of UWB magnetic antennas. Slots tend to have nondispersive quadrupole-type patterns. Any of these antennas can yield acceptable UWB performance for some applications.

The principal disadvantage of small-element magnetic antennas is their difficulty in achieving low-dispersion, omnidirectional patterns. Only the clover-leaf loop antenna (among the magnetic antennas discussed in this section) meets that goal. The omnidirectional pattern of the clover-leaf loop is horizontally polarized, in contrast to the vertically polarized pattern of a typical electric dipole antenna.

6.4 ELECTRICALLY SMALL ANTENNAS

Typical consumer-electronic devices sacrifice antenna performance in favor of smaller, more cosmetic form factors [13]. A foot-diameter AM-band loop antenna would consume a sizeable fraction of the volume of a portable radio, so a smaller loopstick antenna is used instead. An FM-band whip antenna is thought to mar the lines of an automobile, so a discreet but inferior antenna is embedded in the window. A cell phone with a telescoping quarter-wave whip would be awkward, so an inefficient stub is used instead. An antenna that protrudes from a notebook computer is thought unsightly and might be prone to break off, so an embedded antenna is utilized instead. Consumers are generally more concerned with having a compact and aesthetic form factor than with having maximal range or sensitivity. Manufacturers are well aware of this and are happy to trade link margin for a more cosmetically appealing, better-selling consumer-electronic device.

Very few consumers even realize that there are shortcomings in their devices. If a radio station to which they are listening fades into static, how are they to know that with a better antenna, they could be listening to the same station for another half hour down the road? If their cell phone doesn't pick up calls in a particular area, how are they to know that a better antenna would provide them enough link margin to operate despite the fade? How many users of 802.11b wireless networks know that their link is supposed to be able to operate over a range of 1,605ft (503m) in an open field environment? How many users have actually taken an 802.11b wireless network out to an open field and discovered that the maximum range is actually far less [76]?

These performance shortcomings of typical consumer-electronic devices are not particularly evident; nor are they obvious to the casual user. Thus, there tends to be little demand for high-performance antennas and generous amounts of link margin are sacrificed on the altar of aesthetics. This section considers the trade-off between the size and performance of UWB antennas. The aim of this section is to demonstrate that the nature of UWB technology is likely to break this general trend and lead to consumer demand for efficient, high-performance antennas. First, this section examines the performance sacrifices inherent to miniaturizing UWB antennas. Then, this section will present a quick survey of a few interesting ideas for miniaturizing UWB antennas without significantly impacting their performance.

6.4.1 Antenna Scaling

Antennas may be scaled, but when their size is reduced, their frequency response is increased. Halve the size of a 1–5GHz antenna and it becomes a 2–10GHz antenna. More to the point, by shrinking an antenna, its performance below its lower cutoff moves increasingly in band. Figure 6.51 shows the performance of scaled-sector bow tie antennas like those described in Figures 6.6 and 6.7. A 13mm bow tie element yields an acceptable 3–10GHz antenna. Reducing the size by a factor of two yields a 10-dB penalty in performance below 4.5GHz. Significant antenna miniaturization results in substantial performance penalties. UWB links rarely have enough margin to warrant wholesale sacrifice of antenna performance. Thus, the UWB antenna design challenge is to miniaturize an antenna as much as possible without significantly impacting antenna performance.

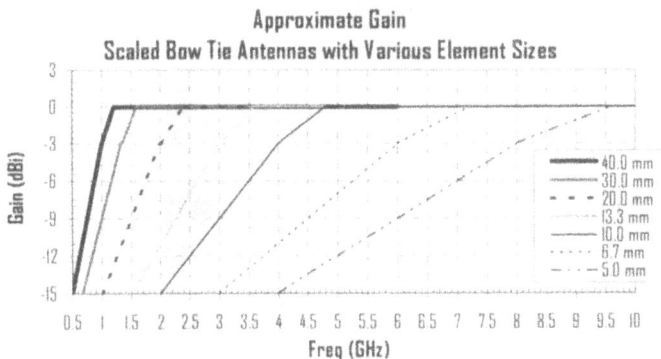

Figure 6.51 Miniaturizing a UWB antenna results in significant performance penalties ([13]; © 2004 CMP Media, LLC).

6.4.2 Dielectric Loading

Dielectric loading is one of the most important techniques for antenna miniaturization. Loading an antenna with dielectric reduces the frequency response of an antenna according to the square of the dielectric constant. For instance, an antenna embedded in TiO_2 with a relative dielectric constant $\varepsilon_r = 100$ will see its frequency response shifted down by a factor of ten. Conversely, an antenna embedded in a $\varepsilon_r = 100$ material will be a factor of ten smaller than its free-space counterpart.

Unfortunately, the dielectric loading also creates an impedance mismatch. The free-space impedance is $Z_S = 377\Omega$. In a medium of relative dielectric constant $\varepsilon_r = 100$, the impedance becomes $Z = Z_S = 37.7\Omega$. The power loss due to this mismatch is: $P_{loss} = (1 - ((Z_S - Z)/(Z_S + Z))^2) = 0.33$ (i.e., about a 4.8-dB hit in performance). Although significant, this is a much smaller performance impact than would be expected from miniaturization alone.

6.4.3 Conducting Enclosure Antenna

If space is limited for an antenna, the enclosure itself may be used as an antenna by driving one-half of an enclosure against the other half as a dipole [77]. A conducting enclosure antenna takes maximal advantage of the available device form factor. Dielectric loading may further improve the performance of such antennas. A similar scheme was used in the original Explorer satellites first launched in 1958 [78]. The frequency response of such an antenna depends upon the details of the enclosure size and shape, as well as the dielectric loading. As always, additional spectral shaping is possible using filtering. Figure 6.52 shows a potential configuration for a conducting enclosure antenna.

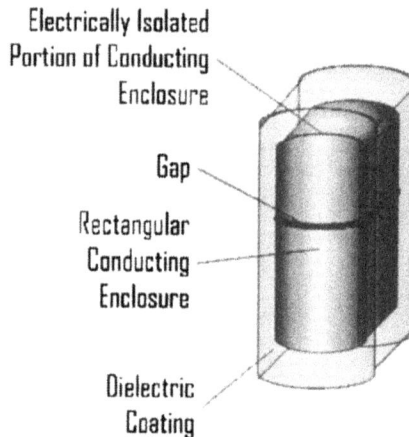

Figure 6.52 A conducting enclosure antenna ([13]; © 2004 CMP Media, LLC).

6.4.4 Electric-Magnetic Antenna

A final technique for miniaturizing UWB antennas follows from Chu's observation that superimposed electric and magnetic dipoles may yield at least a factor of two reduction in the size of an efficient antenna relative to a single small-element antenna alone [79]. Dual electric-magnetic antennas were originally introduced by Wilhelm Runge in the 1930s [see Figure 6.53(a)] [80]. Kraus observed that such a configuration in which the small loop and small dipole are fed in phase yields circular polarization everywhere in the field of view [81]. These antennas were improved upon by Kandoian in the 1940s [see Figure 6.53(b)] [82]. Kandoian described these antennas in the same article in which he introduced the discone antenna [83]. The energy flow around a dual electric and magnetic antenna exhibits fortuitous cancellations, resulting in significant reductions in near-field reactive energy [84]. These properties make dual electric-magnetic antennas well suited for UWB use [85]. In principle, this technique may yield a well-matched UWB antenna with element sizes on the order of 0.07λ. This suggests 3.1–10.6GHz elements with dimensions of 0.28in (7.0mm).

6.4.5 Summary of Electrically Small Antennas

Many consumer-electronic devices sacrifice performance for aesthetics because consumers better appreciate good aesthetics. UWB-enabled devices may be the exception to this general rule. Because UWB links are likely to be very high data rate and very short range, both ends of the link will be under the consumer's control, and a failure to live up to advertised specifications will be immediately evident.

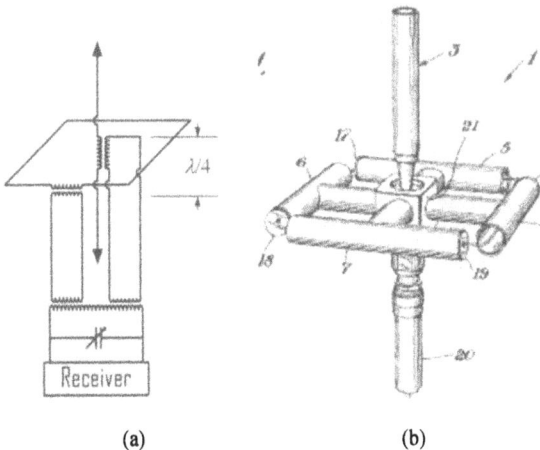

(a)   (b)

Figure 6.53 (a) Runge's electric-magnetic antenna system for polarization diversity (After [80]), and (b) Kandoian's electric-magnetic antenna [82].

Thus, the severe performance degradations introduced by arbitrarily shrinking UWB antennas are unlikely to be accepted by the marketplace. It is essential to get as much performance as possible out of a UWB antenna. Conventional UWB antennas can deliver excellent performance with element sizes on the order of 15 mm (0.60 in). Dielectric loading, conducting enclosure antennas, and dual electric-magnetic element antennas promise to deliver comparable performance in an even smaller package.

6.5 HORN ANTENNAS

Originally introduced by Bose in the 1890s, a horn antenna is a flared or tapered transmission line designed to transmit and receive electromagnetic energy in one or more particular directions. Bose's original conical and pyramidal horn antennas (or "collecting funnels") follow simply from acoustical analogy. For decades after Bose's pioneering work, frequencies remained low and wavelengths long, rendering horn antennas impractical. With advances in microwave RF technology around the time of World War II, frequencies became high enough and wavelengths short enough for reasonable sized horn antennas to be practical again. Katzin, King, and others rediscovered Bose's designs and extended them to even higher levels of performance.

Horn antennas tend to be relatively large, often a wavelength or more in dimension at a typical operating frequency. These large apertures lead horn antennas to be directional with relatively high gains. The usual trade-off between directionality and field of view applies. Horn antennas will be able to transmit and receive signals very well (with high gain) but only over a limited field of view. The higher the gain, the narrower the field of view. Thus, horn antennas are typically used only in point-to-point links where a narrow field of view is not a liability. Alternatively, horn antennas may be used in an array with each horn providing coverage of a particular desired sector.

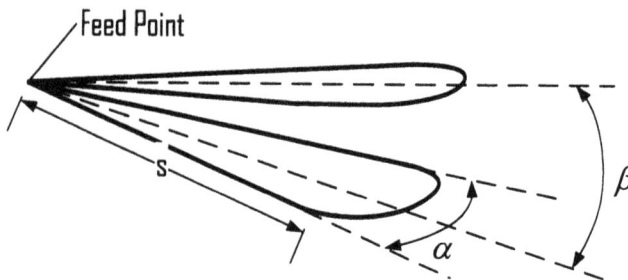

Figure 6.54 A conical horn antenna begins at a feed point and comprises two conducting plates with length s, typically with a fixed plate angle (α) and a fixed plate pitch angle (β).

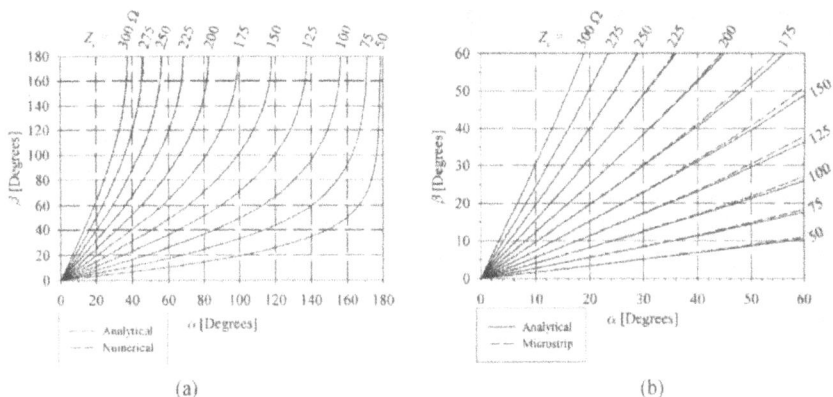

Figure 6.55 (a) Characteristic impedance of a conical plate horn antenna as a function of plate (α) and pitch (β) angles, and (b) close-up for small angles ([87]; © 2004, IEEE; Courtesy: Glenn Smith).

This section discusses conical plate horn antennas, one of the most basic and best understood horn designs. Then, this section addresses proper termination of horns. Planar horn antennas offer a relatively inexpensive alternative to full 3-D horn antenna structures. Finally, this section provides a brief survey of other interesting horn antenna designs and a summary of horn antennas and their properties.

6.5.1 Conical Plate Horn Antenna

Figure 6.54 shows a conical plate horn antenna. These antennas are among the most common of horn antenna designs. A conical plate horn antenna begins at a feed point and comprises parallel conducting plates of length s, typically with a fixed plate angle (α) and a fixed plate pitch angle (β).

There exists an analytical solution for the impedance of a conical plate horn antenna [86]. This solution involves elliptic integrals and some fairly involved mathematics. However, once the math is solved, it is possible to generate a plot showing lines of constant impedance on an α-β plane. Figure 6.55(a) shows how the characteristic impedance of a conical plate horn depends upon the plate (α) and pitch (β) angles [87]. Figure 6.55(b) shows a close-up for smaller angles.

One may add a tapered double ridged wave-guide to a conical plate antenna. A tapered double ridged wave-guide lowers the cutoff frequency of the dominant mode and increases the cutoff frequency of the next dominant mode. The net effect is to increase the bandwidth of the horn antenna. "Standard gain" horn antennas with this feature are available from a variety of manufacturers [88]. Figure 6.56 shows a 200MHz–2GHz standard gain horn antenna of this design.

This standard gain horn has an aperture of 93cm × 73cm (37in × 29in) and is 98cm (39in) long. At 200 MHz, the gain is about 6dBi, increasing to 9dBi at 400MHz. Gain is nominally around 9dBi up through nearly 2.0GHz; however, in places across the 400MHz–1900MHz band, gain varies between 7.5dBi and 10dBi. This variation in gain means antenna calibration is essential for most applications. The performance of a standard gain horn is not sufficiently uniform as to be characterized by a single constant parameter (gain or aperture) across the band. The antenna is well matched, with a VSWR of 1.5:1 or better to about 1.5GHz, peaking just above 2:1 at the high end of the band. Typical half-power beam widths are about 30–40° in the H-plane and approximately 45° in the E-plane.

Figure 6.56 A 200-MHz to 2.0-GHz "standard gain" horn with a tapered double-ridged waveguide.

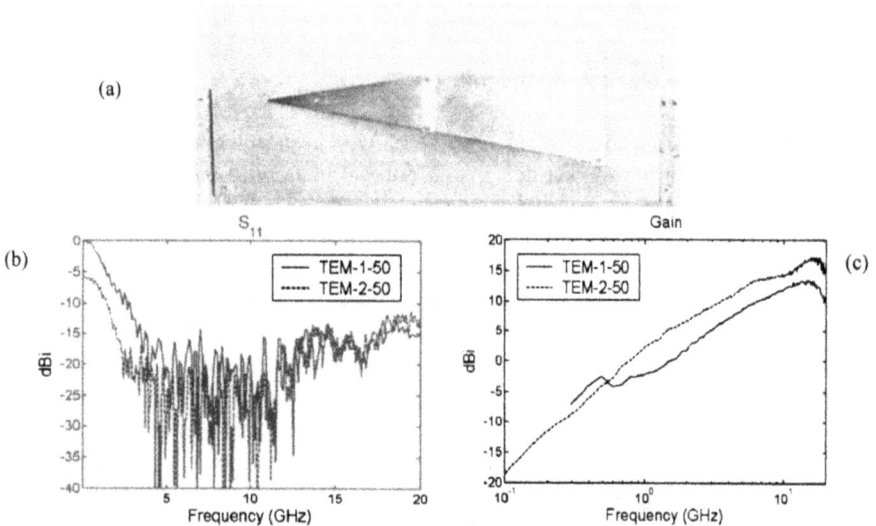

Figure 6.57 (a) A Farr Research TEM 2-50 conical element horn, (b) matching (S_{11} in dB) of the TEM 2-50 horn, and (c) gain (in dBi) of the TEM 2-50 horn ([89]; Courtesy: Farr-Research).

The design plots of Figure 6.55(a, b) also apply to conical horn elements driven against a ground plane, provided impedance is divided in half to account for the ground plane effect. Farr-Research manufactures and sells two antennas of this design [89]. Figure 6.57(a) shows a photo of a Farr-Research Model TEM 2-50 horn antenna. This 50Ω antenna comprises a conical horn element driven against a 51cm × 122cm (20in × 48in) ground plane. Matching is better than S_{11} ~ 10dB above about 1.5GHz. The Farr-Research TEM 2-50 horn is approximately (but not quite) a constant aperture antenna with a gain varying as frequency squared. Figure 6.57(b) shows the matching, and Figure 6.57(c) shows the gain. This antenna is optimized for boresight response and is suitable for use in a point-point link or as a test horn.

6.5.2 Termination of Horn Antennas

The abrupt discontinuity at the end of a typical horn element may be a source of undesired reflections. Furthermore, the end of a horn antenna may also be a source of diffraction. There are a variety of ways to terminate a horn element so as to reduce both reflection and diffraction. One of the most common and effective is a rolled-edge.

Figure 6.58 shows a rolled-edge feed horn. This design was pioneered by Walter Burnside and Chiwei Chuang [90]. A rolled termination can significantly reduce undesired sidelobes and yield more stable pattern and gain performance across ultrawide bandwidths. An elliptical cross-section taper works very well, although other taper geometries are certainly possible.

Figure 6.58 A feed horn with an elliptical rolled-edge termination conformal to shaped absorber material (Courtesy: The ElectroScience Lab, Ohio State University, Columbus, Ohio).

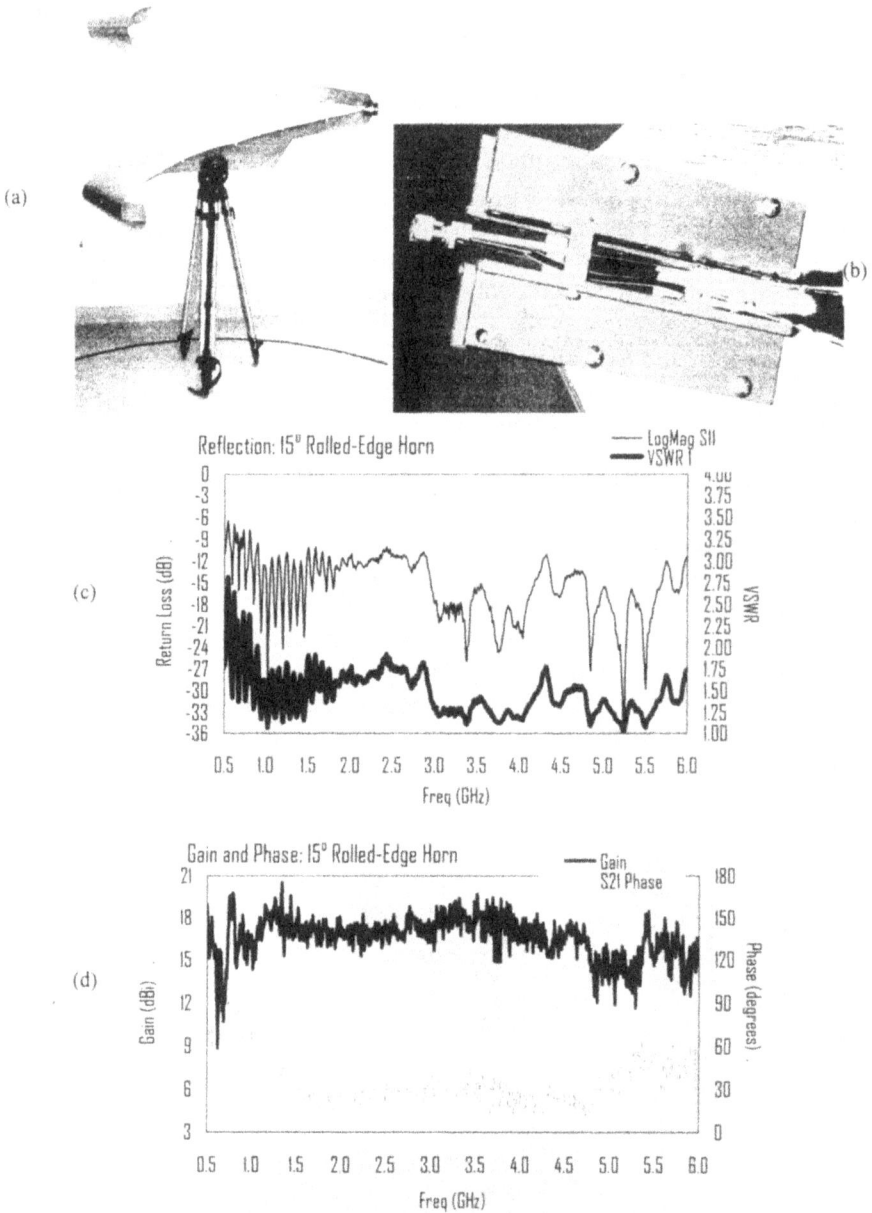

Figure 6.59 (a) A 500-MHz–5.0-GHz elliptical rolled-edge horn antenna, (b) details of the feed region: a PCB-mounted SMA end-launch connector is mechanically isolated from the aluminum plate elements by a thin brass plate transition, (c) matching (both S_{11} and VSWR) for this antenna, and (d) gain and phase behavior of this horn (Courtesy: Time Domain Corporation, Huntsville, Alabama).

Figure 6.59(a) shows a 500MHz–5.0GHz rolled-edge horn antenna. The plate angle is $\alpha = 15°$ and the plate pitch angle $\beta = 15°$. The horn is 152cm (60in) long. The most complicated part of constructing a horn like this is to couple to an appropriate feed. Figure 6.59(b) shows the feed region of this antenna. An SMA end launcher attaches to a PCB feedline. The PCB in turn couples to a parallel strip feed line of relatively thin brass. This strip feed line is flexible enough to be adjusted to optimize the feed, while also mechanically isolating the SMA and PCB from the mechanical stress of being directly attached to the horn elements. The feed mechanism of Figure 6.59(b) helps this horn antenna achieve a match with $S_{11} < -10$dB across much of the operating band. Figure 6.59(c) shows the VSWR and return loss for this antenna. Finally, Figure 6.59(d) shows that this antenna has a relatively flat boresight gain of about +17dBi across the operating band. Figure 6.59(d) also shows a relatively flat and linear phase response, meaning that this antenna is not frequency dispersive. This antenna is well suited for use in a UWB point-point link or as a standard gain horn for making precision field or antenna pattern measurements. Rolled edges do not exhaust techniques for terminating horn elements. Other techniques include serration of an edge or resistive terminations [91].

Excellent performance may be obtained from the horn antennas described in this section. However, these antennas tend to be rather complicated and difficult to build. Section 6.5.3 describes an inexpensive alternative: planar horn antennas.

6.5.3 Planar Horn Antennas

One key principle of small element design is that a planar cross-section of a surface-of-revolution UWB antenna makes a good planar UWB antenna. This same principle applies to horn antennas. A planar cross-section of a 3-D UWB horn antenna is a good starting point for a 2-D, or planar, UWB horn antenna. A planar cross-section of a horn antenna effectively sets the plate angle to $\alpha = 0°$ and relies only on "plate pitch" (β). In practice, this plate pitch is just the taper between opposing antenna elements.

Planar horn antennas are sometimes referred to as "Vivaldi notch" antennas. Traditionally, these antennas have been fed using strip line-to-slot line transitions such as those shown in Figure 3.13(a, b). Optimization of these antennas has been an area of great interest [92]. A Y-Y balun (see Figure 3.14) can also be used to transition between a slot line and a strip line for purposes of feeding a horn. [75]. Taper feeds, like those of Nester's microstrip notch antenna (see Figure 3.10) also show significant promise [93].

Planar horns have been employed in a variety of UWB system implementations. For instance, now defunct Fantasma Network, Inc., used a dual planar horn antenna designed by G. Roberto Aiello and Patricia Foster [94]. Figure 6.60 shows this planar horn antenna system. Depending on the shape and dimensions of the notch and shape of the periphery, this antenna system may be used not only as a directional system but also for omnidirectional applications.

Once one has made a transition from a microstrip or other conventional planar transmission-line geometry to a slot line, the main challenge is to design the radiating elements subject to the dual constraints of good matching and good pattern. The following example demonstrates how an elliptically tapered element horn antenna can yield a compact yet directional antenna.

Figure 6.61(a) shows a medium-gain horn manufactured by Next-RF, Inc. [95]. This antenna consists of a tapered balun feed (similar to Nester's), a right-angle microstrip-to-slot line transition, and an elliptically tapered slot line. The right-angle transition allows this horn to fit within a relatively compact 10cm × 10cm (4in × 4in) form factor. Figure 6.61(b) presents an exploded view of the opposing metalizations. This medium gain horn has good matching, with a VSWR nominally around 2:1. Figure 6.61(c) shows typical matching results. The boresight gain of this antenna varies from about 4dBi at 3.1GHz to about 8dBi at 5GHz. Figure 6.61(d) shows the boresight gain from 2.0GHz to 6.0GHz. Figure 6.61(e) shows a typical 3-D pattern as calculated using an Ansoft HFSS model. This medium-gain antenna was designed to operate with approximately a 90° azimuthal pattern between 3.1GHz and 5.0GHz.

6.5.4 Other Horn Antennas

A horn antenna is a transmission line, tapered and truncated in such a way as to convert guided electromagnetic energy into radiation and vice versa. Not surprisingly, a wide variety of horn antenna geometries are possible.

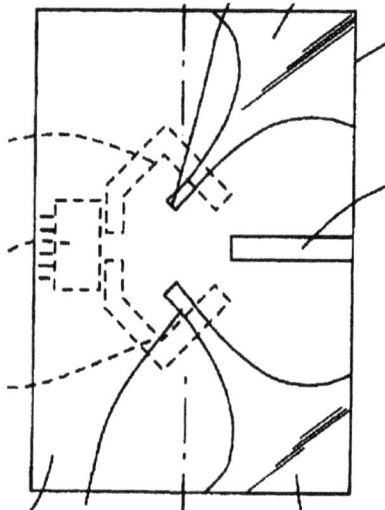

Figure 6.60 A dual feed horn employed by Fantasma Network, Inc. [94].

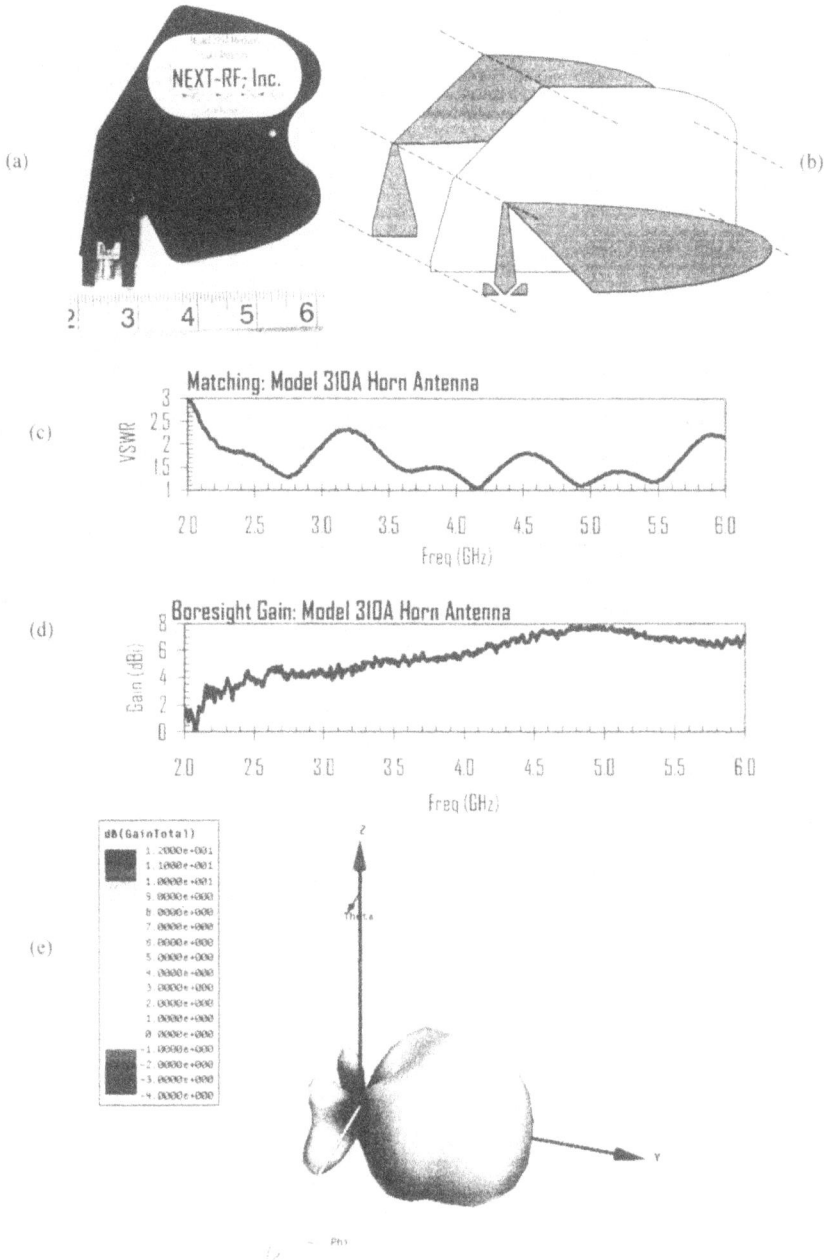

Figure 6.61 (a) A Model 310A medium gain planar horn antenna, (b) exploded view of a Model 310A medium gain planar horn antenna, (c) matching of the 310A planar horn antenna, (d) boresight gain of the 310A planar horn antenna, and (e) typical pattern (Courtesy: Next-RF, Inc., Hunstville, Alabama).

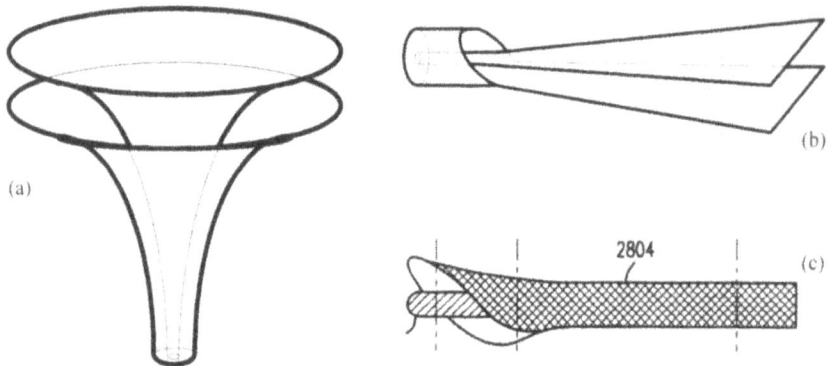

Figure 6.62 (a) Brillouin's omnidirectional coaxial horn antenna (After [96]), (b) Brillouin's directional coaxial horn antenna involves a tapered transition between a coaxial line and parallel plate horn elements (After [96]), and (c) a semi-coaxial horn antenna [97].

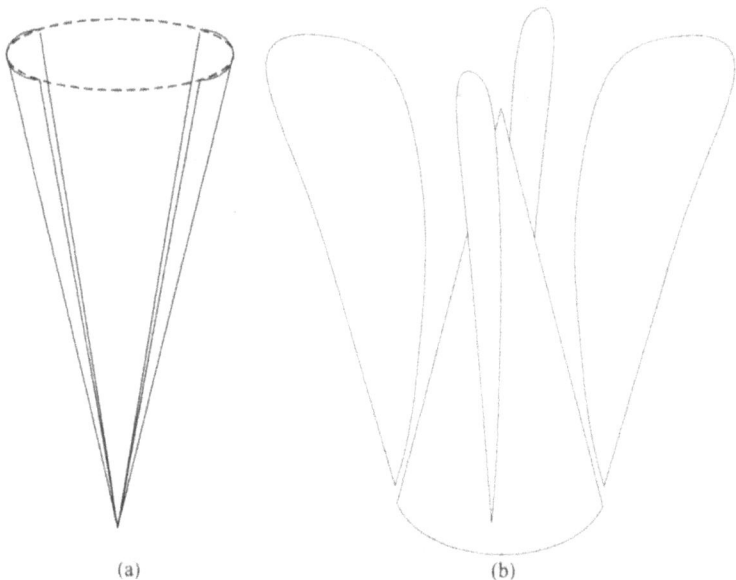

Figure 6.63 (a) The V-conical horn of Shen, King, and Wu (After [98]), and (b) the "polarization diverse" horn of Wicks and Antonik (After [99]).

Given the prevalence of coaxial transmission lines, horn antennas that take a coaxial form factor as a starting point are of particular interest. Brillouin introduced a couple of horn antennas based on coaxial transitions (see Section

3.3.2) [96]. Figure 6.62(a) shows Brillouin's omnidirectional coaxial horn antenna, which flairs out in a controlled fashion to yield an omnidirectional pattern. With appropriate tapering, a high-gain "pancake pattern" (omnidirectional on azimuth but narrow on elevation) is possible. Figure 6.62(b) shows Brillouin's directional coaxial horn antenna. This coaxial horn antenna tapers into conical horn radiating elements. Figure 6.62(c) shows a semicoaxial horn antenna [97]. A semicoaxial horn antenna transitions from a coaxial line to a horn by splitting the outer sheath of a coaxial line into a coaxial arc of expanding radius. Using conformal mapping techniques, one may determine relations for the impedance of the semicoaxial transmission line as a function of inner (cylindrical element) radius, outer (sheath) radius, and sheath arc angle.

Given that a horn antenna is a tapered transmission line, any transmission-line structure can serve as the starting point for a horn architecture. For instance, instead of the traditional conical parallel plate horn structure, one may use arcs on a conical form [98]. Figure 6.63(a) presents a sketch of this "V-conical" horn. Alternatively, a plate element may be driven against a circular ground [99]. This architecture is useful for a multiple-element horn antenna. The resulting "polarization diverse" horn of Figure 6.63(b) may be used to radiate or receive vertical, horizontal, or chiral polarization signals.

A good starting place for horn geometry ideas is a standard reference on transmission lines and conformal mapping [100, 101]. Any transmission line can serve as a horn cross-section. Many of these cross-section options already have tabulated expressions for impedance, although they often involve elliptical or other special functions. By varying the dimension and scaling of the cross-section, one can achieve a desired impedance taper.

6.5.5 Summary of Horn Antennas

This section has reviewed horn antennas. A horn antenna is a flared or tapered transmission line designed to transmit and receive electromagnetic energy in one or more particular directions.

The principal advantage of a horn antenna is its relatively directional high gain behavior. Horn antennas are useful for point-point links and other applications in which the corresponding narrow field of view is not a liability. Alternatively, horn antennas are useful in an array in which each horn provides high-gain coverage in a particular direction, with the array as a whole nevertheless providing coverage across an entire field of view. Typical conical plate horn antennas are "constant aperture" with gain increasing as a function of frequency (like the Farr TEM horn). If appropriate termination is used, a conical plate horn may be "constant gain" [like the elliptical tapered horn of Figure 6.59(a)]. The principal disadvantage of horn antennas is their large size. A planar geometry helps somewhat, but even a modest-gain 3–10GHz antenna has about a 10cm × 10cm (4in × 4in) form factor.

6.6 REFLECTOR ANTENNAS

Hertz introduced reflector antennas in the 1880s. Figure 1.3 shows a sketch of a parabolic reflector Hertz used in his pioneering experiments. A reflector antenna relies on a quasioptical reflection of energy from a reflector. Like horn antennas, reflector antennas tend to be relatively large, directional, high-gain antennas. This section addresses some of the simplest and easy to implement UWB reflector antennas, including planar, corner, and parabolic cylinder reflectors.

6.6.1 Planar Reflector

A planar reflector is at once the simplest and yet most practical UWB reflector design [102]. Few commercial applications can tolerate an antenna protruding from an enclosure. If instead the antenna is conformal to the enclosure, using the enclosure as a back reflector, a more compact antenna structure may be feasible. Of course, no antenna can radiate through a conducting enclosure. The conducting enclosure planar reflector serves to concentrate the radiated and received signals to the antenna side of the planar reflector.

To model this configuration, consider an antenna a distance d from a conducting plane. The effective origin of the reflected signal is an image antenna a distance d below the conducting plane. Figure 6.64 shows the geometry of this configuration. A signal from the image must travel an excess path length:

$$L = 2d \cos \theta \qquad (6.5)$$

Thus, the reflected signal is subject to a delay:

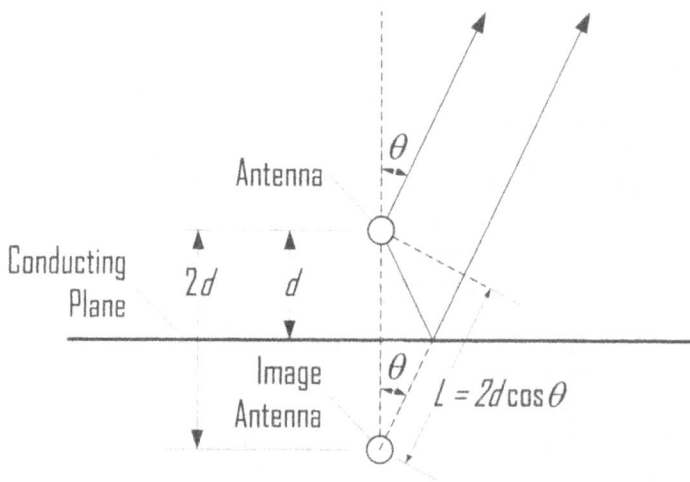

Figure 6.64 Geometry of an antenna and its image in a planar reflector.

$$dt = \frac{2d}{c} \cos \theta \qquad (6.6)$$

A composite signal from a planar reflector is a combination of a direct signal with time dependence $T(t)$ and an inverted reflected signal with time dependence $-T(t - dt)$. To carry the analysis further requires a specific time dependence, so assume a Gaussian W signal—second derivative of a Gaussian signal. In the nomenclature of Section 4.1.3.1, such a signal has a time dependence

$$T_2(t) = \left[1 - 2\left(\pi t f_c\right)^2\right] e^{-(\pi t f_c)^2} \qquad (6.7)$$

where f_c is the center frequency and the time dependence is normalized to unit amplitude. Suppose $f_c = 2.4\text{GHz}$. If there is no delay ($dt = 0$), then the direct and reflected signals suffer total destructive interference as shown in Figure 6.65(a). As the delay increases, the waveform reaches the "destructive interference threshold" of Figure 6.65(b), where the resulting composite waveform has unit amplitude. If $f_c = 2.4\text{GHz}$, this occurs for a delay about $dt = 77$ ps (0.18 period). Smaller delays result in a destructive interference waveform with less than the original unit amplitude. Larger delays result in a constructive interference waveform with more than the original unit amplitude. Figure 6.65(c) shows maximal constructive interference. This happens when the peak of each W waveform aligns with the secondary peak of the other. The resulting maximal constructive interference waveform has an amplitude of 1.45 (+3.2dB relative to the original waveform). If $f_c = 2.4\text{GHz}$, this occurs for a delay about $dt = 160$ ps (0.38 period). As delay continues to increase, the waveform amplitude decreases. For a delay of $dt = 320$ ps (0.77 period), the waveforms become partly resolved again with an amplitude near 1.0 as shown in Figure 6.65(d). The waveforms become "fully resolved" when they no longer interfere. This happens for a delay of about $dt = 760$ps (1.8 period).

Figure 6.65(e) shows where each of these waveforms occurs as a function of reflector spacing and angle off boresight. Figure 6.65(f) provides a close-up. For instance, to achieve a maximally constructive waveform on boresight requires a minimum reflector spacing of $d = 0.19\lambda_c$. For this spacing, the destructive interference threshold occurs for an angle off boresight of 62°, so the −3-dB beamwidth is approximately 124°.

As reflector spacing is increased, the pattern becomes "double humped." For a reflector spacing of $d = 0.39\lambda_c$, the pattern is approximately −3dB down on boresight, peaks at ±60° from boresight and is back down to −3dB at ±77° from boresight. Thus, the total −3-dB beamwidth is approximately 154°. Moving the other direction, as the reflector spacing approaches $d = 0.09\lambda_c$, the boresight waveform again has unit amplitude. Smaller reflector spacings will result in a serious reduction in performance.

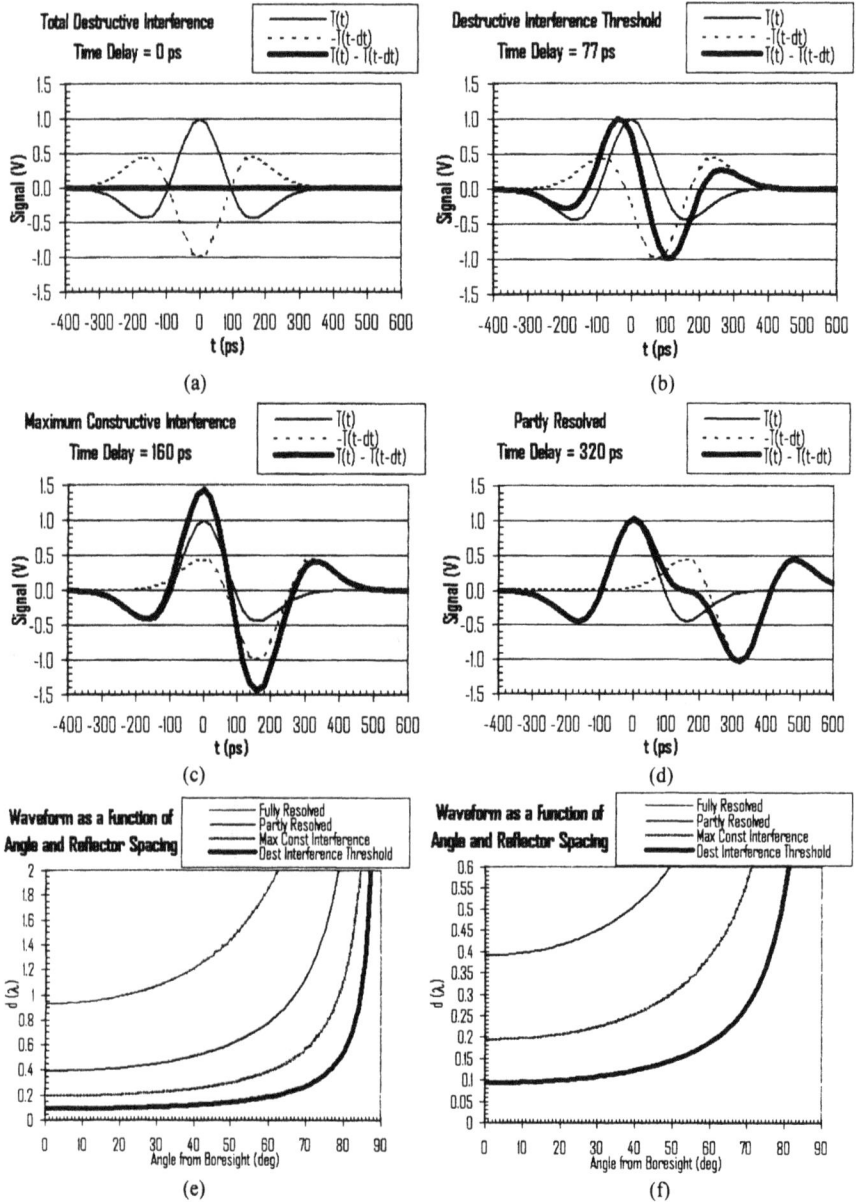

Figure 6.65 (a) Waveform for total destructive interference, (b) waveform for destructive interference threshold, (c) waveform for maximum constructive interference, (d) partly resolved waveform, (e) waveform as a function of reflector spacing and angle off boresight, and (f) close-up of previous figure (Courtesy: Time Domain Corporation, Huntsville, Alabama).

This analytic study assumes a point or a line source. A real antenna actually has a certain physical aperture across which current sources are distributed. Thus although the analytic model provides some guidance, there is no substitute for building and testing a prototype antenna.

I constructed the minireflector antenna of Figure 6.66(a). The dimensions of this antenna were 41.3mm × 69.9mm × 12.7mm (1.625in × 2.75in × 0.5in). At the center frequency of 4GHz, this antenna has a reflector spacing of $d = 0.17\lambda_c$, just enough for about a 3dB gain enhancement on boresight [see Figure 6.66(b)]. In fact, the gain for this antenna is actually slightly more than might be expected for an ideal aperture of the same size. Figure 6.66(c) presents peak power patterns on azimuth as well as on two elevation axes: the frontal elevation (passing through the normal of the antenna plane) and the edge elevation (lying in the antenna plane). The azimuthal pattern is a bit narrower than the model predicts. Finally, Figure 6.66(d) shows that the narrow reflector spacing has severely impacted matching. A planar reflector antenna may be quite compact; however, the price for narrow reflector spacing must be paid for by mismatch.

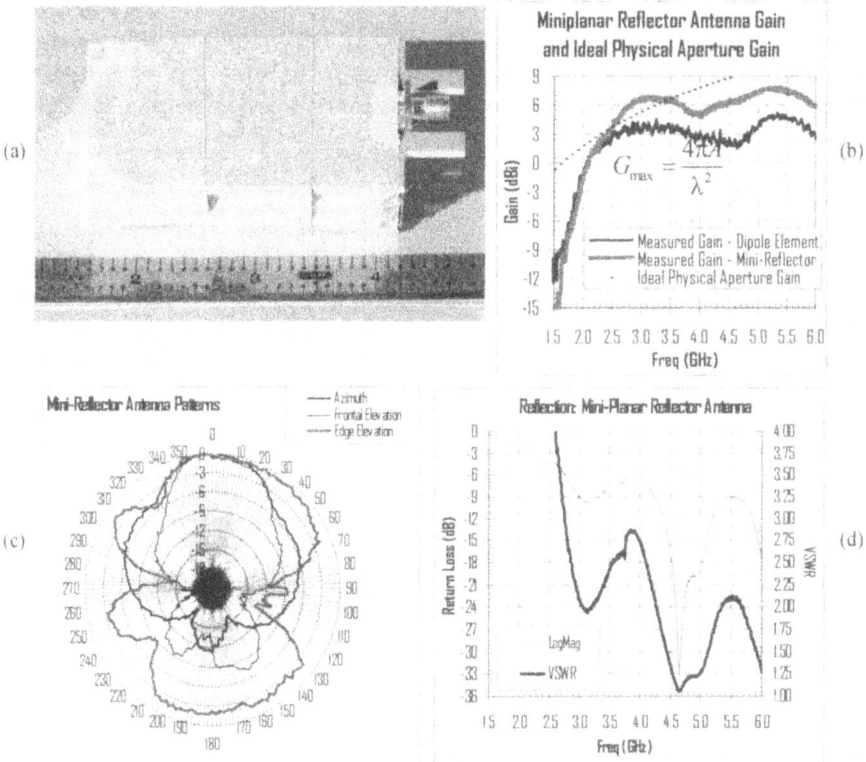

Figure 6.66 (a) Minireflector antenna, (b) boresight gain of a minireflector antenna, (c) peak power patterns for a minireflector antenna, and (d) matching of a minireflector antenna (Courtesy: Time Domain Corporation, Huntsville, Alabama).

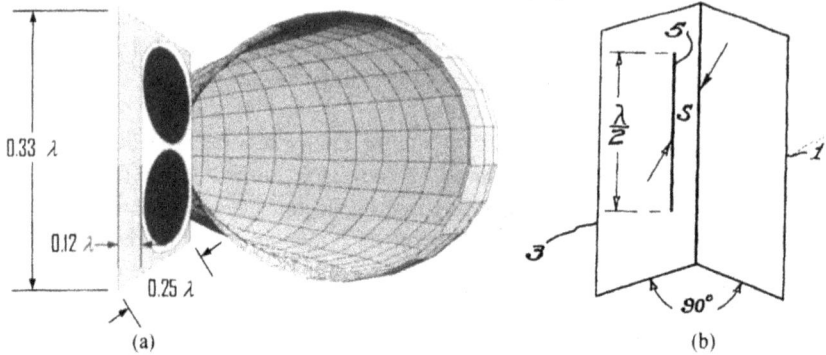

Figure 6.67 (a) Minimum dimension for a planar reflector antenna and approximate pattern ([13]; ©
2004, CMP Media, LLC), and (b) a corner reflector [103].

Figure 6.67(a) shows approximate dimensions and pattern for a planar
reflector antenna. Pattern details will depend on the size of the ground plane. One
can use a ground plane as small as $0.25\lambda \times 0.33\lambda$ and a reflector spacing as small
as about 0.125λ (all with respect to the lower frequency of the operating band). A
planar reflector antenna may also be integrated with a device by using the device
enclosure as the reflecting plane.

6.6.2 Corner Reflectors

Invented by John Kraus in 1942, a corner reflector offers a more directional,
higher-gain antenna system [103]. Figure 6.67(b) shows an example of a corner
reflector. Instead of the 3-dB enhancement of a typical planar reflector, a corner
reflector can offer as much as a 5-dB enhancement.

Consider an antenna a distance d from the corner of a 90° corner reflector, as
shown in Figure 6.68. Here again, the effective origin of the reflected signals may
be thought of in terms of images below the conducting plane. One image lies
behind one side of the reflector, a second lies behind the other side of the
reflector, and a third is formed from two double reflections of signals from each
side. This back image lies directly behind the corner of the reflector. The first and
second images correspond to signals traveling an extra path length of

$$L_1 = \sqrt{2}d\cos(45° - \theta)$$ (6.8)

and

$$L2 = \sqrt{2}d\cos(45° + \theta)$$ (6.9)

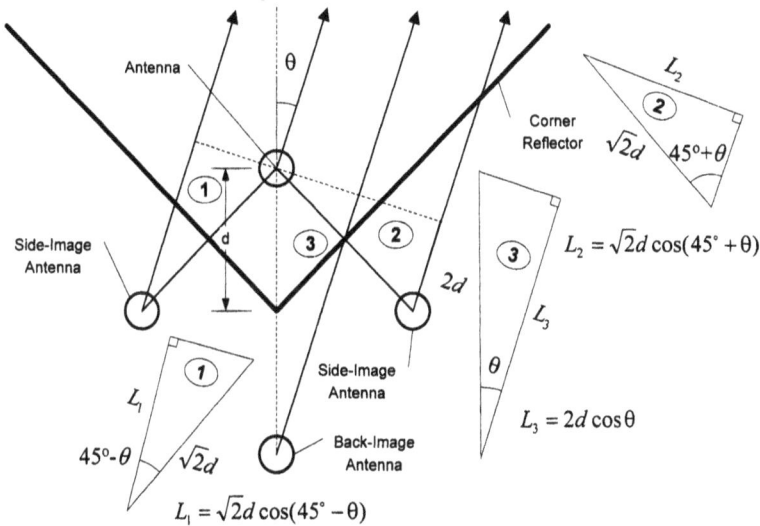

Figure 6.68 Geometry of an antenna and its image in a corner reflector (Courtesy: Time Domain Corporation, Huntsville, Alabama).

respectively. The extra path length from the back image antenna is

$$L_3 = 2d \cos\theta \tag{6.10}$$

The composite waveform is the summation of the direct signal, the two inverted side reflections, and the two colocated double reflections:

$$T_{composite}(t) = T(t) - T_1(t - L_1/c) - T_2(t - L_2/c) + 2\,T_3(t - L_3/c) \tag{6.11}$$

Using the Gaussian W of (6.7), the boresight response is maximized for $d = 0.38\lambda_c$. Figure 6.69(a) shows a diagram of this optimal design, and Figure 6.69(b) shows a corner reflector with a planar dipole antenna. The analytic model predicts the peak amplitude is 2.89 for a 4.6-dB improvement over a single element.

Figure 6.69(c) shows the gain for the corner reflector antenna of Figure 6.69(b) compared to the gain of the dipole element alone. Note that the corner shows roughly a 5-dB improvement over the individual element. Here again, the reflector has a serious impact on impedance and matching. Figure 6.68(d) shows how VSWR peaks as high as 3:1 in band. The azimuthal peak power pattern of Figure 6.69(e) shows a 24°, −3-dB beamwidth, while the elevation peak power pattern of Figure 6.69(f) exhibits a 66° −3-dB beamwidth.

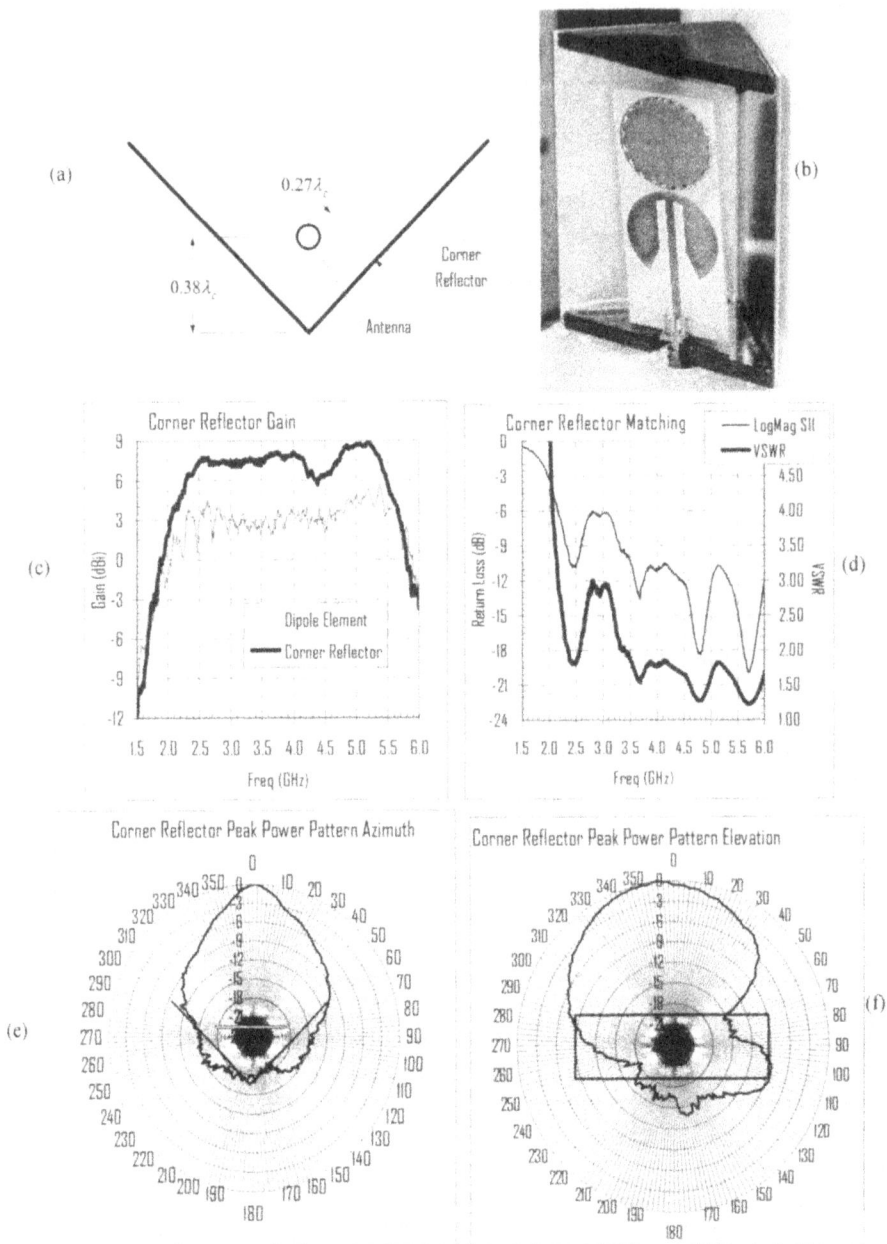

Figure 6.69 (a) Dimensions of a corner reflector optimized for maximal boresight gain, (b) corresponding maximal constructive interference waveform, (c) gain of a corner reflector antenna, (d) matching of a corner reflector antenna, (e) azimuthal pattern of a corner reflector antenna, and (f) elevation pattern of a corner reflector antenna (Courtesy: Time Domain Corporation, Huntsville, Alabama).

6.6.3 Parabolic Cylinder Reflectors

Hertz's parabolic cylinder reflector (see Figure 1.3) can offer additional gain beyond that of a corner reflector, depending on the size of the aperture. Masters applied a similar scheme in conjunction with a diamond dipole element (see Figure 1.18). The geometry of a parabolic reflector is defined by a focal point. Assuming the vertex lies at the origin and the parabola is oriented toward the positive y-axis with a focal point at $y = p$, the equation of the parabola is

$$y = \frac{1}{4p} x^2 \qquad (6.12)$$

Figure 6.70(a) presents the geometry of a parabolic reflector. There are two basic design options for a parabolic cylinder: compact and large.

For a compact parabolic cylinder reflector, choose a focal point at a distance comparable to that for a planar reflector: $p \approx 0.20\lambda - 0.25\lambda$. The criterion for selecting the spacing is to align the direct signal from the dipole at the focal point with the reflected signal from the parabolic cylinder to achieve a favorable constructive superposition. A compact parabolic reflector antenna combines good gain in a small form factor. Nevertheless, this antenna tends to have mediocre matching due to reflection of energy back into the feed antenna.

For a large parabolic cylinder reflector, $p \gg 0.25\lambda$. Aperture blockage is less of a concern than it is with a compact parabolic cylinder reflector, so a larger, more directive feed may be employed. A large parabolic cylinder reflector may have relatively high gain but will naturally be large and bulky. Because the feed can be placed a substantial distance away from the reflector, matching is prone to be better than in a compact parabolic cylinder reflector.

Figure 6.70(b) shows a compact parabolic cylinder reflector with a planar dipole antenna feed. Figure 6.70(c) shows the gain for the parabolic cylinder reflector antenna of Figure 6.70(b) compared to the gain of the dipole element alone. Note that the parabolic cylinder reflector antenna shows roughly a 6-dB improvement over the individual element. This is approximately 1dB better than that for a comparable aperture corner reflector. This parabolic cylinder reflector also has a serious impact on matching. Figure 6.70(d) shows how VSWR peaks as high as 2.5:1 in band. The azimuthal peak power pattern of Figure 6.70(e) shows a 20° wide, –3-dB beamwidth, while the elevation peak power pattern of Figure 6.70(f) exhibits a 60° –3-dB beamwidth. Like corner reflectors, parabolic cylinder reflectors yield a relatively narrow beam on azimuth without much impacting the elevation pattern.

Planar, corner, and parabolic cylinder reflector antennas are three good options for relatively compact yet more directional antennas. Planar reflectors are the easiest to implement and combine modest gain with relatively small size. In

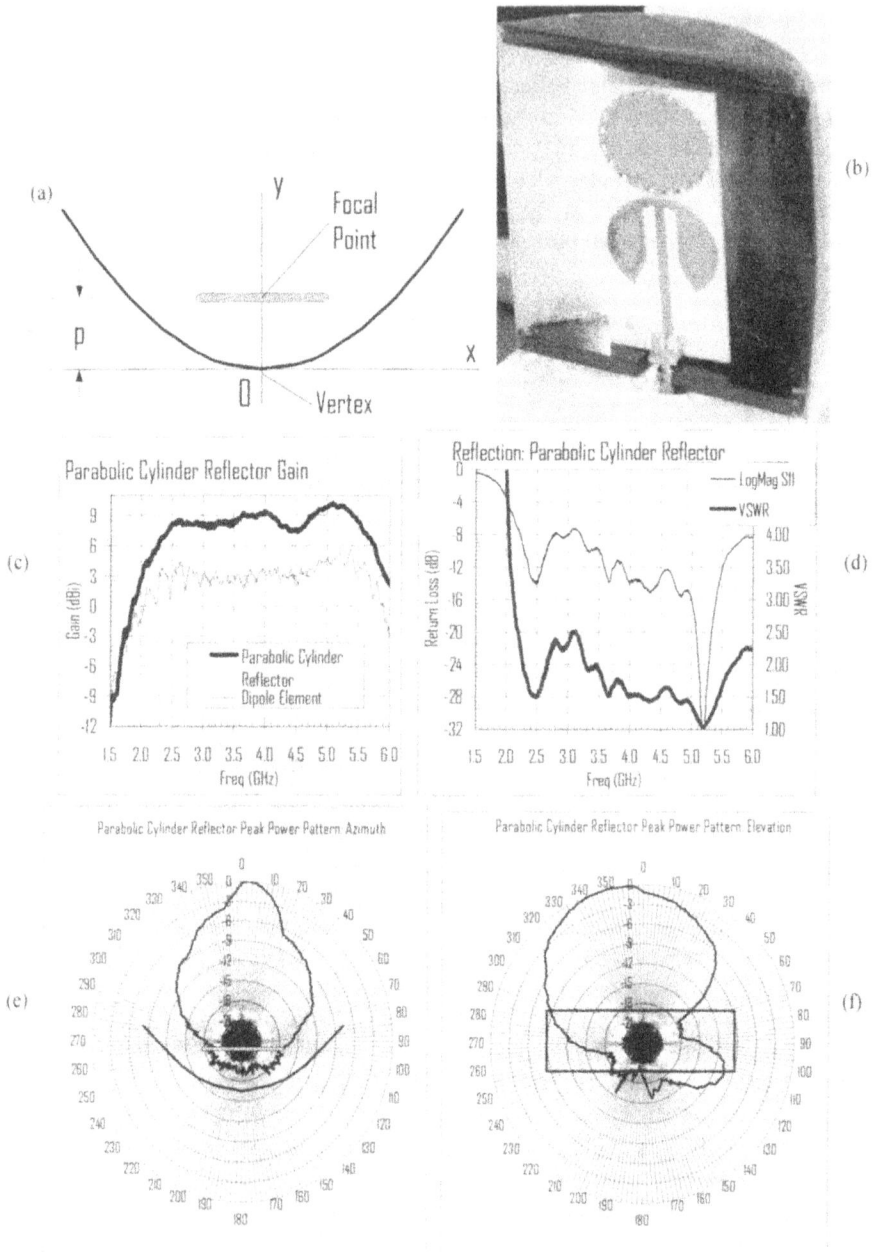

Figure 6.70 (a) Dimensions of a corner reflector optimized for maximal boresight gain.
(b) Corresponding maximal constructive interference waveform. (c) Gain of a corner
reflector antenna. (d) Matching of a corner reflector antenna. (e) Azimuthal pattern of a
corner reflector antenna. (f) Elevation pattern of a corner reflector antenna (Courtesy:
Time Domain Corporation).

some cases, the reflector may be no bigger than a dipole element. Corner and parabolic antennas offer higher gain at the price of a larger form factor. All three compact reflector antennas suffer from poor matching due to reflection of energy back into the feed antenna from the reflector.

6.6.4 Impulse Radiating Antennas

A wide variety of other reflector antennas are suitable for UWB application, including parabolic dish antennas. These are generally well understood and do not require a detailed examination here. One relatively large reflector antenna worthy of a brief mention in this context, however, is the impulse radiating antenna (IRA).

IRAs are parabolic dish reflector antennas fed from traveling wave feed arms. Figure 6.71(a) shows an IRA with feed arms spaced 30° from the vertical, manufactured by Farr Research [104]. These antennas combine high gain with an excellent impulselike time domain response. Figure 6.71(b) shows the gain for this antenna. IRAs approximate constant aperture antennas, so their gain increases with frequency.

6.6.5 Summary of Reflector Antennas

This section has addressed reflector antennas. A reflector antenna relies on a quasioptical reflection of energy from a reflector. Like horn antennas, reflector antennas tend to be relatively large, directional, high-gain antennas. Also, like horn antennas, reflectors can benefit from edge treatment. Figure 6.72 shows a rolled-edge reflector used in a compact range. This edge treatment helps control diffraction and ensure a uniform phase front for precision field-measurement applications.

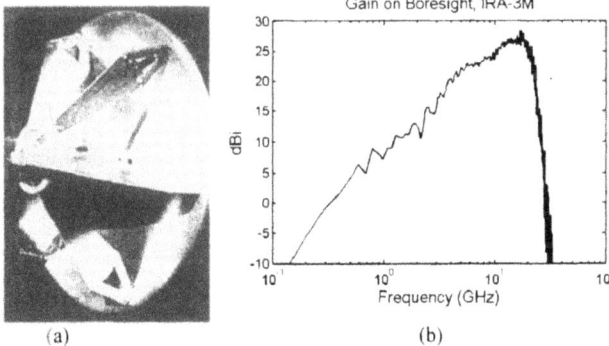

(a)   (b)

Figure 6.71   (a) A Model IRA-3M Farr Research impulse radiating antenna. (b) Gain on boresight for a Farr Research IRA-3M antenna (Courtesy: Farr Research, Inc.).

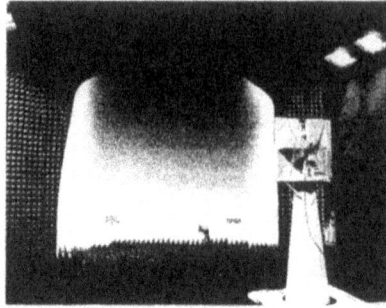

Figure 6.72   Edge treatments are as useful for reflectors as for horns. This rolled-edge reflector is a
key component of a compact range (Courtesy: The ElectroScience Lab, Ohio State
University, Columbus, Ohio).

The main focus of this section has been on compact reflector antennas. Good
performance from a modest-sized antenna is a perennial requirement for
commercial and other practical UWB applications. Planar, corner, and parabolic
cylinder reflector antennas offer designers three relatively simple-to-implement
design alternates when an application calls for a more directional, higher-gain
antenna than a small element.

6.7 SUMMARY

The goal of this chapter was to survey the UWB antenna kingdom. This UWB
antenna taxonomy looked at five important classes of UWB antennas. Frequency-
independent antennas tend to be dispersive and ill suited for many UWB
applications. Small-element electric antennas are the workhorse of UWB
consumer-electronic applications, combining compact size with omnidirectional
field of view. Small-element magnetic antennas are excellent choices for
embedded applications, but many of the more common designs have difficulty
achieving an omnidirectional field of view. If "small" elements are not small
enough, there are a variety of techniques to miniaturize antennas further, including
dielectric loading, using device enclosures as radiating elements, and
superimposing electric and magnetic small-element antennas.

Larger antennas are required where applications call for higher gain and
narrower field of view. Horn antennas are tapered transmission lines that radiate
and receive energy in particular directions. Relatively compact planar horn
antenna designs are also possible. Reflector antennas are also a possibility when
high gain and a narrow field of view are desired.

This completes the taxonomy of UWB antennas. Chapter 7 will move beyond
antennas as discrete components to consider how antennas work as part of a more
comprehensive system to meet overall system goals.

Endnotes

[1] Rumsey, Victor, *Frequency Independent Antennas*, New York: Academic Press, 1965.

[2] Best, Stephen R., "A discussion on the significance of geometry in determining the resonant behavior of fractal and other non-Euclidean wire antennas," IEEE Antenna and Propagation Magazine, Vol. 45, No. 3, June 2003, pp. 9–28.

[3] Kraus, John, *Antennas*, 2nd ed. New York: McGraw-Hill, 1988, p. 346.

[4] Morrow, Jarrett D. and Jeffery T Williams, "Analysis of antennas for ultra-wideband pulse radiation," 2004 IEEE Antennas and Propagation Society International Symposium. Monterey, California. June 20–25, 2004, Vol. 2, pp. 1760–1763.

[5] McNamara, Derek A., D. E. Baker, and L. Botha, "Some design considerations for biconical antennas," 1984 IEEE Antennas and Propagation Society International Symposium, Vol. 22, June 1984, pp. 173–176.

[6] Schelkunoff, S.A. and Harald Friis, *Antennas: Theory and Practice*, New York: John Wiley and Sons, 1952, pp. 314–319.

[7] Baum, Carl E., "Design of a Pulse-Radiating Dipole Antenna as Related to High-Frequency and Low-Frequency Limits," Sensor and Simulation Note #69, January 13, 1969.

[8] Baum, Carl E., "An Equivalent-Charge Method for Defining Geometries of Dipole Antennas," Sensor and Simulation Note #72, January 24, 1969.

[9] Kandoian, A. G., "Three New Antenna Types and Their Applications," Proceedings of the IRE, February 1946, pp. 70W–75W.

[10] Kandoian, A. G., "Broad band antenna," U.S. Patent 2,368,663, 1945.

[11] Snyder, Keith A., and Gary L. Peisley, "Compact omnidirectional antenna," U.S. Patent 5,140,334, August 18, 1992.

[12] Brown, G. H. and O. M. Woodward, "Experimentally Determined Radiation Characteristics of Conical and Triangular Antennas," RCA Review, Vol. 13, December 1952, pp. 425–452.

[13] Schantz, Hans, "Size vs. Performance: The UWB Antenna Dilemma," CommsDesign 2004, San Francisco, California, March 2004. The bow tie results of Figures 6.5–6.7 and other figures in this chapter (as individually noted) were originally presented in this paper. Copyright © 2004 by CMP Media LLC, 600 Harrison Street, San Francisco, CA 94107, USA. Reprinted from the Communications Design Conference 2004 with permission. Section 6.4 of the present chapter is largely derived from this paper.

[14] Masters, Robert W., "Antenna," U.S. Patent 2,430,353, November 4, 1947.

[15] Fullerton, Larry, and Mark Barnes, "Pulse-responsive dipole antenna," U.S. Patent 6,606,051, August 12, 2003.

[16] Schantz, Hans and Larry Fullerton, "The diamond dipole: A Gaussian impulse antenna," 2001 IEEE Antennas and Propagation Society International Symposium. Boston, Massachusetts. July 8–13, 2001, Vol. 4, pp. 100–103.

[17] Kiminami, K., A. Hirata, and T. Shiozawa, "Double-sided printed bow-tie antenna for UWB communications," IEEE Antennas and Wireless Propagation Letters, Vol. 3, 2004, pp. 152–153.

[18] Yazdandoost, Kamya Yekeh and Ryuji Kohno, "Bow-tie antenna for UWB communication frequency," 2004 IEEE Antennas and Propagation Society International Symposium. Monterey, California. June 20–25, 2004, Vol. 3, pp. 2520–2523.

[19] Smith, Les, Tim Starkie, and Jack Lang, "Novel UWB antennas — theory and simulation," 2004 International Workshop on Ultra Wideband Systems. Joint with 2004 Conference on Ultrawideband Systems and Technologies. Kyoto, Japan. May 18–21, 2004, pp. 304–306.

[20] Smith, Les, Tim Starkie, and Jack Lang, "Measurements of Artimi's antenna designs," 2004 International Workshop on Ultra Wideband Systems. Joint with 2004 Conference on Ultrawideband Systems and Technologies. Kyoto, Japan. May 18-21, 2004, pp. 299–303.

[21] Kwon, Do-Hoon and Yongjin Kim, "CPW-fed planar ultra-wideband antenna with hexagonal radiating elements," 2004 IEEE Antennas and Propagation Society International Symposium, Monterey, California. June 20–25, 2004, Vol. 3, pp. 2947–2950.

[22] Lindenblad, Nils E., "Wide band antenna," U.S. Patent 2,239,724, April 29, 1941.

[23] Lindenblad, Nils E., et al., RCA Review, April 1939.

[24] Stöhr, Walter, "Broadband ellipsoidal dipole antenna," U.S. Patent 3,364,491, January 16, 1968.

[25] I have had good experience with spheres from the J.G. Braun Company of Chicago, Illinois. At the time of this writing, J. G. Braun's selection of architectural and ornamental metallic balls and spheres was available at www.jgbraun.com/balls.html.

[26] Lee, K. S. H., "Electrically Small Ellipsoidal Antennas," Sensor and Simulations Note #193, February 1974.

[27] Kraus, John, Antennas, 2nd ed., New York: McGraw-Hill, 1988, pp. 62–64. Note that J. G. Braun also offers hemispheres suitable for constructing dish dipoles.

[28] Lalezari, Farzin, Charles Gilbert, and John Rogers, "Broadband notch antenna," U.S., Patent 4,843,403, June 27, 1989.

[29] Ihara, Taisuke, Koichi Tsunekawa, and Makoto Kijima, "Broadband antenna using a semicircular radiator," U.S. Patent 5,872,546, February 16, 1999.

[30] Anob, P. V., K. P. Ray, Girish Kumar, "Wideband orthogonal square monopole antennas with semi-circular Base," 2001 IEEE Antennas and Propagation Society International Symposium. Boston, Massachusetts. July 8–13, 2001, Vol. 3, pp. 294–297.

[31] Suh, Seong-Youp, Warren L. Stutzman, and William Davis, "Multi-broadband monopole disc antennas," 2003 IEEE Antennas and Propagation Society International Symposium. Columbus, Ohio. June 22–27, 2003, Vol. 3, pp. 616–619.

[32] Yang, Taeyoung and William Davis, "Planar half-disk antenna structures for ultra-wideband communications," 2004 IEEE Antennas and Propagation Society International Symposium. Monterey, California. June 20–25, 2004, Vol. 3, pp. 2508–2511.

[33] Kwon, Do-Hoon; Kim, Yongjin; Hasegawa, Minoru; and Shimamori, Takao, "A small ceramic chip antenna for ultra-wideband systems," 2004 International Workshop on Ultra Wideband Systems. Joint with 2004 Conference on Ultrawideband Systems and Technologies. Kyoto, Japan. May 18–21, 2004, pp. 307–311.

[34] Thomas, Mike and Ronald I. Wolfson, "Wideband arrayable planar radiator," U.S. Patent 5,319,377, June 7, 1994.

[35] For more information on the two penny dipole, see www.timederivative.com/PUBs-2cent-antenna.pdf.

[36] Honda, S., M. Ito, H. Seki, and Y. Jinbo, "A disk monopole antenna with 1:8 impedance bandwidth and omnidirectional radiation pattern," Proc. ISAP'92, Sapporo, Japan, 1992, pp. 1145–1148.

[37] Hammoud, P.P. and Colomel, F., "Matching the input impedance of a broadband disc monopole," Electronics Letters, Vol. 29, February 1993, pp. 406–407.

[38] Yang, Ning, Zhi Ning Chen, and Xuan Hui Wu, "Study of circular planar monopole for radio systems," Asia-Pacific Microwave Conference, Vol. 3, pp. 1628–1631, November 2003.

[39] Agrawall, N. P., et al., "Wide-band planar monopole antennas," IEEE Transactions on Antennas and Propagation, Vol. 46, No. 2, February 1998, pp. 294–295.

[40] Nobuhiro, Kuga, and Hiroyuki Arai, "Wideband elliptical disk monopole antenna," 2003 IEEE Antennas and Propagation Society International Symposium. Columbus, Ohio. June 22–27, 2003, Vol. 3, pp. 624–627.

[41] Song, C. T. P., Peter S. Hall, H. Ghafouri-Shiraz, "Multiband multiple ring monopole antennas," IEEE Transactions on Antennas and Propagation, Vol. 51 No. 4, April 2003, pp. 722–729.

[42] Schantz, Hans, "Planar elliptical element ultra-wideband dipole antennas," 2002 IEEE Antennas and Propagation Society International Symposium, San Antonio, Texas, June 16–21, 2002, Vol. 3, pp. 16–21.

[43] Ye, Quibo and Wilfred R. Lauber, "Microstrip ultra-wideband dipole antenna simulation by FDTD," 2003 IEEE Antennas and Propagation Society International Symposium, Columbus, Ohio, June 22–27, 2003, Vol. 3, pp. 620–623.

[44] Specifications for the PPL Model 4050B pulser are available at www.picosecond.com.

[45] Specifications for the Farr Research Model TEM-2-50 TEM horn are available at www.farr-research.com.

[46] Schantz, Hans, "Bottom fed planar elliptical UWB antennas," 2003 IEEE Conference on Ultra Wideband Systems and Technologies. Reston, Virginia. November 16–19, 2003, pp. 219–223.

[47] Schantz, Hans, "Apparatus for establishing signal coupling between a signal line and an antenna structure," U.S. Patent 6,512,488, January 28, 2003.

[48] Schantz, Hans, "Apparatus for establishing signal coupling between a signal line and an antenna structure," U.S. Patent 6,642,903, November 4, 2003.

[49] Suh, Seong-Youp, Warren Stutzman William Davis, Alan Waltho, and Jeffrey Schiffer, "A novel broadband antenna, the low profile dipole planar inverted cone antenna (LPdiPICA)," 2004 IEEE Antennas and Propagation Society International Symposium, Monterey, California. June 20–25, 2004, Vol. 1, pp. 775–778.

[50] Suh, Seong-Youp, and Warren L. Stutzman, "Planar wideband antennas" U.S. Patent Publication 2003/0210207 A1, November 13, 2003.

[51] Zollinger, Ernst, "Planare breitbandantenne," German Patent DE 197 29 664 A 1, February 18, 1999.

[52] Schantz, Hans, "UWB magnetic antennas," 2003 IEEE Antennas and Propagation Society International Symposium. Columbus, Ohio. June 22–27, 2003, Vol. 3, pp. 604–607.

[53] Harmuth, Henning F., "Frequency independent shielded loop antenna," U.S. Patent 4,506,267, March 19, 1985.

[54] Harmuth, Henning F., *Nonsinusoidal Waves for Radar and Radio Communication*, New York: Academic Press, 1981, pp. 108–110.

[55] Harmuth, Henning F. and S. Ding-rong, "Large-current, short-length radiator for nonsinusoidal waves," IEEE International Symposium on Electromagnetic Compatibility, 1983 pp. 453–456.

[56] *FCC Filing of Aether Wire re 98-153, March 27, 2001* See also: www.aetherwire.com.

[57] Pochanin, Gennadiy P., "Large Current Radiator for the Short Electromagnetic Pulses Radiation," *Ultra-Wideband, Short-Pulse Electromagnetics*, Vol. 4, Heyman et al. eds. New York: Plenum Publishers, 1999, p. 150.

[58] Farr, Everett G., et al. "A Two-Channel Balanaced-Dipole Antenna (BDA) with Reversible Antenna Pattern Operating at 50 Ohms," Sensor and Simulation Notes #441, Air Force Research Laboratory, December 1999.

[59] Barnes, Mark, private communication with the author, 2000.

[60] Collin, Robert E., *Field Theory of Guided Waves*, New York: McGraw-Hill, 1960, p. 272.

[61] Turner, Edwin M. and William P. Turner, "Scimitar antenna," U.S. Patent 3,015,101, December 26, 1961.

[62] Partridge, R. E., "Combined E and B-dot Sensor," Sensor and Simulation Note #3, Air Force Research Lab Directed Energy Directorate, February 1964.

[63] Schantz, Hans, "Single element antenna apparatus," U.S. Patent 6,437,756, August 20, 2002.

[64] Barnes, Mark, et al., "Impulse radar antenna array and method," U.S. Patent 6,667,724, December 23, 2003.

[65] Schantz, Hans, "Planar loop antenna," U.S. Patent 6,593,886, July 15, 2003.

[66] Harmuth, Henning, *Antennas and Waveguides for Nonsinusoidal Waves*, New York, Academic Press, 1984, pp. 98–99.

[67] Barnes, Mark, "Ultra-wideband magnetic antenna," U.S. Patent 6,091,374, July 18, 2000.

[68] Barnes, Mark, "Ultra-wideband magnetic antenna," U.S. Patent 6,400,329, June 4, 2002.

[69] Barnes, Mark, "Ultra-wideband magnetic antenna," U.S. Patent 6,621,462, September 16, 2003.

[70] Hannigan, Andrew B., Steve R. Pencock, and Peter Shepard, "Broadband Operation of Tapered Inset Dielectric Guide and Bowtie Slot Antennas," in *Ultra-Wideband, Short-Pulse Electromagnetics*, Vol. 5, P. Smith and S. Cloude, eds., New York: Kluwer Academic/Plenum Publishers, 2002, pp. 319–326.

[71] Suh, Young Hoon and Ikmo Park, "A broadband eccentric annular slot antenna," 2001 IEEE Antennas and Propagation Society International Symposium. Boston, Massachusetts. July 8–13, 2001, Vol. 3, pp. 294–297.

[72] Chen, Horng-Dean, "Broadband CPW-fed square slot antennas with a widened tuning stub" IEEE Transactions on Antennas and Propagation, Vol. 51, No. 8, August 2003, pp. 1982–1986.

[73] McCorkle, John, "Electrically small planar UWB antenna apparatus and related system," U.S. Patent 6,590,545, July 8, 2003.

[74] Powell, Johnna and Anantha Chandrakasan, "Differential and single ended elliptical antennas for 3.1-10.6GHz ultra-wideband communication," 2004 IEEE Antennas and Propagation Society International Symposium, Monterey, California. June 20–25, 2004, Vol. 3, pp. 2935–2938.

[75] Raman, Sanjay, and Gabriel M. Rebeiz, "Single- and dual-polarized millimeter-wave slot-ring antennas," IEEE Transactions on Antennas and Propagation, Vol. 44, No. 11, November 1996, pp. 1438–1444.

[76] About 225 ft in an admittedly unscientific experiment performed by the author.

[77] Schantz, Hans, "Nano-antenna apparatus and method," U.S. Patent Pending.

[78] Harper, Warren, and Tom Barr in private communication with the author, 2004.

[79] Chu, L.J., "Physical limitations of omni-directional antennas," Journal of Applied Physics, Vol. 19, 1948, pp. 1163–1175.

[80] Runge, Wilhelm, "Polarization diversity reception," U.S. Patent 1,892,221, December 27, 1932.

[81] Kraus, John, *Antennas*, 2nd ed. New York: McGraw-Hill, 1988, p. 264. See Problem 6-9.

[82] Kandoian, Armig, "Antenna unit," U.S. Patent 2,465,379, March 29, 1949.

[83] Kandoian, Armig, "Three New Antenna Types and Their Applications," Proceedings of the IRE, February 1946, pp. 70W–75W.

[84] Schantz, Hans, "The Energy Flow and Frequency Spectrum about Electric and Magnetic Dipoles," Ph.D. dissertation, University of Texas, Austin, Austin, Texas, 1995.

[85] Schantz, Hans, "UWB loop antenna and electric magnetic antenna system," U.S. Patent Pending.

[86] Yang, F.C. and K. S. H. Lee, "Impedance of a two conical plate transmission line," Sensor and Simulation Note #221, November, 1976.

[87] Lee, R.T., and Glenn S. Smith, "On the characteristic impedance of the TEM horn antenna," IEEE Transactions on Antennas and Propagation, Vol. 52, No. 1, January 2004, pp. 315–318.

[88] For instance, ETS-Lindgren. See www.ets-lindgren.com.

[89] For more information on Farr Research, Inc. and its antenna offerings see www.farr-research.com.

[90] Burnside, Walter D. and Chiwei Chuang, "An aperture matched horn design," IEEE Transactions on Antennas and Propagation, Vol. 30, July 1982, pp. 790–796.

[91] Chang, Li-Chung T. and Walter D. Burnside, "An ultrawide-bandwidth tapered resistive TEM horn antenna" IEEE Transactions on Antennas and Propagation, Vol. 48, No. 12, December 2000, pp. 1848–1857.

[92] Shin, Joon and D. H. Schaubert, "A parameter study of stripline-fed Vivaldi notch-antenna arrays," IEEE Transactions on Antennas and Propagation, Vol. 47, No. 5, May 1999, pp. 879–886.

[93] Nester, William, "Microstrip notch antenna," U.S. Patent 4,500,887, Feb. 19, 1985.

[94] Aiello, G. Roberto and Patricia Foster, "Antenna comprising two wideband notch region on one coplanar substrate," U.S. Patent 6,292,153, September 18, 2001.

[95] Schantz, Hans, "Improved system and method for directional transmission and reception of ultra-wideband signals," U.S. Patent Pending.

[96] Brillouin, Leon N., "Broad band antenna," U.S. Patent 2,454,766, November 30, 1948.

[97] Schantz, Hans, "Semi-coaxial horn antenna," U.S. Patent 6,538,615, March 25, 2003.

[98]   Shen, Hao-Ming, R. W. P. King, and Tai Tsun Wu, "V-conical antenna," IEEE Transactions on Antennas and Propagation, Vol. 36 , No. 11 , November 1988, pp. 1519–1525.

[99]   Wicks, Michael C. and Paul Antonik, "Polarization Diverse Ultra-Wideband Antenna Technology," in *Ultra-Wideband, Short-Pulse Electromagnetics*, Vol. 1, H. Bertoni, L. Carin, L. Felsen, eds., New York: Plenum Press, 1993, pp. 177–187.

[100]   Collin, Robert E., *Op. Cit.*, see in particular Chapter 4.

[101]   Smythe, William R., *Static and Dynamic Electricity*, 3rd ed., New York: McGraw Hill, 1968. See in particular Chapter 4 and Chapter 11.

[102]   Hallquist, Richard, private communication, 2000. Much of the work of this and the following section is originally due to Richard Hallquist.

[103]   Kraus, John D., "Corner reflector antenna," U. S. Patent 2,270,314, January 20, 1942.

[104]   Additional specifications and ordering information for Farr Research IRAs are available at www.farr-research.com.

Chapter 7

UWB Antennas in Systems

> *The antenna is the connecting link between free-space and the transmitter or receiver. As such, it plays an essential part in determining the characteristics of the system in which it is used.*

Henry Jasik, 1961

Thus far, this book has considered UWB antennas as an individual component characterized by certain properties and parameters. UWB antennas do not exist in a vacuum, however. They are part of an overall RF system designed to meet the requirements of specific applications. A UWB antenna may work together with other components of a system to meet overall system goals. A poorly designed antenna will detract from overall system performance. The aim of this chapter is to explore how antenna-system interactions can improve system performance.

The shortcomings and deficiencies of a traditional narrowband antenna operating at a particular frequency may often be accommodated by an easily designed matching filter. A poorly matched UWB antenna may benefit somewhat from a matching network, but a better approach is to design in a good match from first principles. Chapter 3 addresses this important topic. While a good impedance match is crucial, an important corollary is "over what frequencies?" This chapter addresses additional techniques for spectral control using UWB antennas, including use of UWB antennas to provide spectral filtering.

Antenna efficiency is also an important system-level concern. Inefficient antenna operation impacts system performance and cannot easily be remedied elsewhere in the system. Thus, this chapter will discuss UWB antenna efficiency and present a technique for assessing UWB antenna efficiency.

Field of view and directivity considerations also play an important role in determining overall system performance. A directional, high-gain antenna provides a more robust link but suffers from a restricted field of view. This chapter considers how to balance the trade-off between directionality and field of view. This chapter further considers some techniques for direction finding (DF) in UWB systems using relatively compact antenna systems.

7.1 ANTENNA SPECTRAL CONTROL

Antenna spectral control has its roots in the early days of radio. The antennas of the early radio pioneers were often electrically small, tuned to radiate the higher-order harmonics of a damped sinusoid excitation. Thus, these antennas behaved as high pass filters, preferentially radiating higher frequencies and rejecting lower frequencies.

The radiated spectrum from a UWB system is the product of the applied excitation spectrum and the antenna spectral response. Control of a UWB antenna spectral response contributes to the overall system response. First, this section addresses antenna scaling for gross spectral control. Then, this section considers antenna filtering for fine spectral control. UWB antennas may be modified to include a band stop filter response, a low pass filter response, or a high pass filter response.

7.1.1 Antenna Scaling

Just as a desired impedance can be designed into an antenna, so also can a desired frequency range. The simplest such manipulation is varying the scale of an antenna. For instance, planar elliptical dipoles offer a S_{11} on the order of –20dB across a 3:1 frequency range (see Section 6.2.4.2). The minor axis is approximately 0.14λ at the lower end of the operating band. Thus, a 1–3GHz antenna will have approximately 42.4mm (1.67in) elements, a 2–6GHz antenna will be half the size (one-fourth the area) with about 21.2mm (0.83in) elements, and a 3–9GHz antenna will be one third the size (one-ninth the area) with approximately 14.1mm (0.56in) elements. The antenna size may be scaled to select any particular 3:1 range of desired frequencies. Figure 7.1 shows three versions of these antennas scaled for various frequency ranges.

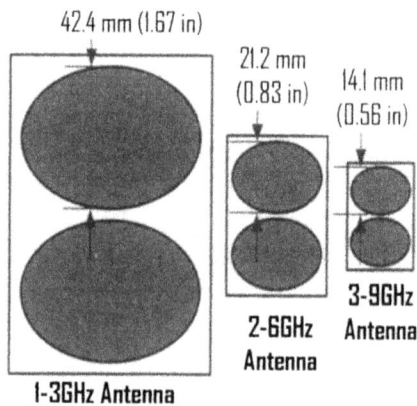

Figure 7.1   Antenna scaling adjusts the frequency range of an antenna ([1]; © 2003, IEEE).

The feed region poses the only significant difficulty to adjusting an antenna's response by scaling. For instance in the case of the elliptical dipoles of Figure 7.1, the feed impedance depends on the gap between the elements. To maintain the same feed impedance requires the same gap spacing.

Element scaling only allows selection of a particular band of operation. More elaborate spectral shaping requires more sophisticated techniques. In-line filters are a traditional means for controlling RF spectral response. In many cases, however, an antenna can be designed so as to implement desired spectral behavior. These techniques are the subject of the following section.

7.1.2 Antenna Filtering

UWB systems use vast expanses of bandwidth in their operation, leaving them vulnerable to whatever narrowband sources of interference might exist in their environment. If narrowband sources of interference occupy known fixed frequencies, it is possible to use RF filtering techniques to block these frequencies.

For instance, it may be desirable to collocate a UWB receiver and a relatively high-power narrowband system. The high-power narrowband signals will be prone to cause interference to the collocated UWB receiver. Alternatively, perhaps it is desired to collocate a UWB transmitter and a sensitive narrowband receiver. Then the UWB system may be prone to cause interference with the narrowband receiver under some circumstances. In either case, it is desirable to have a UWB antenna system that is not sensitive to one or more narrowbands of interest.

The traditional in-line filter is a lumped element or transmission line structure in the front end of an RF device. When the spectral response of a UWB antenna is combined with the spectral response of an in-line notch filter, the resulting composite spectral response exhibits the expected nulls, or notches. Figure 7.2 illustrates this behavior.

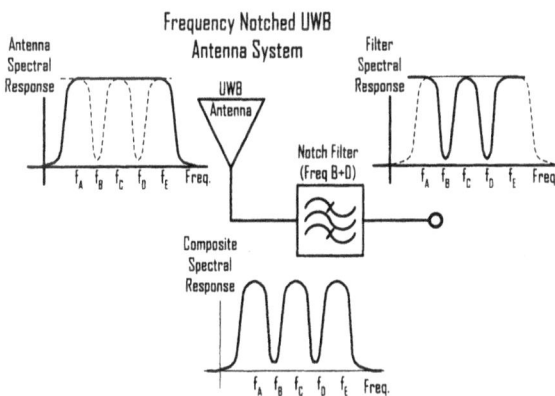

Figure 7.2 Application of a notch filter in line with a UWB antenna can yield a composite spectral response with frequency notches ([1], © 2003, IEEE).

In the example of Figure 7.2, a UWB antenna element has a frequency response sensitive across a UWB range of frequencies from f_A to f_E. A frequency notch filter passes a UWB range of frequencies from f_A to f_E with the exception of those frequencies in the vicinity of f_B and f_D. The resulting spectral response of a frequency notched UWB antenna system is sensitive to a UWB range of frequencies from f_A to f_E with the exception of those frequencies in the vicinity of f_B and f_D. Frequencies in the vicinity of f_B and f_D are "notched out" of the overall spectral response, and a frequency-notched UWB antenna system may be coupled via an output to an appropriate transmitter, receiver, or transceiver.

Note that the UWB antenna element and frequency notch filter are distinct entities separated by a transmission line. This means that the cost and complexity of a frequency-notched UWB antenna system will be greater than the cost of implementing a nonnotched UWB antenna system.

7.1.2.1 Resonant Structure Filtering

An alternate technique to achieve a desired antenna spectral response is to incorporate narrowband resonant structures in an otherwise UWB antenna element so as to create a frequency-notched UWB antenna element. Figure 7.3 shows the general concept.

A UWB antenna element with a spectral response covering a desired range of frequencies serves as a starting point. Then, narrowband resonant structures with response at a particular frequency modify the UWB element. These narrowband resonant structures act as wave traps, capturing particular frequency components and reflecting them away. In Figure 7.3, an otherwise UWB element incorporates two narrowband resonant structures that ring at frequencies f_B and f_D. Frequencies in the vicinity of f_B and f_D are notched out of the overall spectral response.

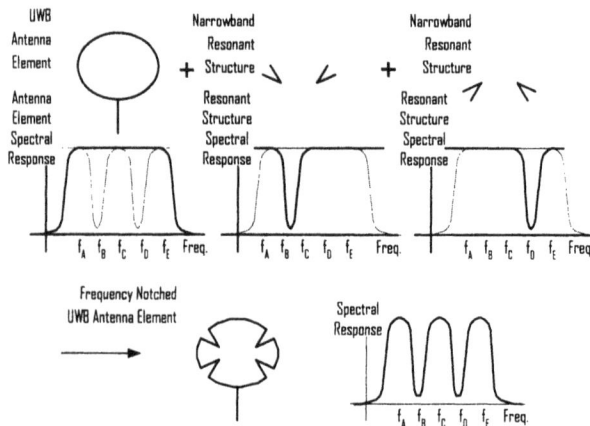

Figure 7.3   Application of a notch filter in line with a UWB antenna can yield a composite spectral response with frequency notches ([1]; © 2003, IEEE).

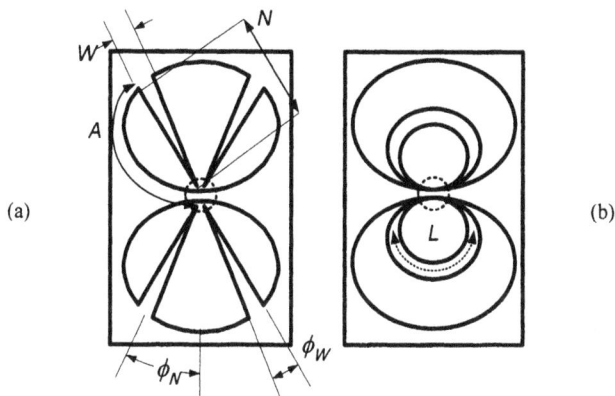

Figure 7.4 (a) A triangular notch UWB antenna is insensitive to frequencies where $A + N$ is a multiple of a half wavelength. (b) An elliptical notch UWB antenna is insensitive to frequencies where L is a multiple of a half wavelength ([1]; © 2003, IEEE).

There is a wide variety of ways to implement a narrowband resonant structure. For instance, lumped elements can yield a narrowband resonance response. This technique has the disadvantage of requiring additional components and complexity but the advantage of being able to implement a diplexing filter that resistively terminates undesired frequency components. An alternate technique is to implement a narrowband resonance in the structure of the antenna itself. This technique has the advantage of low cost and ease of implementation but results in a reactive filter that reflects undesired frequency components.

One way to implement a frequency-notched UWB antenna is to incorporate a half-wave resonant structure in an antenna. Various implementations are possible, two of which are shown in Figure 7.4. A "triangular notch" UWB antenna element has a frequency notch where the arc length A and the notch length N add up to form a half-wavelength resonant structure. Figure 7.4(a) shows a triangular notch antenna. When the $A + N$ path is a half-wavelength at a particular frequency, a destructive interference takes place rendering the antenna nonresponsive at that frequency.

Similarly, an "elliptical notch" UWB antenna element has a frequency notch where the arc length L forms a half-wavelength resonant structure. When the L path is a half-wavelength at a particular frequency, a destructive interference takes place rendering the antenna nonresponsive at that frequency. Figure 7.4(b) shows an elliptical notch antenna.

To evaluate this concept, my colleagues and I created a test matrix of various triangular notch and elliptical notch dipoles [1]. Figure 7.5(a) shows a matrix of triangle notch antennas. The triangular notch antennas are described by the notch width angle ϕ_W and the notch angle ϕ_N. For instance, antenna 1030 has notch width angle $\phi_W = 10°$ and notch angle $\phi_N = 30°$.

The theoretical notch frequency is determined by the frequency at which $A + N$ is a half-wavelength for the triangular notch dipoles. For the elliptical notch dipoles, the theoretical notch frequency is the frequency at which L is a half-wavelength. Although the antennas were constructed on a Rogers RO-4003 dielectric with $\varepsilon_r = 3.38$, free-space propagation was assumed since the electromagnetic energy propagates predominantly in the free-space around the antennas. Although most of the triangle notches coupled back to the feed region forming a closed loop, two of the antennas in the matrix had more shallow notches. The shallow notches should also show a resonance where $A + N$ is a half-wavelength, but the resulting resonance was at too high a frequency to spot in measurements and so was omitted.

Figure 7.5(c) shows boresight gain for the triangular notch antennas of Figure 7.5(a). Detailed results are in Table 7.1. Fractional bandwidths of the nulls ranged from 3% to 13%. For notches located at the same angle, angular notch width is correlated with frequency notch width. The location of the notch was within 10% of the theoretically predicted value.

Figure 7.5(b) shows elliptical notch antennas. The elliptical notch antennas are described as a percentage of the outer circumference. For example, C70 has an elliptical arc length L equal to 70% of the outer circumference of the element. Figure 7.5(d) shows boresight gain for the elliptical notch antennas of Figure 7.5(b). Fractional bandwidths of the notches were around 25%. As with triangular notches, varying the width of the elliptical notches would probably effect the width of the notch, although this hypothesis was not tested. Again, the location of the notch was within about 10% of the theoretically predicted value.

Table 7.1

Performance of Selected Frequency Notch Antennas

Antenna	Notch Frequency (GHz):			Null BW (%)	Null Depth (dB)
	Measured	Theory	Delta (%)		
1015	2.44	2.52	−3.3	2.8	9.9
1030	2.48	2.64	−6.1	6.3	15.1
1530	2.50	2.73	−8.4	8.1	17.9
530	2.64	2.56	3.2	3.4	12.2
1060	2.71	2.92	−7.2	13.2	24.8
C90	2.20	2.37	−7.4	24.1	29.4
C80	2.61	2.72	−3.8	25.1	22.5
C70	3.08	2.99	2.9	25.5	23.3
C60	3.79	4.23	−10.4	25.0	17.5

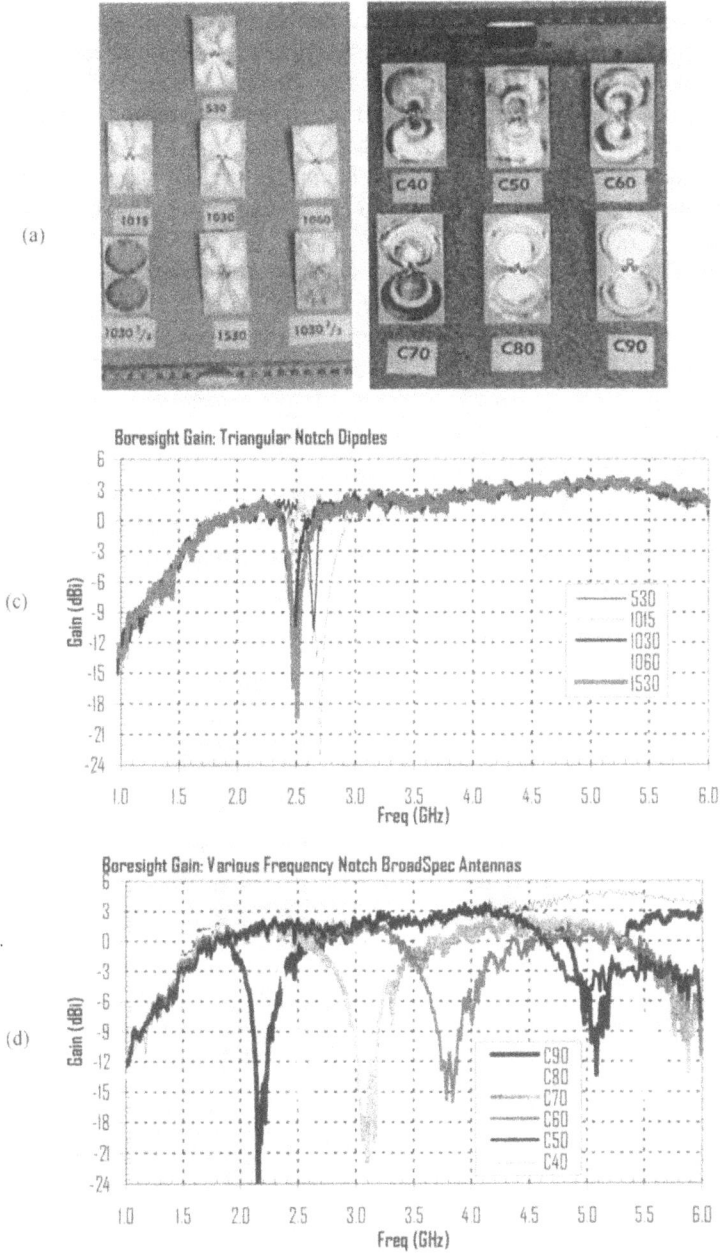

Figure 7.5 (a) A matrix of triangular notch UWB antennas, (b) a matrix of elliptical notch UWB antennas, (c) gain of various triangular notch UWB antennas, and (d) gain of various elliptical notch UWB antennas ([1]; © 2003, IEEE).

Additional details of antenna frequency notching are presented elsewhere [2]. In general, performance of a UWB system deteriorates with an increasing number of spectral notches, but a system designer may be able to accept one or perhaps two as part of an overall compatibility and interference strategy. There are many ways in which one can introduce a narrowband resonance in a UWB antenna so as to create a desired frequency-notching behavior. A variety of researchers have proposed alternate frequency notch UWB antennas using narrowband resonant structures.

Figure 7.6(a) shows a tapered horn antenna with integrated quarter-wave stubs similar to the antenna proposed by Ick-Jae Yoon et al. [3]. The authors obtained about a 256Ω impedance in band which suggests about a 2.6-dB deep notch. This antenna rejects the 5.05–5.93GHz band. Figure 7.6(b) presents a coplanar counterpoise monopole element with resonant stubs suggested by Hyungkuk Yoon et al. [4]. This antenna offers similar performance: about a 3-dB rejection. Figure 7.6(c) shows the square element frequency notch slot of Aaron Kirchoff and Hao Ling [5]. This monopole antenna achieved rejection on the order of 20dB. Figure 7.6(d) shows a CPW-fed bow tie slot antenna devised by Yongjin Kim and Do-Hoon Kwon [6]. This antenna employs a resonant quarter-wave slot at 5.25GHz to achieve a 9-dB reduction in gain around that frequency.

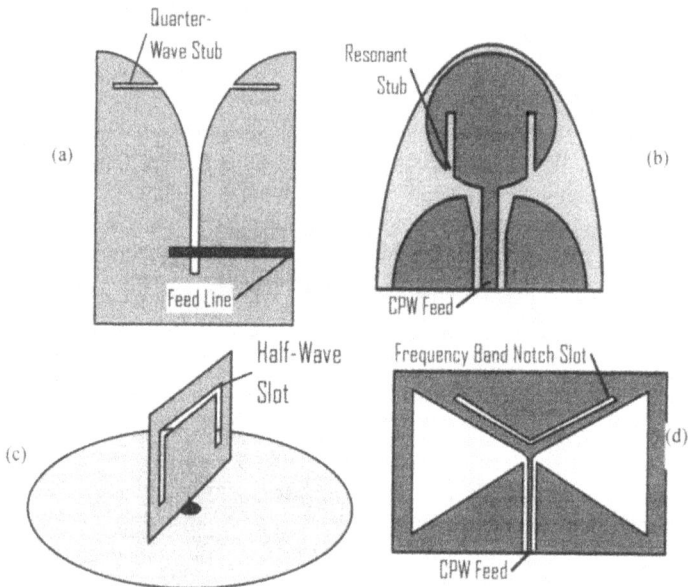

Figure 7.6   (a) Frequency notched tapered horn of Yoon et al. (After [3]), (b) frequency notched co-planar counterpoise monopole element of Yoon et al. (After [4]), (c) frequency notched square monopole element of Kirchoff and Ling (After [5]), and (d) frequency notched bowtie slot antenna of Kim and Kwon (After [6]).

7.1.2.2 Stepped-Impedance Line Filtering

The narrowband resonant structures of the previous section provide excellent frequency notch filtering responses with rejection of as much as 20dB. The techniques of Section 7.1.2.1 work well as band stop filters but do not readily provide low pass or other filter responses. Stepped-impedance line filtering opens new opportunities for spectral control of antennas [7].

A high-impedance transmission line (for instance, a wide slot line) electrically equates to a series inductor. A low-impedance transmission line (for instance, a narrow slot line) electrically equates to a parallel capacitor. By combining the electrical equivalents of series inductors and shunt capacitors, a low pass filter can be created. The range between a high and a low impedance achievable with a particular transmission-line geometry determines the efficacy of filtering. A large range in impedance allows for compact, electrically short filtering. In the context of transmission lines, stepped-impedance filtering is well understood and often discussed in textbooks [8]. This technique has not apparently been applied to antennas until recently.

One possible reason is that it remains difficult to achieve a very low impedance using a traditional slot line. If, however, the slot line is formed by metallizations on opposite sides of a dielectric, the slot width can be reduced to zero or even made to overlap. This yields a very low impedance. Opposing metallization slot lines allows one to benefit from the advantages of a slot line transmission line while also being able to achieve a low-impedance transmission line suitable for use as a stepped-impedance transmission line filter. For instance, consider the slot element of Figure 7.7(a). The wide and narrow slot line sections electrically equate to a four-element filter comprising two inductors and two parallel capacitors, shown in Figure 7.7(b). This circuit constitutes a fourth-order low pass filter.

Table 7.2

Example Sixth-Order Stepped-Impedance Low Pass Filter

Phase Angle at 5.9GHz (°)	$Z(\Omega)$
6.9	8.62
36.8	376.7
11.4	4.8
78.8	376.7
14.1	52.4
92.7	376.7

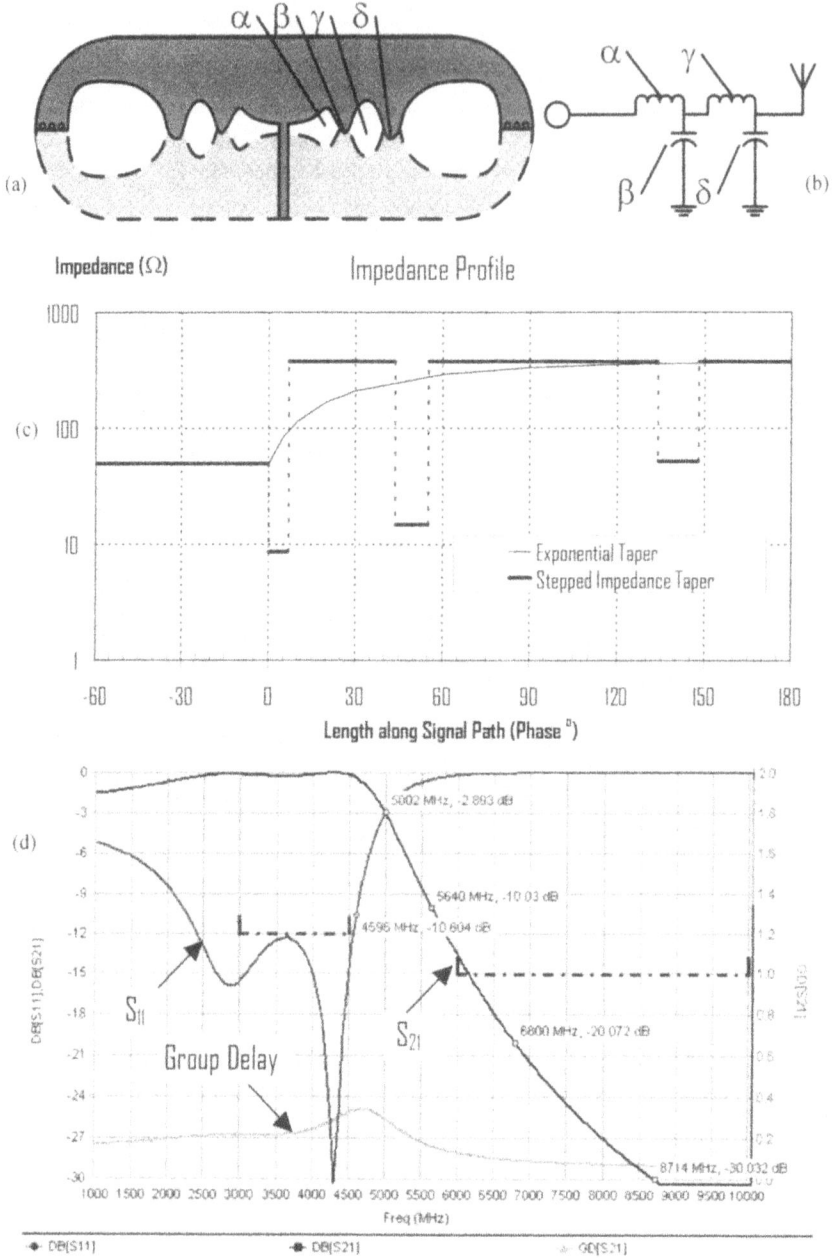

Figure 7.7   (a) A stepped impedance UWB slot antenna with a four pole low pass response, (b) equivalent circuit, (c) an impedance profile to yield a six pole low pass filter response, and (d) an Eagleware Genesys simulation of a six pole stepped impedance antenna low pass filter response (Courtesy: Next-RF, Inc., Huntsville, Alabama).

Figure 7.7(c) shows an impedance taper for a potential sixth-order low pass filter implementation. Table 7.2 presents the exact parameters. Impedance is stepped along 241° at 5,900MHz. This yields an electrical length of 34mm in free-space, less in a dielectric. The stepped-impedance filter of Figure 7.7(c) (and Table 7.2) has the low pass spectral response of Figure 7.7(d). Return loss, or S_{11}, is -12dB or better in the pass band from 3,000–4,500MHz. The 3-dB point is 5,000MHz, and the through response, or S_{21}, is down -10dB by 5,640MHz. This excess out-of-band rejection does not need to be provided by a front-end RF filter. The group delay is not excessive.

Of course, one may also inductively or capacitively load an antenna using discrete components, but using variations in the antenna transmission line or other variations in antenna metallization is preferred for reasons of cost and simplicity. The antenna essentially represents a fixed cost, and the techniques discussed in this section allow for clever modifications of the antenna to replace or eliminate discrete components in a matching filter, reducing overall system cost, manufacturing cost, system size, and system complexity.

Antennas are inherently high pass filters. Using stepped-impedance filter techniques, one may insert a low pass response into an antenna. Although the series inductor and shunt capacitor behavior of stepped impedance filters is ideal for low pass filter responses, these techniques may be applied to other filters as well. In particular, stepped impedance techniques can help a UWB antenna improve matching by allowing construction of a broadband impedance matching network.

7.1.3 Antennas and Spectral Control

Antenna spectral control is an important and often overlooked aspect of UWB system design. Designers are accustomed to working with narrowband systems in which the flaws of an antenna may be rectified by an easily implemented matching network. UWB matching networks are a last resort, not first aid. Ideally, a UWB antenna should be designed so as to cooperate with the RF front end to achieve a desired spectral result. Figure 7.8 illustrates this interplay.

The Golden Rule of UWB antenna design is

$$P_{TX}(f)\, G_{TX}(f) = \text{Mask}(f) \tag{7.1}$$

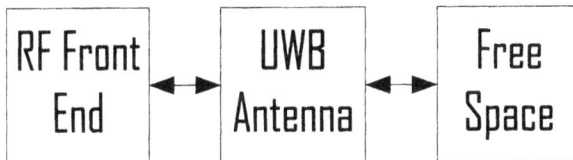

Figure 7.8   RF front ends and UWB antennas work together to define the overall system response.

The transmit power $P_{TX}(f)$ and the antenna gain $G_{TX}(f)$ at any frequency (f) cooperate to meet a desired mask. This "holistic" point of view is essential to a good UWB system design [9, 10].

For instance, radiated power must be down −10dB below 3.1GHz for an indoor UWB device to meet the FCC mask. Thus, an antenna that is perfectly matched at 3.1GHz represents poor design practice. Instead, a designer should incorporate as much of the antenna high pass response as possible to meet the mask. This not only places less demand on the RF front end (resulting in a simpler, less-expensive design) but also involves shifting the antenna response up in frequency (scaling the antenna down in size).

This section has presented two powerful techniques for UWB antenna spectral control: resonant structure filtering and stepped-impedance filtering. These methods offer an unprecedented amount of control to a UWB system designer. Rather than having to impose a complicated matching network between an RF front end and an antenna, in many cases the spectral properties of the antenna may be deliberately modified and controlled. This allows a UWB system designer to eliminate or reduce significantly the complexity of a matching network, thus reducing parts count, component cost, and manufacturing cost.

7.2 ANTENNA EFFICIENCY

In many areas of RF design, efficiency is an important but not critical consideration, and an inefficiency in one area may be accommodated elsewhere in the system. One would like to have an efficient amplifier for instance, but a certain degree of inefficiency may be accommodated by a larger battery or a shorter battery lifetime without directly impacting overall system performance. Efficiency, in this context, represents just one more factor to be weighed in the overall design process. In a system limited by thermal noise, antenna inefficiencies, however, cannot be similarly compensated for without directly impacting critical system parameters like range or data rate. There may be more leeway for interference limited systems.

Fortunately, a typical UWB antenna of approximately quarter-wavelength scale is highly efficient. If a designer uses a high-quality dielectric substrate, losses will be minimal. Nevertheless, the inevitable pressure toward ever smaller and more compact antennas means that eventually a designer will need to weigh the trade-off between antenna size and efficiency. Thus, antenna efficiency is an important consideration in UWB antenna design.

This section discusses the theory behind antenna efficiency and the controversy underlying the definition of antenna efficiency. Then, it reviews experimental techniques for evaluating the efficiency of UWB antennas. One technique of particular interest is the use of a spherical Wheeler Cap to assess the efficiency of UWB antennas.

7.2.1 Efficiency Theory

The radiation efficiency of an antenna is the ratio of the total power radiated by the antenna to the net power accepted by the antenna at its terminals during the radiation process.

IEEE Standard Test Procedures for Antennas
IEEE Std 149-1979, p. 112

Efficiency is defined as the ratio of radiated power (P_{rad}) to input power for an antenna (P_{in}):

$$\eta \equiv \frac{P_{rad}}{P_{in}} \tag{7.2}$$

Of course, one must either integrate power over the duration of the radiation process to obtain the total energy or average the power over an appropriate period. For a harmonic or narrowband system, one may use the root mean square (rms) power; for a UWB or short-pulse radiating antenna, one may calculate efficiency as a function of frequency or determine the efficiency relative to a certain excitation of interest.

A more subtle complication lies in the definition of input power. The *IEEE Standard Test Procedures for Antennas* defines input power as "the net power accepted by the antenna at its terminals during the radiation process" [11]. This is a reasonable assumption to make in the limit of a narrowband antenna. In the narrowband case, one assumes that the antenna is operating at a resonance and that the reactance is negligible. In principle, one could devise a narrowband matching network to eliminate any mismatch, so there would be no mismatch reflected power ($P_{reflected}$). Under these assumptions, all the applied power is accepted:

$$\eta_{IEEE} \equiv \frac{P_{rad}}{P_{in}} = \frac{P_{rad}}{P_{applied}} = \frac{P_{rad}}{P_{accepted}} \tag{7.3}$$

There is no distinction between the applied power at the antenna terminals ($P_{applied}$) and the accepted power at the antenna terminals ($P_{accepted}$). Any accepted power is either radiated away (P_{rad}) or consumed in ohmic or dielectric losses (P_{loss}):

$$P_{in} = P_{applied} = P_{accepted} = P_{rad} + P_{loss} \tag{7.4}$$

This traditional approach to antenna efficiency ignores the power lost due to mismatch reflection. Even in the case of a narrowband antenna at resonance, if the radiation resistance deviates from the source impedance, some of the applied power will be lost to mismatch reflection. Certainly one can devise a matching network that will offer good performance over a narrow bandwidth. But to devise a matching network that will match an arbitrary impedance over an ultrawide bandwidth is a daunting task. Further, in the case of a UWB, or "a-resonant," antenna, mismatch reflection losses may be the most significant factor impairing antenna performance in an overall system. This is certainly the case for a UWB antenna that does not employ resistive loading. In the UWB limit, a better expression for the power budget of an antenna is

$$P_{in} = P_{applied} = P_{accepted} + P_{reflected} = P_{reflected} + P_{rad} + P_{loss} \qquad (7.5)$$

and a better definition for antenna efficiency is

$$\eta_{UWB} \equiv \frac{P_{rad}}{P_{in}} = \frac{P_{rad}}{P_{applied}} = \frac{P_{rad}}{P_{reflected} + P_{rad} + P_{loss}} \neq \frac{P_{rad}}{P_{accepted}} \qquad (7.6)$$

Because mismatch reflected energy is typically the single most significant loss term in the overall power budget, it makes sense to use a definition of efficiency that takes this into account.

7.2.2 Efficiency Measurement

A variety of techniques have been proposed over the years for the measurement of antenna efficiency. This section begins with a survey of efficiency measurement techniques. These include the gain/directivity method, the radiometric method, and the Wheeler Cap method. Then, this section will present a modification of the traditional Wheeler Cap method.

7.2.2.1 Survey of Efficiency Measurement Methods

Perhaps the most conceptually simple approach is the gain/directivity method. An ideal, lossless antenna has gain (G) equal to directivity (D). If half the applied power is dissipated in losses, then the gain is half the directivity. The ratio of these two quantities yields the antenna efficiency [12]:

$$\eta_{G/D} = \frac{G}{D} \qquad (7.7)$$

The difficulty with this method is that it requires complete knowledge of the pattern, not only in the principal planes, but in every direction around the antenna. The directivity is defined as the ratio of the maximum radiation intensity in any direction $[U_{max}(\theta, \phi)]$ to the average radiation intensity (U_{avg}):

$$D = \frac{U_{max}(\theta, \phi)}{U_{avg}} \tag{7.8}$$

The radiation intensity must be measured at a sufficiently dense collection of points to evaluate the average accurately. This is a difficult experimental challenge, although there are commercially available systems for measuring full 3-D patterns [13]. The gain/directivity method lends itself well to determining efficiency from the results of a computational model. Some software packages (including Ansoft's HFSS) calculate total accepted and radiated power, as well as efficiency [14]. The results from computational modeling can be compared to efficiency measurements obtained by alternate methods.

An alternate technique is the radiometric method [15]. Resistive and dielectric losses of an antenna look like a noise source at the ambient temperature, so the ratio of the noise power when the antenna is directed at a warm load (such as an anechoic chamber at room temperature) to the noise power when one directs the antenna at clear sky is a measure of efficiency. System noise may be corrected by using a known high-efficiency antenna (such as a standard gain horn) as a reference. This technique has been applied to measure the efficiency of printed antennas [16].

The radiometric method is challenging to apply to UWB antennas because the bandwidth of these antennas leaves them vulnerable to nonthermal ambient energy from the plethora of narrowband sources in the environment. Also, special precautions must be taken with a low-gain antenna to ensure that its pattern is actually directed to the clear sky. This can include the use of a reflector. The complexity of applying this method makes it a less desirable option.

Perhaps the most common technique for measuring antenna efficiency is the Wheeler Cap method. This technique, originated by H. A. Wheeler in the 1950s [17], has been revisited many times since then [18, 19]. The basic idea is to begin by measuring the reflection coefficient of an antenna in free-space (S_{11FS}). Then the reflection coefficient is measured with the antenna placed in a Wheeler Cap (S_{11WC}). A Wheeler Cap is a closed conducting shell (or cap if the antenna is a monopole) with radius $r \approx \lambda/(2\pi)$ at the frequency of interest (see Figure 7.9). Ideally, the Wheeler Cap should be spherical with the antenna centered at the origin, although good results may also be obtained from cylindrical geometries. This places the closed conducting surface at the radial distance where the near, or reactive, fields are comparable in magnitude to the far, or radiation, fields. This is where the radiated energy decouples from the near fields. The effect is to inhibit radiation without significantly perturbing the near, or reactive, fields about the antenna. Assuming the antenna in free-space may be modeled as a loss resistance (R_{loss}) in series with a radiation resistance (R_{rad}), in the Wheeler Cap, the radiation resistance is effectively turned off, so the antenna now exhibits only the loss resistance.

The Wheeler Cap method is inherently narrowband since the size of the Wheeler Cap follows directly from the frequency of interest. Ronald Johnston and John McRory have suggested an alternate approach [20]. These investigators used a variable waveguide as their Wheeler Cap and did a traditional one-port reflection measurement, as well as a two-port transmission measurement. Sliding shorts were used to fine-tune the waveguide and adjust its impedance for optimal performance. The authors report that one such waveguide Wheeler Cap worked well from 650MHz to 1GHz, a 42% fractional bandwidth. See Figure 7.10 for a conceptual diagram.

The waveguide Wheeler Cap has the advantage of allowing operation over an ultrawide range of frequencies. However, it requires significant tuning and tweaking to optimize performance at each frequency and to gather enough data points (at least three) to infer the antenna's actual free-space impedance.

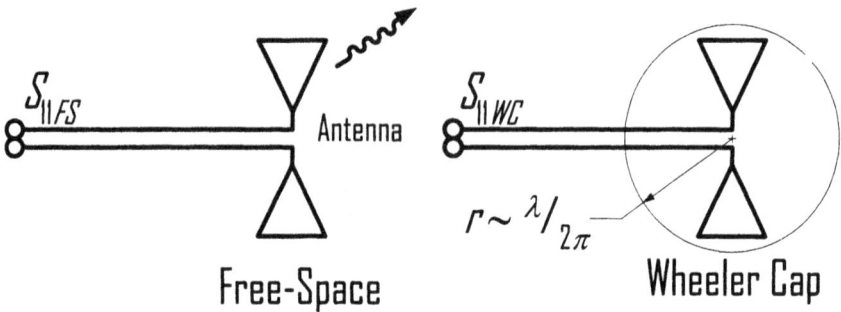

Figure 7.9   The Wheeler Cap method requires measurement of the scattering parameters of the antenna in free-space, and inside a closed conducting shell, or "Wheeler Cap." The Wheeler Cap is sized to coincide with the radian sphere, the distance away from a small dipole antenna at which the near, or reactive, fields and the far, or radiation, fields are comparable in magnitude. This has the effect of shorting out the radiation fields and inhibiting radiation without significantly perturbing the reactive field in the immediate vicinity of the antenna feed ([21]; © 2002, IEEE).

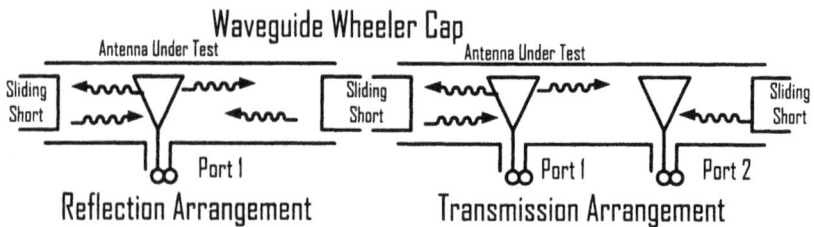

Figure 7.10   The Waveguide Wheeler Cap Method of Johnston and McRory. The reflection arrangement is similar to the standard Wheeler Cap, except that the sliding shorts allow it to be tuned over a broad range of frequencies. The transmission arrangement allows a true measurement of antenna TX/RX efficiency, but requires tuning by adjusting the sliding shorts for each measurement (After [10]).

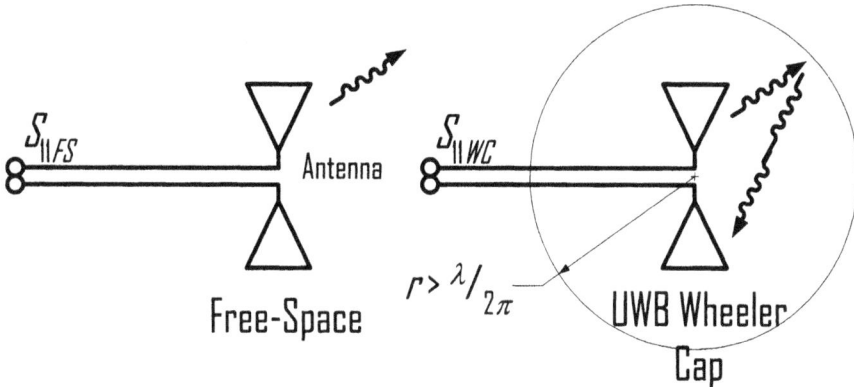

Figure 7.11   A UWB Wheeler Cap, sized to be larger than the radian sphere dimension of the traditional Wheeler Cap. Instead of inhibiting radiation, a UWB Wheeler Cap allows the antenna to transmit and receive the reflected signal ([21]; © 2002, IEEE).

7.2.2.2 UWB Wheeler Cap Theory

Recently, however, a new method has been developed for evaluating UWB antenna efficiency [21]. Instead of a closed spherical shell of radius $r \approx \lambda/(2\pi)$ at the frequency of interest, the "UWB Wheeler Cap" method uses a much larger spherical shell. Rather than inhibiting radiation from the antenna, the UWB Wheeler Cap allows the antenna to radiate freely, then receive its own transmitted, reflected signal (see Figure 7.11).

First, consider the power budget for a transmit antenna. A fraction of the incident energy is dissipated in losses ($\ell \equiv P_{loss}/P_{in}$), a fraction is reflected away due to mismatch ($m \equiv P_{reflected}/P_{in}$), and a fraction is radiated ($\eta \equiv P_{loss}/P_{in}$). If averaged over a suitable time interval, applying conservation of energy yields:

$$\ell + m + \eta = 1 \tag{7.9}$$

With the antenna under test in the center of the spherical UWB Wheeler Cap, the geometry imposes a nearly ideal time reversal of the radiated fields. The antenna thus receives the radiated signal with negligible structural scattering. From thermodynamic considerations, the antenna-mode scattering term will be identical to the mismatch fraction ($m = |S_{11FS}|^2$) [22]. By reciprocity, the receive and transmit efficiencies (η) must be the same. Assuming all of the power radiated at the transmit antenna is available for reception at the receive antenna, the power budget is as shown in Figure 7.12(a).

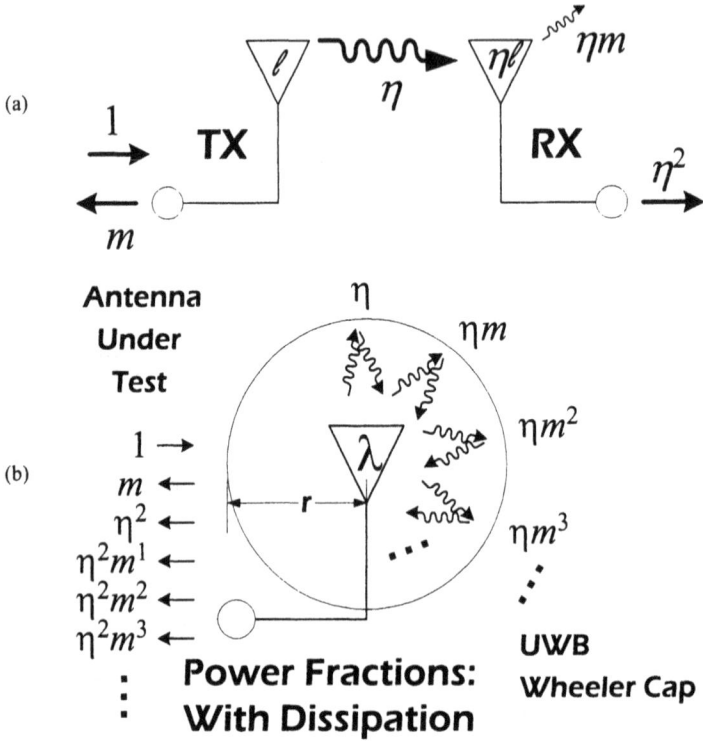

Figure 7.12 (a) Power budget for a TX-RX pair if all the transmitted power is available at the receive antenna, and (b) reflection components for a UWB Wheeler Cap ([21]; © 2002, IEEE).

The scattering coefficient inside the UWB Wheeler Cap follows from summing the components shown in Figure 7.12(b):

$$|S_{11WC}|^2 = m + \eta^2 + \eta^2 m^1 + \eta^2 m^2 + \eta^2 m^3 + \cdots$$

$$= |S_{11FS}|^2 + \eta^2 \sum_{n=0}^{\infty} |S_{11FS}|^{2n} = |S_{11FS}|^2 + \eta^2 \frac{1}{1 - |S_{11FS}|^2} \qquad (7.10)$$

which solves to yield the following frequency domain result for the radiation efficiency:

$$\eta_{FD} = \sqrt{\left(|S_{11WC}|^2 - |S_{11FS}|^2\right)\left(1 - |S_{11FS}|^2\right)} \qquad (7.11)$$

This approach assumes that the various reflections inside the UWB Wheeler Cap are orthogonal to each other. This is a valid assumption provided the characteristic time required by the antenna to radiate or receive a signal is less than the time delay between transmission and reflection.

In fact, if the antenna response is sufficiently short in time and if the UWB Wheeler Cap is large enough to provide sufficient delay to the radiated-reflected-received signal, then the prompt mismatch reflection can be completely time-gated out. Further, the first radiated-reflected-received signal can be isolated and its spectral energy content analyzed: a method dubbed the "time-gated efficiency." In practice, this involves measuring the reflection coefficient, time-gated to capture only the first transmitted-reflected-received impulse ($|S_{11TG}|$). The power content fraction is the product of the transmit and the receive efficiencies; so:

$$\eta_{TG} = \sqrt{|S_{11TG}|^2} = |S_{11TG}| \qquad (7.12)$$

The following section will apply the time gate and frequency domain methods for analyzing results from a UWB Wheeler Cap.

7.2.2.3 UWB Wheeler Cap Results

This section presents experimental results from a UWB Wheeler Cap for two antennas: BroadSpec 101 and BroadSpec 102. A BroadSpec 101 antenna is a center-fed planar elliptical dipole that fits within a boundary sphere of radius $r = 3.9$cm. A "BroadSpec 102" antenna is a center-fed planar elliptical dipole that fits within a boundary sphere of radius $r = 3.8$cm. Recall from Section 5.5.1, the boundary sphere is the smallest sphere within which an antenna can be enclosed. Its radius is the characteristic size of the antenna, and the longest dimension of the antenna is the diameter of the boundary sphere. The boundary sphere is necessarily much smaller than the UWB Wheeler Cap. Additionally, an Ansoft HFSS model of the BroadSpec 102 antenna was developed for comparison purposes.

To measure the efficiency of these antennas, two UWB Wheeler Caps were constructed. Figure 7.13(a) shows the first "small" UWB Wheeler Cap: a mated pair of 30cm (12in) aluminum hemispheres. Figure 7.13(b) shows the second "large" UWB Wheeler Cap. The large UWB Wheeler Cap used 61-cm (24-in) aluminum hemispheres. Conducting tape was used to connect the two hemispheres of the small sphere. A more elaborate mated flange arrangement was also tried on the large sphere without a noticeable improvement in performance [see Figure 7.13(b)]. The antenna under test was located in the center of the sphere. An HP 8753D 6-GHz network analyzer (with time domain analysis option) was used to gather the data.

Typical time domain results are shown in Figure 7.13(c). When the smaller (30-cm diameter) sphere is used, the time of flight for a radiated-reflected-received signal is about 1ns. Since the antenna requires about 1ns to finish radiating an impulse, it is difficult to isolate the prompt mismatch reflection from the radiated-reflected-received signal in the smaller sphere. When the larger (61-cm diameter) sphere is used, the time of flight for a radiated-reflected-received signal is about 2ns, so these signals can be fully resolved. An "early" time gate (–0.5ns–1.5ns) was used with the larger sphere to isolate the prompt mismatch reflection. A "late" time gate (1.5ns–3.5ns) was used to isolate the radiated-reflected-received signal.

For late times, a persistent high-frequency ringing is visible. This is due to excitation of spherical modes and their subsequent ring down. Much of this energy is eventually captured by the antenna at the center of the sphere, but because of this effect, the time-gated efficiency measurement seriously underestimates efficiency at higher frequencies.

Figure 7.13(d) shows reflection data for the BroadSpec 102 antenna in the large UWB Wheeler Cap. The free-space S_{11} is compared to the frequency response of the early time gate. Good agreement is obtained, indicating that the early gate isolates the same prompt reflection as is observed in the free-space measurement. The Ansoft HFSS model shows similar agreement, although the antenna resonances are at a slightly lower frequency. The HFSS model does not include the details of the connecting structure present in the physical implementation of the antenna. Finally, Figure 7.13(d) shows the S_{11} of the antenna in the Wheeler cap. An ideal, lossless antenna in an ideal, lossless Wheeler cap would show a perfect reflection and an S_{11} of 0dB for all frequencies. The experimental set up as realized exhibits numerous narrow spikes. These are narrowband resonant modes of the spherical UWB Wheeler Cap itself. Note that the density of these spherical resonant modes increases with increasing frequency. This effect limits the usefulness of the UWB Wheeler Cap at higher frequencies. Since the frequency range of 30kHz to 6GHz was sampled in 1,601 steps, each data point is about 3.75MHz apart.

Figure 7.13(e) provides the results of an analysis of the reflection data. A "smoothed peak" treatment was applied to the frequency domain efficiency result. The maximum efficiency observed in a 60-MHz span around the frequency in question is taken as the smoothed peak result. This helps avoid the effect of the narrowband resonances and yield an easier-to-follow curve. From about 1.0GHz to 2.5GHz, the time-gated efficiency of (7.12) closely matches the frequency domain efficiency of (7.11). Above about 2.5GHz, the mode density in the spherical UWB Wheeler Cap becomes so large that significant high-frequency energy is captured by resonant modes. Much of this energy is later released as the sphere rings down. Thus, the high-frequency energy content is not evident in the time-gated efficiency measurement but does show up in the frequency domain efficiency. Below about 1.0GHz, it is likely that currents are being excited on the feed cable leading to the antenna in the center of the sphere.

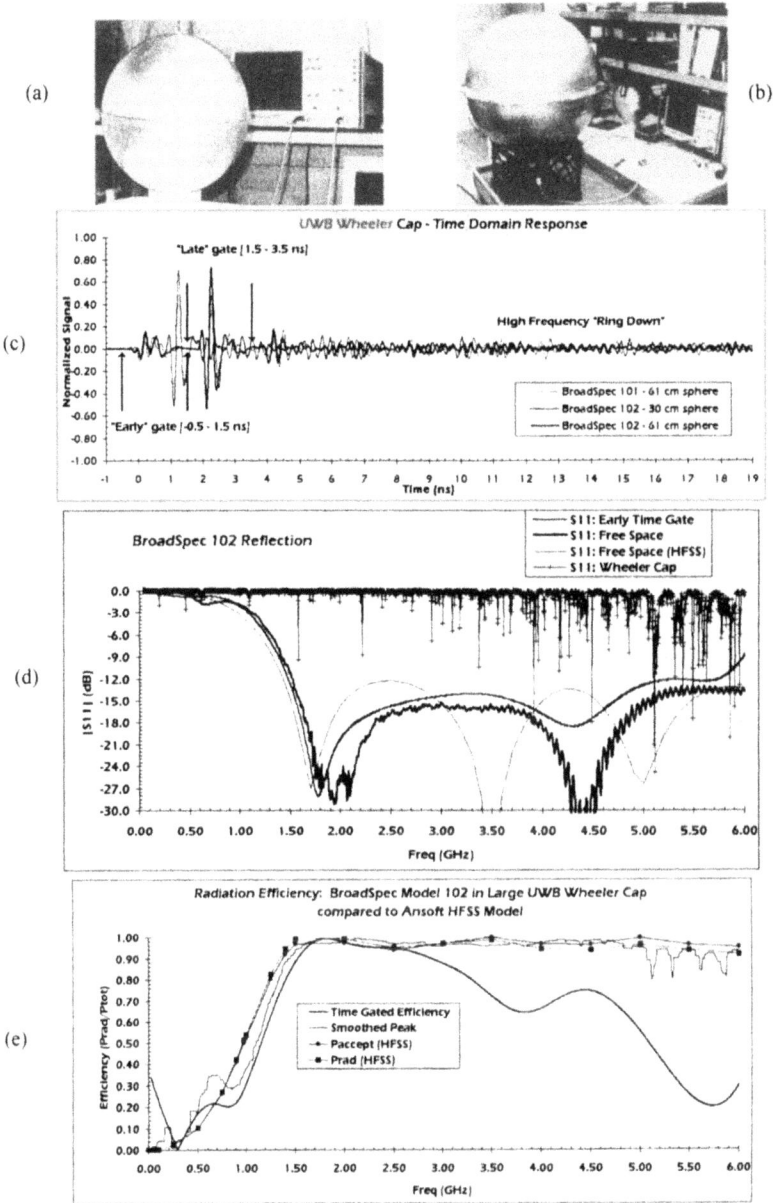

Figure 7.13 (a) A 30-cm (1-ft)-diameter UWB Wheeler Cap, (b) a 61-cm (2-ft)-diameter UWB Wheeler Cap, (c) time domain response of various antennas in UWB Wheeler Caps, (d) S_{11} response of various antennas in UWB Wheeler Caps, and (e) efficiency results for various antennas in UWB Wheeler Caps ([21]; © 2002, IEEE).

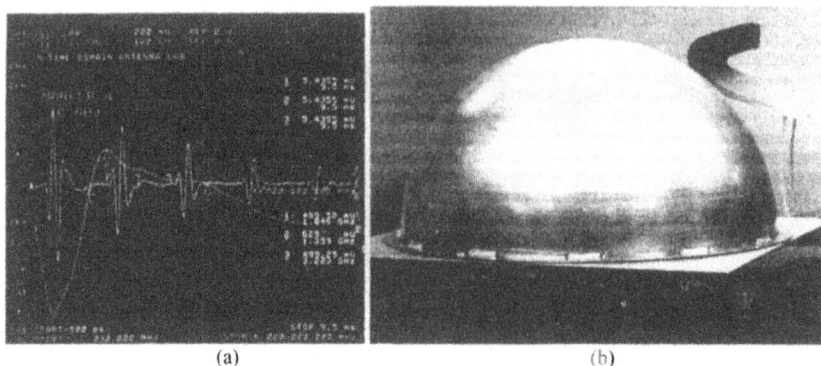

(a)   (b)

Figure 7.14 (a) Time domain and frequency domain efficiency results for an experimental monopole in a UWB Wheeler Cap, and (b) 61-cm (2-ft)-diameter hemispherical Wheeler Cap.

The results for the BroadSpec 102 antenna may be compared to those predicted by an Ansoft HFSS model [see Figure 7.13(e)]. The predicted radiated power is only in rough agreement with measured results. Note that the Ansoft HFSS results lack the increase in efficiency observed in the experimental results around 600–700 MHz. This observed increase in radiation efficiency is probably due to cable currents on the feed cable leading to the antenna at the center of the sphere.

The accepted energy and the radiated energy calculated by the HFSS model are essentially the same within the 4% or so margin of error expected in the calculation (in fact, the calculated radiated power slightly exceeds the calculated accepted power, yielding an efficiency slightly better than 1.00). The equivalence of accepted to radiated power is also evident in the results. If the traditional definition of efficiency as the ratio of radiated to accepted power were used, the BroadSpec 102 antenna would be regarded as an efficient radiator down to a few 10s of megahertz. At 50 MHz, for instance, the radius of the antenna's boundary sphere is less than 1% of wavelength. Clearly, however, a more reasonable definition of antenna efficiency is the ratio of radiated to applied power, as advocated in Section 7.2.1.

The results shown in Figure 7.14 provide an example of what one can achieve if the goal of a small UWB antenna is pursued at the expense of good matching. In the case of this experimental monopole evaluated in the hemispherical UWB Wheeler Cap of Figure 7.14(b), an efficiency only slightly better than 50% is obtained over an inverse fractional bandwidth of $Q = 0.843$. This element fits within a boundary hemisphere of radius 3.58 cm, so $r_{\lambda L} = 0.125$. The return loss from this element is only about −3dB, so it is not a practical real-world element. Note again that this is smaller than allowed by the theoretical limit proposed by McLean (Figure 5.36).

Taking the frequency at which efficiency is 50% as the lowest operating frequency, the model 101 antenna works down to f_L = 1.073GHz, and the model 102 antenna works down to f_L = 1.139GHz. Figure 7.14(a) shows the linear magnitude of S_{11} which by (7.11) is also the efficiency. In units of wavelengths at the lowest operating frequency, the boundary spheres containing the antennas are $r_{\lambda L}$ = 0.140 and $r_{\lambda L}$ = 0.144 for the 101 and 102 antennas, respectively. The Ansoft HFSS model predicts that the 102 antenna will work down to f_L = 0.980GHz for a boundary sphere in units of wavelength of $r_{\lambda L}$ = 0.126. These results are summarized in Table 7.3.

Table 7.3

Size and Lower Operating Frequency for Selected Antennas

Antenna	r (cm)	f_L (GHz)	λ_L(cm)	$r_{\lambda L}$
101	3.9	1.073	27.95	0.140
102	3.8	1.139	26.32	0.144
102 (HFSS)	3.8	0.980	30.59	0.126
Experimental monopole	3.58	1.046	28.66	0.125

In summary, the traditional definition of antenna efficiency (as the ratio of radiated to accepted power) should be modified. Mismatch reflection is as much a system inefficiency as ohmic or dielectric losses. A better definition for antenna efficiency is the ratio of radiated to applied power. Further, this section has surveyed traditional antenna efficiency-measurement techniques and presented a variation of the Wheeler Cap technique that can be applied to evaluate UWB antenna efficiency. The UWB Wheeler Cap is a closed spherical reflector centered on the antenna under test. The sphere is sized to be much larger than the radian sphere $[r = \lambda/(2\pi)]$ at the lowest frequency of interest. Efficiency may be calculated from the free-space and Wheeler Cap reflection coefficients. If the sphere is sufficiently large, it may be possible to time-gate the prompt reflection and to isolate the spectral power content of the initial radiated-reflected-received signal.

7.3 ANTENNA DIRECTIVITY

The demands of a particular application impose certain specifications on an antenna. One of the most fundamental specifications is field of view. Is the device intended for an ad hoc network? Or is the device in question to operate in a point-point link? High-gain antennas are typically bulky and have a narrow field of view. Omnidirectional antennas are typically smaller and by definition have a wide field of view. This section examines the impact of antenna directivity on overall system performance and discusses how small-aperture direction-finding techniques offer a potential solution to the problem of obtaining directivity from a compact antenna system [23].

7.3.1 Omni Versus Directional

One dilemma facing designers of broadband and UWB systems is the trade-off between antenna size and gain. Higher antenna gain means a more robust link that can send more data further and faster. But a high-gain antenna is large and bulky. Furthermore, the field of view of a high-gain antenna is relatively narrow, making a single high-gain antenna inappropriate for use in an ad hoc network. Multiple high-gain antennas might provide sufficient field of view coverage, but the size of the overall array is likely to be prohibitively large.

Even in the case of an ad hoc system where a large field of view is essential, a certain degree of directionality is advantageous. Consider the ad hoc network of four transceivers and two links in Figure 7.15(a). The transmitters radiate energy omnidirectionally, allowing their receivers to pick up the energy and close the link. The side effect of that omnidirectionality is that the receivers are victimized by interference from the nearby link.

There are several ways to deal with the impact of this interference. Orthogonal coding can reduce but cannot eliminate mutual interference. Time-division schemes eliminate interference at the cost of lower data throughput. Ideally, one would like to have a directional high-gain antenna on each transceiver aligned with the corresponding transceiver at the other side of the link as in Figure 7.15(b). The narrow field of view and relatively bulky size of directional antennas make this an impractical idea. The challenge is to find a way to obtain directional information from a relatively small and compact form factor antenna system.

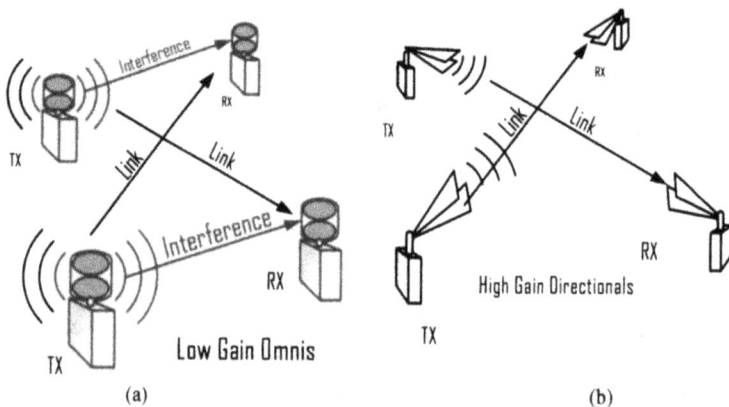

Figure 7.15 (a) An ad hoc network requires low gain omnidirectional antennas but results in significant interference, and (b) high gain directional antennas are preferred but have a limited field of view and are relatively bulky ([23]; © 2004, IEEE).

7.3.2 Amplitude Comparison Direction Finding

Amplitude comparison is a technique for radio direction finding (DF) [24]. It involves a pair of vertically oriented loops with orthogonal axes, and a vertical whip or "sense" antenna. Excellent discussions are available in the literature [25, 26]. Figure 7.16(a) depicts a typical DF receiver (RX).

Each of the two vertical loops has the typical dipole "doughnut" antenna pattern with nulls lying in the horizontal or azimuthal plane. Figure 7.16(b) shows these antenna patterns. Mathematically, the pattern function $P(\theta, \phi)$ is given by

$$P(\theta,\phi)=\begin{cases} \sin^2\phi & Loop\#1 \\ \cos^2\phi & Loop\#2 \end{cases} \tag{7.13}$$

A vertical loop will be maximally sensitive to signals in the plane of the loop, and minimally sensitive to signals incident along the axis of the loop. A sense antenna has a uniform omnidirectional pattern.

The two loop antenna signals may be summed to create a virtual loop antenna oriented in any direction in the plane. A goniometer or other variable summing amplifier may be used. Figure 7.17 shows the effective virtual loop pattern. The virtual loop pattern may be oriented to either maximize or minimize the received signal. Alternatively, the angle of arrival (ϕ) of an incident signal may be found from the amplitude of the first loop signal (A_1) and the amplitude of the second loop signal (A_2):

$$\phi = \tan^{-1}\frac{A_1}{A_2} \tag{7.14}$$

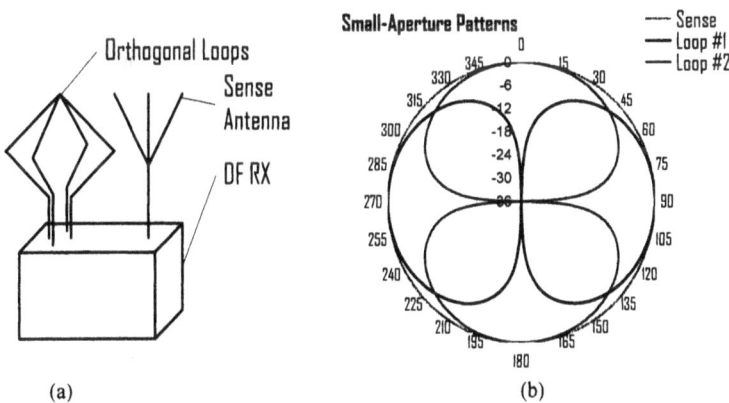

Figure 7.16 (a) A small aperture amplitude comparison DF receiver. (b) Element patterns from a small aperture DF receiver ([23]; © 2004, IEEE).

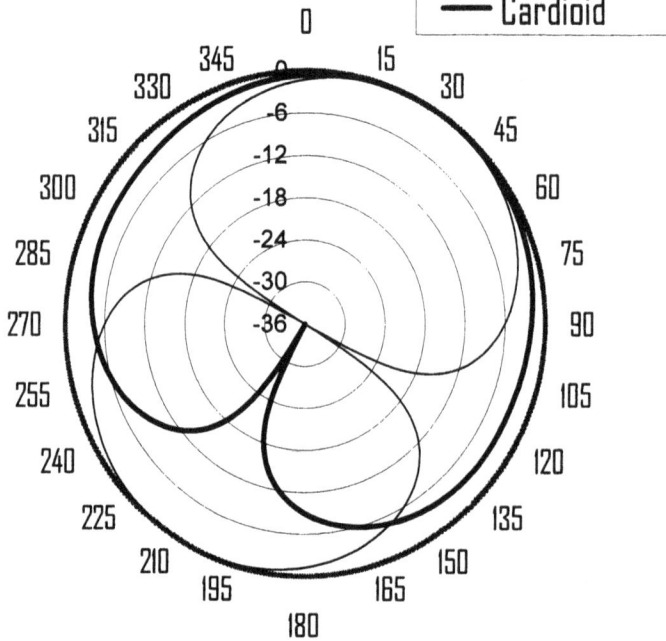

Figure 7.17 Virtual loop, sense, and cardioid patterns. Radial scale in decibels ([23]; © 2004, IEEE).

This virtual pattern has two nulls and two maxima in the horizontal plane. Thus, it is subject to a "front-back" ambiguity: there is no way to identify unambiguously whether a signal arrives from the front or the back. Similarly, the solution obtained from (7.13) is also subject to an ambiguity. Because the amplitudes are by definition positive, (7.13) will only yield solutions in a 180° range.

The sense antenna resolves this ambiguity. The two lobes of the virtual loop pattern exhibit a 180° phase difference. A "cardioid," or heart-shaped, pattern follows from summing the virtual loop response with the response of the sense antenna. Because the addition is constructive for one lobe and destructive for the other, the resulting combined sense and loop pattern has a single sharp null. This null may be aligned with an incoming signal in order to identify unambiguously the signal's angle of arrival. Figure 7.17 also shows the cardioid pattern formed by summing the omnidirectional sense antenna pattern with the virtual loop pattern. Alternatively, the sense antenna yields a phase reference that provides a sign for the amplitudes in (7.13) and makes (7.13) applicable for a full 360° range.

7.3.3 Small-Aperture UWB Direction Finding

Using two orthogonal broadband or UWB planar loop antennas, the angle of arrival of an incoming signal may be obtained by the amplitude comparison method already described. Figure 7.18(a) presents such a receiver. As noted above, the traditional narrowband amplitude comparison direction-finding method suffers from a front-back ambiguity and cannot resolve whether a signal was incident in the forward or reverse direction. Unlike the narrowband case, a broadband or UWB direction-finding receiver does not require a sense antenna to resolve the ambiguity.

In a broadband or UWB system, waveform polarity may be used to resolve the front-back ambiguity. The 180° phase difference in a narrowband direction-finding receiver manifests itself as a waveform inversion in the broadband limit. Thus, on opposite sides of the loop, the antenna will receive an inverted signal. By detecting whether signals from particular antennas are upright or inverted, the quadrant from which the signal arrived may be unambiguously identified.

Start with the "first" quadrant (x, y both positive). Suppose received signals from both antennas are upright. In the "second" quadrant (x negative, y positive), the signal from the first antenna is now inverted while the signal from the second antenna is still upright. In the "third" quadrant (x, y both negative), both signals are inverted. In the "fourth" quadrant (x positive, y negative), the signal from the first antenna is upright while the signal from the second antenna is inverted. See Figure 7.18(b).

A particular asymmetric coding sequence may be used to identify an absolute reference point for a binary phase-shift keyed (or BPSK) modulated sequence. For instance, suppose an upright waveform is denoted by a "1" and an inverted waveform is denoted by a "0." Then a training sequence such as 11011 might be sent as a header of a packet. If 00100 is received instead, the receiver knows that the packet is inverted. Thus, a receiver can identify exactly and unambiguously which signals are upright and which are inverted.

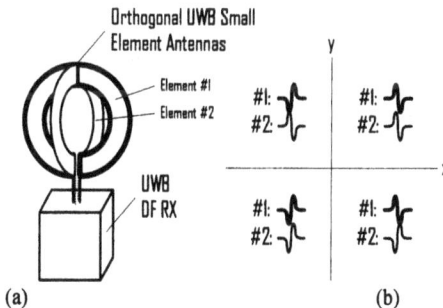

Figure 7.18 (a) Conceptual diagram of a small-aperture UWB direction finding radio, and (b) example pattern of waveform inversions used to identify quadrant of arrival ([23]; © 2004, IEEE).

This broadband or UWB direction-finding technique offers a variety of significant advantages. First, by relying on waveform inversion, a sense antenna can be avoided in many circumstances of interest. Thus, direction finding is possible with only a two-channel receiver instead of a three-channel receiver. Even if a third "sense" channel is required in some cases, the inversion detection technique described here yields a more robust form of angle-of-arrival measurement. Second, the technique of this section can use electrically small elements with dipolelike patterns. Thus, a direction-finding array may be made exceptionally small, limited only by the signal-to-noise ratio obtainable from small antenna elements. Finally, signals from two small elements may be summed to create a virtual pattern that may be aligned to either enhance or null out particular signals. The following section looks at these and additional applications in greater detail.

7.3.4 Applications

The broadband or UWB direction-finding technique introduced in this section has utility in a wide variety of applications. One key application is use in a "spatial rake receiver." Other applications include interference mitigation, radar probing of an environment, and positioning or tracking systems.

7.3.4.1 Spatial Rake Receiver

A rake receiver (or more specifically, a "temporal rake receiver") collects and coherently adds energy arriving at different times so as to optimize a received signal. The direction-finding technique described here may be used to identify and characterize a variety of multipath components incident on a receiver. This enables a "spatial rake receiver," one that collects, rejects, and otherwise combines multipath components so as to optimize a received signal.

For example, consider the propagation environment shown in Figure 7.19(a). A transmitter (TX) communicates with a receiver (RX) via four paths: a direct path (#1) and three indirect paths (#2–#4) that bounce off a reflecting object. The received signals are shown in Figure 7.19(b). These signals are composed of four wavelets, one for each path. These wavelets may be inverted by a combination of the propagation environment and the antenna responses. An omnidirectional or sense antenna receives Signal #0. A loop in the x-plane receives Signal #1 and a loop in the y-plane receives Signal #2.

Signal #0 exhibits the inversions due to the propagation path, allowing them to be distinguished from the inversions due to the function of the angle-of-arrival antenna system. Combining amplitude comparison (to determine the angle) and inversion analysis (to resolve the front-back ambiguity), the angle of arrival of each multipath component may be determined. Then, these individual wavelets arriving from different directions may be summed to yield an optimized signal. This process may be called a "spatial rake."

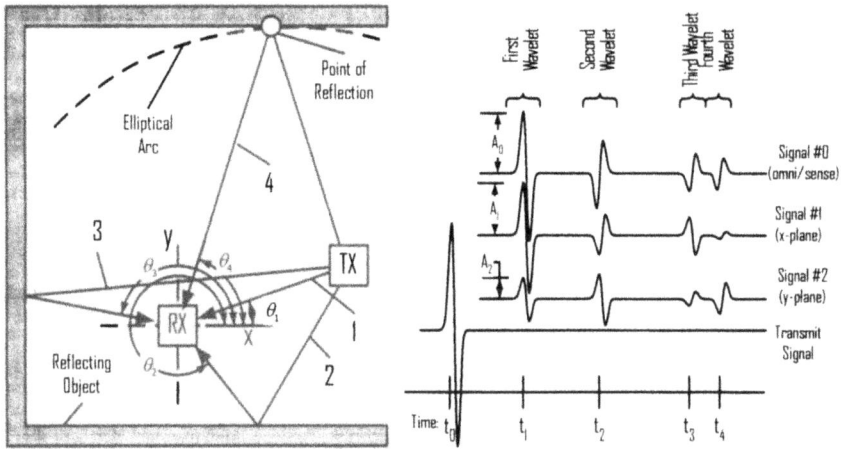

Figure 7.19   (a) A propagation environment with four signal propagation paths, (b) time domain
signals as transmitted and as received by a sense antenna (Signal #0), and two orthogonal
small-element antennas (Signals #1 and #2) ([23]; © 2004, IEEE).

7.3.4.2 Location Awareness

Another application is in a UWB positioning, or locating, system. The methods of
this chapter allow a transponder-type UWB positioning system to determine both
range and bearing using a significantly smaller aperture than a traditional time-
difference of arrival (TDOA) angle-of-arrival (AoA) system would require.
Whereas a traditional TDOA AoA system might require two antennas to be
separated by feet, the proposed system allows an aperture no larger than a single
antenna to make an AoA measurement.

Knowing the time of flight of a direct signal allows calculation of the range
between a transmitter and a receiver. If a receiver also measures an angle of
arrival, then both range and bearing follow. Unlike traditional UWB location
systems that rely on multilateration from a collection of ranges, the system
described in this chapter enables a localized location awareness that allows an
individual receiver to locate a transmitter without requiring a complicated network
of receivers to collect, share, and analyze range data. Alternatively, adding
bearing information into a multilateration calculation makes such a calculation
more robust and accurate.

Further, knowing the time of arrival of a single-bounce multipath signal
establishes the location of a reflector on an ellipse whose foci are located at the
transmitter and the receiver. If the receiver can determine the angle of arrival, then
the exact point of reflection is known. This enables a bistatic radar probe of the
environment surrounding the transmitter and receiver so that one may compile a

radar map of the surroundings or identify intruders or other changes in the environment.

7.3.4.3 Interference Mitigation

The concepts of this section may also be applied toward the problem of interference rejection. The signals of the two orthogonal antennas may be combined so as to null out an undesired interfering signal, in effect rotating a virtual antenna so as to align a null of a virtual antenna with the direction from which an undesired signal is incident. In addition, the two orthogonal elements can be used to null out a transmitted signal in a particular direction.

7.3.5 Conclusion

In short, directionality in a UWB or broadband system does not necessarily require electrically large antennas or arrays. This section has described how a small two-element array can use waveform inversion to obtain unambiguous angle-of-arrival information for broadband or UWB signals. Even in a multipath environment, waveform inversion and a sense antenna signal can yield valuable information about the angle of arrival. This multipath information enables a spatial rake receiver, one that coherently adds signals arriving from different directions. In addition, angle-of-arrival information is useful for location awareness in positioning as well as radar systems. Also, the directivity of a two-element array is useful for interference mitigation in both reception and transmission. Additional information on small-aperture direction finding for UWB systems is available elsewhere [27].

7.4 UWB ANTENNAS IN SYSTEMS

This chapter examined antenna-system interactions in three ways. First, spectral control of an antenna is an important technique to help a UWB system meet spectral goals. This chapter discussed using narrowband resonant structures and stepped-impedance filtering to help antennas meet spectral goals. Second, antenna efficiency can have a serious impact on overall system performance. This chapter explained how mismatch loss is typically the most serious UWB antenna loss term and presented a technique for assessing UWB antenna efficiency. Finally, this chapter discussed the trade-off between field of view, gain, and antenna size and presented a potential solution to the problem of achieving directionality from a compact antenna system.

Endnotes

[1]   Schantz, Hans, Glenn Wolenec, Edward (Mike) Myszka, III, "Frequency notched UWB antennas," 2003 IEEE Conference on Ultra Wideband Systems and Technologies. Reston, Virginia. November 16-19, 2003, pp. 214-218. Much of the material from the present section is reprinted from this paper.

[2]   Schantz, Hans, and Glenn Wolenec, "Ultra-wideband antenna having frequency selectivity," U.S. Patent 6,774,859, August 10, 2004.

[3]   Yoon, Ick-Jae, Hyungrak Kim, Kihun Chang, Young Joong Yoon, and Young-Hwan Kim, "Ultra-wideband tapered slot antenna with band-stop characteristic," 2004 IEEE Antennas and Propagation Society International Symposium, Monterey, California. June 20-25, 2004, Vol. 2, pp. 1780-1783.

[4]   Yoon, Hyungrak, Hyungrak Kim, Kihun Chang, Young Joong Yoon, and Young-Hwan Kim, "A study on the UWB antenna with band-rejection characteristic," 2004 IEEE Antennas and Propagation Society International Symposium, Monterey, California. June 20-25, 2004, Vol. 2, pp. 1784-1787.

[5]   Kerkhoff, Aaron, and Hao Ling, "A parametric study of band-notched UWB planar monopole antennas," 2004 IEEE Antennas and Propagation Society International Symposium, Monterey, California. June 20-25, 2004, Vol. 2, pp. 1768-1771.

[6]   Kim, Yongjin and Do-Hoon Kwon, "Planar ultra wide band slot antenna with frequency band notch function," 2004 IEEE Antennas and Propagation Society International Symposium, Monterey, California. June 20-25, 2004, Vol. 2, pp. 1788-1791.

[7]   Schantz, Hans, "Spectral controlled antenna apparatus and method," U.S. Patent Pending.

[8]   Pozar, David M., *Microwave Engineering*, 2nd ed., New York: John Wiley & Sons, 1998, pp. 470-473.

[9]   Chen, Zhi Ning, Xuan Hui Wu, Ning Yang, and M. Y. W. Chia, "Design considerations for antennas in UWB wireless communication systems," 2003 IEEE Antennas and Propagation Society International Symposium. Columbus, Ohio. June 22-27, 2003, Vol. 1, pp. 822-825.

[10] Schantz, Hans; "An introduction to ultra-wideband antennas," 2003 IEEE Conference on Ultra Wideband Systems and Technologies. Reston, Virginia. November 16-19, 2003, pp. 1-9.

[11] *IEEE Standard Test Procedures for Antennas* (IEEE Std 149-1979), (New York: IEEE, 1979) p. 112. "The radiation efficiency of an antenna is the ratio of the total power radiated by the antenna to the net power accepted by the antenna at its terminals during the radiation process."

[12] *IEEE Standard Test Procedures for Antennas*, Op. Cit., p. 113.

[13] For instance, Satimo's "Stargate" system. See www.satimo.com.

[14] For instance, Ansoft's HFSS. See www.ansoft.com.

[15] Ashkenazy, J., E. Levine, and D. Treves, "Radiometric measurement of antenna efficiency," Electron. Lett., vol. 21, no. 3, Jan. 31, 1985.

[16] Pozar, David M. and Barry Kaufman, "Comparison of three methods for the measurement of printed antenna efficiency," IEEE Transactions on Antennas and Propagation Vol. 36, No. 1, January 1988, pp. 136-139.

[17] Wheeler, H. A. "The Radiansphere Around a Small Antenna," Proceedings of the IRE, Vol. 47, August 1959.

[18] Johnston, Ronald H. and John G. McRory, "An improved small antenna radiation-efficiency measurement method," IEEE Antennas and Propagation Magazine, Vol. 40, No. 5, October 1998 pp. 40-48. See also Pozar et al. [16].

[19] Agahi, D., and W. Domino, "Efficiency Measurements of Portable-Handset Antennas Using the Wheeler Cap," Applied Microwave and Wireless, June 2000. This article is available online at: www.amwireless.com/archives/2000/v12n6/pg34.pdf

[20] Ronald H Johnston and John G. McRory, Op. Cit.

[21] Schantz, Hans, "Radiation efficiency of UWB antennas," 2002 IEEE Conference on Ultra Wideband Systems and Technologies. Washington, D.C. May 21-23, 2002, pp. 351-355.

[22] Schantz, Hans, "*Maxwell's demon and reciprocity,*" 2003 URSI National Radio Science Meeting. Columbus, Ohio. June 22-27, 2003.

[23] Schantz, Hans, "Smart antennas for spatial RAKE UWB systems," 2004 IEEE Antennas and Propagation Society International Symposium, Monterey, California. June 20-25, 2004, Vol. 3, pp. 2524-2527. Much of the material in this section was adapted from this conference paper.

[24] Carr, Joseph J., *Joe Carr's Loop Antenna Book*, Reynoldsburg, Ohio: Universal Radio Research, 1999, p. 107.

[25] Jenkins, Herndon H. *Small-Aperture Radio Direction-Finding,* Norwood, MA: Artech House, 1991, pp. 12-14.

[26] Moell, Joseph D. et al, *Transmitter Hunting: Radio Direction Finding Simplified,* New York: McGraw Hill, 1987, pp. 2, 28, 237, 248.

[27] Schantz, Hans, "System and method for ascertaining angle of arrival of an electromagnetic signal," U.S. Patent Publication 2004/0239562 A1 Dec. 2, 2004.

Chapter 8

Conclusion

Over the last few years much has been published about the principles and applications of electromagnetic waves with large relative bandwidth, or nonsinusoidal waves for short. The next step is the development of the technology for the implementation of these applications. It is generally agreed that the antennas pose the most difficult technological problem...

Henning F. Harmuth, 1984

When Harmuth wrote these words in 1984, there was some justification for his conclusion that antennas were the most difficult aspect of implementing UWB technology [1]. Existing UWB antenna designs were obscure and forgotten. The work of earlier generations of investigators lay hidden like needles in haystacks, overlooked snippets lost in the vast expanse of dusty, decades-old journals stacked in dimly lit lonely aisles in the basements of technical libraries. There having been no significant interest in UWB technology, there was no comprehensive body of knowledge, no community of experience, no standard treatises and texts to capture preexisting work and serve as a stepping off point for further exploration. Innovators of necessity worked in a vacuum, struggling to reinvent antennas known to earlier generations.

Today, the surge of interest in UWB means that a community of experience is coming into being. A body of knowledge is rapidly coalescing in conference proceedings and journal articles. The aim of the present book has been to fill the UWB antenna information void with a reasonably comprehensive summary of the existing art and an overview of the technical tools needed to develop new and truly innovative UWB antenna designs. The reader will have to judge how well this work met its goals. This final chapter serves as a brief summary and review of this book's lessons. In addition, this final chapter provides a few more philosophic concluding thoughts about UWB antenna technology in particular and electromagnetics in general.

8.1 SUMMARY

Through a myriad of examples, Chapter 1 demonstrated how the history of UWB antennas is replete with examples of designers reinventing preexisting designs. Carter reinvented Lodge's biconicals, King and Kaitzen rediscovered Bose's horns, and any number of moderns (present company included!) have resurrected the pioneering designs of Masters, Lamberty, Stöhr, Thomas, and others. The lesson of Chapter 1 is that a prudent designer should be familiar with the history of the art so as to spend time implementing useful existing designs and looking for new solutions rather than reinventing old ones. The historical introduction of Chapter 1 further provided a wealth of examples to serve as a basis for additional discussion and analysis and to provide an overview to orient readers to what was to come.

Chapter 2 described UWB antennas as transducers—as black boxes that couple between signals in transmission lines and signals in free-space. For many applications, UWB antennas can be put to good use with only a basic understanding of such key concepts as bandwidth, dispersion, pattern, directivity, gain, aperture, field of view, polarization, and matching. These concepts are already familiar to those with antenna experience. The main wrinkle introduced in the UWB context is the fact that certain frequency-dependent parameters may no longer be considered constants for antennas with such broad bandwidths. In some cases, this matters little. One of the most important classes of UWB antennas is those with constant gain. A nondispersive constant-gain antenna has a fixed pattern that does not change significantly with frequency. The performance of such an antenna can be readily analyzed using the traditional Friis's Law approach and integrating over frequency. Constant-aperture and other UWB antennas may require somewhat more sophisticated analyses that explicitly take into account the frequency dependence of the antenna's key parameters.

Chapter 3 looked at UWB antennas as transformers providing an impedance match between a transmission line and free-space. Unlike narrowband antennas whose impedance flaws may be readily corrected using a simple external matching network, UWB antennas need to have good matching built in from the beginning. Designers of broadband antennas have long realized that a gradual taper between a transmission line and the radiating elements of an antenna meets this need. Unlike narrowband antennas, UWB antenna impedance is subject to the designer's control. Although in the narrowband case, a monopole has half the impedance of a dipole, in the UWB limit, a monopole and a dipole using the same element may share the same impedance. Designing a well-matched UWB antenna is an exercise in designing a gradual transition between a transmission line and free-space with a characteristic impedance that varies smoothly and continuously.

Chapter 4 analyzed UWB antennas as radiators and receivers of electromagnetic fields. First, Chapter 4 considered time domain signal analysis as a complement to traditional frequency domain analysis. Maxwell's equations, written as the time-dependent Coulomb and Biot-Savart laws, allow an analysis of the time domain fields from known current geometries. A few simple approximations allow a quick, qualitative prediction of antenna behavior. Finally, Hertzian electric and magnetic dipoles present a theoretical model fundamental to the understanding and analysis of antennas.

Chapter 5 considered UWB antennas as energy converters. A study of energy around antennas leads to a variety of important principles. First, excessive reactive energy around an antenna tends to narrow bandwidth and impact matching. Thus, "fatter is better." Bulbous antennas tend to break up current concentrations that lead to excessive reactive energy. Even planar bulbous antennas can do an excellent job of reducing reactive energy and implementing a UWB antenna response. Second, energy considerations lead to fundamental limits governing antenna performance. Caution must be used in applying resonant concepts such as the quality factor, or Q, to a-resonant or UWB antennas. This chapter used a combination of theoretical analysis and an analytic example to demonstrate that UWB antennas as small as about 0.10λ at the low end of the operating band are possible.

Chapter 6 presented a taxonomy of UWB antennas. There exists a vast antenna menagerie of excellent UWB designs. Small-element electric antennas are a good choice for many applications because they combine a compact size with an omnidirectional pattern. A variety of techniques exist to make antennas even smaller. Magnetic antennas are a good choice for embedded applications, though many of the more compact designs have difficulty achieving omnidirectional coverage. Larger horn and reflector antennas are useful where high gain or directive performance is needed.

Chapter 7 discussed three important aspects of how UWB antennas behave as part of an overall customer solution. First, an antenna can be an important tool to meet overall system spectral goals by implementing notch or other filter characteristics. Second, antenna efficiency is a key part of overall system performance. Mismatch loss is the most significant inefficiency facing most well-designed antennas. Antenna efficiency may be assessed using a UWB Wheeler Cap. Finally, Chapter 7 discussed a possible solution to the problem of obtaining antenna directionality from a relatively compact antenna system.

8.2 CONCLUDING THOUGHTS

There are those who argue that wisdom and wide-eyed wonder are mutually exclusive. There are those who believe that the unknown and the mysterious are divine and that bringing things once unknown "down" to the level of human understanding leaves them sullied and unappreciated. Certainly, the light of scientific inquiry dispels mystery and transforms what appears to be magic into mundane technology. Scientific investigation is the art by which the mysterious is transformed into human knowledge, and engineering is the art by which human knowledge is applied for the material benefit of humankind.

Replacing mystery with understanding in no way diminishes the marvel of the world around us. UWB antennas are no less marvelous for an understanding of the basic principles that govern their performance. The sky is no less blue for knowledge of Rayleigh scattering, an airplane is no less adventuresome for a familiarity with aerodynamics, and Yosemite is no less majestic for exposure to the principles of orogeny and glaciation. If anything, knowledge of a subject merely adds richness and detail to our appreciation of it. And always, somewhere out there, lies the frontier of human knowledge where there are new truths to be uncovered and a deeper understanding to be found.

Despite more than a century of progress in advancing the radio and electromagnetic arts, we remain hard-pressed to provide a satisfactory answer to the questions Heaviside alluded to in the epigraph for Chapter 4. Thanks to Maxwell, we know the fundamental equations to describe electricity and magnetism. Thanks to Poynting and Heaviside we know how to mathematically describe the process by which electromagnetic energy is stored, somehow, in so-called free-space. Thanks to Hertz, we know we can generate and detect Maxwell's theoretical abstraction called an electromagnetic wave. Thanks to Lodge, Marconi, Bose, and a host of successors up to the present day we know seemingly endless ways to put electromagnetic behavior to work for human benefit—not only in radio communications but also in the vast and ubiquitous expanse of electronic technology. And still, as Heaviside observed in the nineteenth century, this progress and understanding comes "in spite of the abstract nature of the theory, involving quantities unknown to us at present."

What is an electric field? In many respects, our knowledge of electromagnetics is akin to the science of chemistry before the atomic theory was verified. We have the math to mix electronic components together in the right portions to generate useful results, but we really don't understand what we are doing on a fundamental level.

UWB antenna design is one of those rare niches in engineering in which Nature and Nature's truths lay buried just below the surface on the frontiers of our knowledge, awaiting excavation by intrepid explorers. UWB antenna design truly does combine the science and wisdom of electromagnetics with the creativity, pleasure, and wide-eyed wonder of an art.

Endnote

[1]   Harmuth, Henning F., *Antennas and Waveguides for Nonsinusoidal Waves*, New York: Academic Press, 1984, p. xi.

Appendix

Energy around Time-Harmonic Dipoles

The time average of a function, u, is denoted by triangular brackets and defined as follows:

$$\langle u \rangle = \frac{1}{\tau} \int_0^\tau u \, dt \qquad (A.1)$$

where $\tau = \frac{1}{f} = \frac{2\pi}{\omega}$ is the period. Time-average analysis is simplified by neglecting the effects of retardation ($t \to t - \frac{r}{c}$). This neglect makes calculations easier but is part of the way in which time average analysis tends to obscure the physical behavior of the electromagnetic fields. Table A.1 presents a variety of time functions relevant to the present analysis of the harmonic dipole and shows the results of the time-averaging process.

The radial component of the time-average harmonic Poynting vector is:

$$S_r = \frac{p_0^2 \sin^2 \theta}{\varepsilon_0 (4\pi)^2} \left(\frac{\omega^4}{2r^2 c^3} \right) \qquad (A.2)$$

and the angular component of the time average harmonic Poynting vector is:

$$S_\theta = 0 \qquad (A.3)$$

The time average time domain harmonic energy density becomes:

$$\langle u_E \rangle = \frac{p_0^2}{32\pi^2 \varepsilon_0 r^2} \left[4\cos^2 \theta \left(\frac{1}{2r^4} + \frac{\omega^2}{2c^2 r^2} \right) + \sin^2 \theta \left(\frac{1}{2r^4} - \frac{\omega^2}{2c^2 r^2} + \frac{\omega^4}{2c^4} \right) \right] \qquad (A.4)$$

311

Table A.1

Harmonic Time Domain Functions

Time Domain Functions	Time Average Values of Time Harmonic Functions
$T^2 = \sin^2 \omega t = \frac{1}{2}(1 - \cos 2\omega t)$	$\langle T^2 \rangle = \langle \frac{1}{2}(1 - \cos 2\omega t) \rangle = \frac{1}{2}$
$T\dot{T} = \omega \sin \omega t \cos \omega t = \frac{1}{2}\omega \sin 2\omega t$	$\langle T\dot{T} \rangle = \langle \frac{1}{2}\omega \sin 2\omega t \rangle = 0$
$\dot{T}^2 = \omega^2 \cos^2 \omega t = \frac{1}{2}\omega^2(1 + \cos 2\omega t)$	$\langle \dot{T}^2 \rangle = \langle \frac{1}{2}\omega^2(1 + \cos 2\omega t) \rangle = \frac{1}{2}\omega^2$
$\dot{T}^2 + 2T\ddot{T} = \omega^2(\cos^2 \omega t - 2\sin^2 \omega t)$	$\langle T\ddot{T} \rangle = -\frac{1}{2}\omega^2$
$\quad = \frac{1}{2}\omega^2(1 + \cos 2\omega t) - \omega^2(1 - \cos 2\omega t)$	$\langle \dot{T}^2 + 2T\ddot{T} \rangle = \langle -\omega^2(\frac{1}{2} + \frac{3}{2}\cos 2\omega t) \rangle$
$\quad = -\omega^2(\frac{1}{2} + \frac{3}{2}\cos 2\omega t)$	$\quad = -\frac{1}{2}\omega^2$
$\dot{T}\ddot{T} = -\omega^3 \sin \omega t \cos \omega t = -\frac{1}{2}\omega^3 \sin 2\omega t$	$\langle \dot{T}\ddot{T} \rangle = \langle -\frac{1}{2}\omega^3 \sin 2\omega t \rangle = 0$
$\ddot{T}^2 = \omega^4 \sin^2 \omega t = \frac{1}{2}\omega^4(1 - \cos 2\omega t)$	$\langle \ddot{T}^2 \rangle = \langle \frac{1}{2}\omega^4(1 - \cos 2\omega t) \rangle = \frac{1}{2}\omega^4$

for the electric field energy, and

$$\langle u_H \rangle = \frac{p_0 \mu_0}{32\pi^2 r^2}\left(\frac{\omega^2}{2r^2} + \frac{\omega^4}{2c^2}\right)\sin^2 \theta \qquad (A.5)$$

for the magnetic field energy. Expressing these results in terms of wave number $\left(k = \frac{2\pi}{\lambda} = \frac{\omega}{c}\right)$, the electric energy density is

$$\langle u_E \rangle = \frac{p_0^2 k^3}{64\pi^2 \varepsilon_0}\left[4\cos^2 \theta\left(\frac{1}{k^3 r^6} + \frac{1}{kr^4}\right) + \sin^2 \theta\left(\frac{1}{k^3 r^6} - \frac{1}{kr^4} + \frac{k}{r^2}\right)\right] \quad (A.6)$$

and the magnetic energy density is

$$\langle u_H \rangle = \frac{p_0 \mu_0 c^2}{64\pi^2 r^2}\left(\frac{k^2}{r^2} + k^4\right)\sin^2 \theta \qquad (A.7)$$

Noting that $\mu_0 c^2 = \frac{1}{\varepsilon_0}$, the propagating, or radiation, portion of the energy density is

$$\langle u_E \rangle_{rad} = \frac{p_0 k^4 \sin^2 \theta}{64 \pi^2 \varepsilon_0 r^2} \quad (A.8)$$

for the electric energy and:

$$\langle u_H \rangle_{rad} = \frac{p_0 k^4 \sin^2 \theta}{64 \pi^2 \varepsilon_0 r^2} \quad (A.9)$$

for the magnetic energy. The radiation energy is equally divided between electric and magnetic energy.

The traditional division of energy into propagating radiation energy and fixed reactive energy has a significant conceptual difficulty. Assume the electric field can be broken up into a reactive and radiation term:

$$\mathbf{E} = \mathbf{E}_{reac} + \mathbf{E}_{rad} \quad (A.10)$$

Then, the electric field energy is

$$
\begin{aligned}
u_E &= \frac{\varepsilon_0}{2} |\mathbf{E}|^2 = \frac{\varepsilon_0}{2} |\mathbf{E}_{reac} + \mathbf{E}_{rad}|^2 \\
&= \frac{\varepsilon_0}{2} |\mathbf{E}_{rad}|^2 + \varepsilon_0 \mathbf{E}_{reac} \cdot \mathbf{E}_{rad} + \frac{\varepsilon_0}{2} |\mathbf{E}_{reac}|^2 \\
&= u_{rad} + \varepsilon_0 \mathbf{E}_{reac} \cdot \mathbf{E}_{rad} + u_{reac}
\end{aligned}
\quad (A.11)
$$

Lumping the cross-term into the reactive field energy leads to the unsatisfying result that the reactive field energy depends on the radiation energy. The cross-term has time dependence $\dddot{T}T$. This averages to zero for the time-average harmonic case, and thus may be ignored for purposes of the present discussion.

The reactive electric energy is

$$\langle u_E \rangle_{reac} = \frac{p_0^2 k^3}{64 \pi^2 \varepsilon_0} \left[4 \cos^2 \theta \left(\frac{1}{k^3 r^6} + \frac{1}{kr^4} \right) + \sin^2 \theta \left(\frac{1}{k^3 r^6} - \frac{1}{kr^4} \right) \right] \quad (A.12)$$

and the reactive magnetic energy is

$$\langle u_H \rangle_{reac} = \frac{p_0 k^3}{64 \pi^2 \varepsilon_0} \left(\frac{1}{kr^4} \right) \sin^2 \theta \quad (A.13)$$

Integrating over the 4π steradian shell yields relations for the radial dependence of the electric energy density:

$$\langle U_E \rangle_{reac} = \int_V \langle u_E \rangle_{reac} dV = \int_0^{2\pi} \int_0^\pi \int_R^\infty \langle u_E \rangle_{reac} r^2 \sin\theta \, dr \, d\theta \, d\phi$$

$$= \frac{p_0^2 k^3}{64\pi^2 \varepsilon_0} \int_0^{2\pi} \int_0^\pi \int_R^\infty \left[\begin{array}{l} 4\cos^2\theta \sin\theta \left(\dfrac{1}{k^3 r^4} + \dfrac{1}{kr^2} \right) \\ + \sin^3\theta \left(\dfrac{1}{k^3 r^4} - \dfrac{1}{kr^2} \right) \end{array} \right] \sin\theta \, dr \, d\theta \, d\phi$$

$$= \frac{p_0^2 k^3}{32\pi\varepsilon_0} \int_0^\pi \int_R^\infty \left[\begin{array}{l} 4\cos^2\theta \sin\theta \left(\dfrac{1}{k^3 r^4} + \dfrac{1}{kr^2} \right) \\ + \sin^3\theta \left(\dfrac{1}{k^3 r^4} - \dfrac{1}{kr^2} \right) \end{array} \right] dr \, d\theta$$

$$= \frac{p_0^2 k^3}{32\pi\varepsilon_0} \int_R^\infty \frac{8}{3}\left(\frac{1}{k^3 r^4} + \frac{1}{kr^2} \right) + \frac{4}{3}\left(\frac{1}{k^3 r^4} - \frac{1}{kr^2} \right) dr$$

$$= \frac{p_0^2 k^3}{8\pi\varepsilon_0} \int_R^\infty \left(\frac{1}{k^3 r^4} + \frac{1}{3kr^2} \right) dr = \frac{p_0^2 k^3}{8\pi\varepsilon_0} \left[-\frac{1}{3k^3 r^3} - \frac{1}{3kr} \right]_R^\infty$$

$$= \frac{p_0^2 k^3}{24\pi\varepsilon_0} \left(\frac{1}{k^3 R^3} + \frac{1}{kR} \right) \tag{A.14}$$

The time-average electric radiation field is

$$\mathbf{E}_{rad} = \frac{p_0 \omega^2 \sin\omega t \sin\theta}{4\pi\varepsilon_0 c^2 r} \hat{\boldsymbol{\theta}} \tag{A.15}$$

and the time-average magnetic radiation field is

$$\mathbf{H}_{rad} = \frac{p_0 \omega^2 \sin\omega t \sin\theta}{4\pi c r} \hat{\boldsymbol{\phi}} \tag{A.16}$$

Thus, the total time-average radiated power is

$$P_{rad} = \left\langle \oint_{\sigma} \left(\mathbf{E}_{rad} \times \mathbf{H}_{rad} \cdot \hat{\mathbf{r}} \right) d\sigma \right\rangle$$

$$= \left\langle \oint_{\sigma} \left(\frac{p_0 \omega^2 \sin \omega t \sin \theta}{4\pi\varepsilon_0 c^2 r} \frac{p_0 \omega^2 \sin \omega t \sin \theta}{4\pi c r} \right) d\sigma \right\rangle$$

$$= \int_0^{2\pi} \int_0^{\pi} \left(\frac{p_0^2 \omega^4 \sin^2 \theta}{32\pi^2 \varepsilon_0 c^3 r^2} \right) r^2 \sin \theta \, d\theta \, d\phi$$

$$= \frac{8\pi}{3} \frac{p_0^2 \omega^4}{32\pi^2 \varepsilon_0 c^3} = \frac{p_0^2 \omega \, k^3}{12\pi\varepsilon_0}$$

(A.17)

The relations of this Appendix are useful for setting fundamental limits on antenna performance, a topic that is addressed in Section 5.5.

About the Author

Dr. Hans Schantz is a consulting physicist and engineer for Next-RF, Inc. His specialties include time-domain electromagnetics and high-performance antennas, particularly ultrawideband (UWB) ones. A popular and engaging lecturer, his short course on UWB antennas is a regular feature at the IEEE Antenna and Propagation Society International Symposium. He is also a senior member of the IEEE. For more information on Dr. Schantz's UWB work, please refer to his web site at www.UWBAntenna.com.

Dr. Schantz has published work in the American Journal of Physics, the IEEE Aerospace and Electronic Systems Magazine, and the IEEE Antennas and Propagation Magazine. He has written over a dozen conference papers. He has ten U. S. patents and several patent applications pending to his credit. Dr. Schantz is also the chief scientist for the Q-Track Corporation, a start-up company that is developing near-field electromagnetic ranging (NFER™) technology. This technology enables very precise positioning using narrowband, low-frequency techniques. For more information, please see www.q-track.com.

Dr. Schantz holds a B. S. I. E. in industrial engineering and a B. S. in the honors curriculum in physics, both from Purdue University. While a student, he worked at IBM in Manassas, Virginia, and at the Lawrence Livermore National Laboratory in Livermore, California. He earned his Ph.D. in theoretical physics from the University of Texas, at Austin, studying under E. C. G. Sudarshan and John A. Wheeler. Dr. Schantz's doctoral dissertation discussed the flow of electromagnetic energy around electric and magnetic dipoles and the process by which reactive energy becomes radiation energy.

Dr. Schantz lives in Huntsville, Alabama, with his wife, Barbara, and his newborn twin daughters, Greta and Cora. When not engaged in professional work, his favorite hobbies of late include diapering, burping, feeding, and, most of all, sleeping.

Index

Defense Advanced Research Projects
    Agency, 31
Density functions, for UWB parameters, 55
DF see direction finding
Diamond dipole antennas, 19, 24, 115,
    208–210
Dielectric constant, 70, 246
Differential annular slot antenna, 243
Differential ended, 91
Differrential feed, 234
Differentiation of radiated impulses,
    98, 128
Diffraction, at end of horns, 251
Ding-Rong, S., 272
Diplexing filter, 68, 279
Dipole
    antenna see specific kind ideal (or
        Hertzian)
    dc and quasi-dc excitation, 133–134
    electric, defined and fields, 124–126
    energy density around, 152–154
    energy-flow streamlines, 175
    field impedance of, 129–130
    magnetic, defined and fields, 126–127
    Larmor's radiation formula for, 156
    mode of spherical discharge, 196
    pattern, directivity, aperture, 50, 299
    time harmonic fields and energy, 129,
        161–166, 311–315
    transient excitations, 156–160,
        167–173
    optimal shapes, 175–176
    thin wire, 66, 117–118
Dirac delta function, 96
Direction finding (DF), 275, 299–303
Directional antennas, 50, 248–268, 298
Directionality, 33
Directivity (D), 50–51, 297
Discone antenna, 18–19, 24, 25, 205
Discontinuities, affecting antenna
        performance, 23
    concentration of reactive energy, 174
    in balanced and unbalanced lines, 85
    reflections from, 23
    transmission line, 75
Dispersion, 42–47
    from path length, delays and phase
        offsets, 116
    loop antenna, 239

monoloop antenna, 236
    of planar loop antenna, 176
    reduction in clover-leaf loop, 177
Distortion, 116, 258, 263
Distributed source antenna, 121
Domino, W., 306
Double differentiation, 128
Double ridged horn, 249
Duality principle, 126
Duncan, J. W., 89, 93

E

Earth ground, 8, 11
Edge current concentration, 114
Edge signals, 98, 101–104
Effective antenna height, 54–55
Effective Isotropic Radiated Power (EIRP),
    52,55
Efficiency, 52, 68, 275, 286–297
    limit, energy-flow based, 191
    of spherical discharge, 195
    volumetric, invalidity of, 174–175
EIRP, see Effective Isotropic Radiated
    Power
Electric dipole, see dipole, ideal, electric
Electrically large antennas vs small-
    aperture, 304
Electrically small antenna, 23, 59, 244, 247
    resistive loading, 68
Electric-magnetic antenna, 247
Electrobit, Ltd., semicircular dipole, 215
Electromagnetic cause-and-effect, 152
The ElectroScience Lab, Ohio State
        University, 20, 251, 267–268
Ellipsoidal antenna, 22, 25, 213
Elliptic planar dipole, 22, 25, 44–45,
        218–224
Emerson, Darrel T., 5, 10, 35
Endloading, spherical and square plate, 6
Energetics of time domain excitations, 172
Energy, 29, 33, 139, 143–177, 190,
        311–315
Energy flow, 143–177
    based antenna limit, 190–195
Equivalence of frequency and time domain,
    37
Explorer satellite 1958, 246
Exponential impedance taper, 80–82, 84–85
Exponentially decaying dipole, 154

For further information on these and other Artech House titles,
including previously considered out-of-print books now available through our
In-Print-Forever® (IPF®) program, contact:

Artech House   Artech House
685 Canton Street   46 Gillingham Street
Norwood, MA 02062   London SW1V 1AH UK
Phone: 781-769-9750   Phone: +44 (0)20 7596-8750
Fax: 781-769-6334   Fax: +44 (0)20 7630 0166
e-mail: artech@artechhouse.com   e-mail: artech-uk@artechhouse.com

Find us on the World Wide Web at:
www.artechhouse.com